# Electroactive Polymers for Robotic Applications

Kwang J. Kim and Satoshi Tadokoro (Eds.)

# Electroactive Polymers for Robotic Applications

## Artificial Muscles and Sensors

Springer

Kwang J. Kim, PhD
Mechanical Engineering Department
(MS312)
University of Nevada
Reno, NV 89557
USA

Satoshi Tadokoro, Dr. Eng.
Graduate School of Information
Sciences
Tohoku University
Sendai
Japan

British Library Cataloguing in Publication Data
Electroactive polymers for robotic applications :
artificial muscles and sensors
1.Actuators 2.Detectors 3.Robots - Control systems
4.Conducting polymers
I.Kim, Kwang Jin, 1949- II.Tadokoro, Satoshi
629.8'933
ISBN-13: 9781846283710
ISBN-10: 184628371X

Library of Congress Control Number: 2006938344

ISBN 978-1-84628-371-0 e-ISBN 978-1-84628-372-7 Printed on acid-free paper



9 8 7 6 5 4 3 2 1

Springer Science+Business Media
springer.com

# Preface

The focus of this book is on electroactive polymer (EAP) actuators and sensors. The book covers the introductory chemistry, physics, and modeling of EAP technologies and is structured around the demonstration of EAPs in robotic applications. The EAP field is experiencing interest due to the ability to build improved polymeric materials and modern digital electronics. To develop robust robotic devices actuated by EAP, it is necessary for engineers to understand their fundamental physics and chemistry.

We are grateful to all contributing authors for their efforts. It has been a great pleasure to work with them. Also, the authors wish to thank Anthony Doyle and Kate Brown of Springer-Verlag, London, and Deniz Dogruer of the University of Nevada-Reno, for their assistance and support in producing the book. One of us (KJK) expresses his thanks to Drs. Junku Yuh and George Lee of the U.S. National Science Foundation (NSF), Drs. Tom McKenna and Harold Bright of the Office of Naval Research (ONR), Dr. Promode Bandyopadhyay of Naval Undersea Warfare Center, and Dr. Kumar Krishen of NASA Johnson Space Center (JSC) for their encouragement.

Kwang J. Kim
University of Nevada, Reno
Reno, Nevada USA

Satoshi Tadokoro
Tohoku University
Sendai, Japan

# Contents

# Contents

## List of Contributors

**K. Asaka**
Research Institute for Cell Engineering, National Institute of AIST, 1-8-31 Midorigaoka, Ikeda, Osaka 563-8577, Japan and Bio-Mimetic Control Research Center, RIKEN
e-mail: asaka-kinji@aist.go.jp

**H.R. Choi**
School of Mechanical Engineering, Sungkyunkwan University, 300 Chunchun-dong, Jangan-gu, Suwon, Kyunggi-do 440-746, South Korea
e-mail: hrchoi@me.skku.ac.kr

**D. Dogruer**
Active Materials and Processing Laboratory, Mechanical Engineering Department (MS 312), University of Nevada, Reno, Nevada 89557, U.S.A.
e-mail: kwangkim@unr.edu

**K.M. Jung**
School of Mechanical Engineering, College of Engineering, Sungkyunkwan University, Suwon 440-746, Korea
e-mail: jungkmok@me.skku.ac.kr

**N. Kamamichi**
Department of Mechanical and Control Engineering, Tokyo Institute of Technology 2-12-1 Oh-okayama, Meguro-ku, Tokyo, 152-8552, Japan
e-mail: nkama@ac.ctrl.titech.ac.jp

**K.J. Kim**
Active Materials and Processing Laboratory, Mechanical Engineering Department (MS 312), University of Nevada, Reno, Nevada 89557, U.S.A.
e-mail: kwangkim@unr.edu

**M. Konyo**
Robot Informatics Laboratory, Graduate School of Information Science, Tohoku University, 6-6-01 Aramaki Aza Aoba, Aoba-ku, Sendai 980-8579. Japan
e-mail: konyo@rm.is.tohoku.ac.jp

**J.C. Koo**
School of Mechanical Engineering, Sungkyunkwan University, 300 Chunchun-dong, Jangan-gu, Suwon, Kyunggi-do 440-746, South Korea
e-mail: jckoo@me.skku.ac.kr

**Y.K. Lee**
School of Chemical Engineering,
Sungkyunkwan University, 300
Chunchun-dong, Jangan-gu, Suwon,
Kyunggi-do 440-746, South Korea
e-mail: yklee@skku.edu

**Z.W. Luo**
Bio-Mimetic Control Research Center,
RIKEN 2271-130 Anagahora,
Shimoshidami, Moriyama-ku, Nagoya
463-0003, Japan
e-mail: luo@bmc.riken.jp

**J.D. Madden**
Molecular Mechatronics Lab,
Advanced Materials & Process
Engineering Laboratory and
Department of Electrical & Computer
Engineering, University of British
Columbia, Vancouver, British
Columbia V6T 1Z4, Canada
e-mail: jmadden@ece.ubc.ca

**T. Mukai**
Bio-Mimetic Control Research Center,
RIKEN, 2271-130 Anagahora,
Shimoshidami, Moriyama, Nagoya
463-0003, Japan
e-mail: mukai@bmc.riken.jp

**Y. Nakabo**
Bio-Mimetic Control Research Center,
RIKEN, 2271-130 Anagahora,
Shimoshidami, Moriyama, Nagoya
463-0003, Japan and Intelligent
Systems Institute, National Institute of
AIST, 1-1-1 Umezono, Tsukuba,
Ibaraki 305-8568, Japan
e-mail: nakabo-yoshihiro@aist.go.jp

**J.D. Nam**
Department of Polymer Science and
Engineering, Sungkyunkwan
University, 300 Chunchun-dong,
Jangan-gu, Suwon, Kyunggi-do 440-
746, South Korea
e-mail: jdnam@skku.edu

**R. Samatham**
Active Materials and Processing
Laboratory, Mechanical Engineering
Department (MS 312), University of
Nevada, Reno, Nevada 89557, U.S.A.
e-mail: kwangkim@unr.edu

**J. Su**
Advanced Materials and Processing
Branch Langley Research Center
National Aeronautics and Space
Administration (NASA)
Hampton, Virginia 23681, U.S.A.
e-mail:ji.su-1@nasa.gov

**S. Tadokoro**
Graduate School of Information
Sciences, Tohoku University, 6-6-01
Aramaki Aza Aoba, Aoba-ku, Sendai
980-8579, Japan
e-mail: tadokoro@rm.is.tohoku.ac.jp

**M. Yamakita**
Department of Mechanical and Control
Engineering, Tokyo Institute of
Technology, 2-12-1 Oh-okayama,
Meguro-ku, Tokyo 152-8552, Japan
e-mail: yamakita@ctrl.titech.ac.jp

**W. Yim**
Department of Mechanical
Engineering, University of Nevada,
Las Vegas, 4505 Maryland Parkway,
Las Vegas, Nevada 89154-4027,
U.S.A.
e-mail: wy@me.unlv.edu

# 1

# Active Polymers: An Overview

R. Samatham[1], K.J. Kim[1], D. Dogruer[1], H.R. Choi[2], M. Konyo[3], J. D. Madden[4], Y. Nakabo[5], J.-D. Nam[6], J. Su[7], S. Tadokoro[8], W. Yim[9], M. Yamakita[10]

[1] Active Materials and Processing Laboratory, Mechanical Engineering Department (MS 312), University of Nevada, Reno, Nevada 89557, U.S.A. (kwangkim@unr.edu)
[2] School of Mechanical Engineering, Sungkyunkwan University, 300 Chunchun-dong, Jangan-gu, Suwon, Kyunggi-do 440-746, South Korea
[3] Robot Informatics Laboratory, Graduate School of Information Sciences, Tohoku University, Sendai 980-8579, Japan
[4] Molecular Mechanics Group, Department of Mechanical Engineering, University of British Columbia, Vancouver BC V6T 1Z4, Canada
[5] Bio-Mimetic Control Research Center, RIKEN, 2271-130 Anagahora, Shimoshidami, Moriyama, Nagoya, 463-0003 JAPAN and Intelligent Systems Institute, National Institute of AIST, 1-1-1 Umezono, Tsukuba, Ibaraki 305-8568 Japan
[6] Department of Polymer Science and Engineering, Sungkyunkwan University, 300 Chunchun-dong, Jangan-gu, Suwon, Kyunggi-do 440-746, South Korea
[7] Advanced Materials and Processing Branch, NASA Langley Research Center, Hampton, VA 23681, U.S.A.
[8] Graduate School of Information Sciences, Tohoku University, 6-6-01 Aramaki Aza Aoba, Aoba-ku, Sendai 980-8579, Japan
[9] Department of Mechanical Engineering, University of Nevada, Las Vegas, 4505 Maryland Parkway, Las Vegas, Nevada 89154-4027, U.S.A.
[10] Department of Mechanical and Control Engineering, Tokyo Institute of Technology, 2-12-1 Oh-okayama, Meguro-ku, Tokyo, 152-8552, Japan

## 1.1 Introduction

In this time of technological advancements, conventional materials such as metals and alloys are being replaced by polymers in such fields as automobiles, aerospace, household goods, and electronics. Due to the tremendous advances in polymeric materials technology, various processing techniques have been developed that enable the production of polymers with tailor-made properties (mechanical, electrical, etc). Polymers enable new designs to be developed that are cost-effective with small size and weights [1].

Polymers have attractive properties compared to inorganic materials. They are lightweight, inexpensive, fracture tolerant, pliable, and easily processed and manufactured. They can be configured into complex shapes and their properties can be tailored according to demand [2]. With the rapid advances in materials used in science and technology, various materials with intelligence embedded at the molecular level are being developed at a fast pace. These intelligent materials can sense variations in the environment, process the information, and respond accordingly. Shape-memory alloys, piezoelectric materials, *etc.* fall in this category of intelligent materials [3]. Polymers that respond to external stimuli by changing shape or size have been known and studied for several decades. They respond to stimuli such as an electrical field, pH, a magnetic field, and light [2]. These intelligent polymers can collectively be called *active polymers*.

One of the significant applications of these active polymers is found in biomimetics—the practice of taking ideas and concepts from nature and implementing them in engineering and design. Various machines that imitate birds, fish, insects and even plants have been developed. With the increased emphasis on "green" technological solutions to contemporary problems, scientists started exploring the ultimate resource—nature—for solutions that have become highly optimized during the millions of years of evolution [4]. Throughout history, humans have attempted to mimic biological creatures in appearance, functionality, intelligence of operation, and their thinking process. Currently, various biomimetic fields are attempting to do the same thing, including artificial intelligence, artificial vision, artificial muscles, and many other avenues [5]. It has been the dream of robotic engineers to develop autonomous, legged robots with mission-handling capabilities. But the development of these robots has been limited by the complex actuation and control and power technology that are incomparable to simple systems in the natural world. As humans have developed in biomimetic fields, biology has provided efficient solutions for the design of locomotion and control systems [6]. Active polymers with characteristics similar to biological muscles hold tremendous promise for the development of biomimetics. These polymers have characteristics similar to biological muscles such as resilience, large actuation, and damage tolerance. They are more flexible than conventional motors and can act as vibration and shock dampers; the polymers are similar in aesthetic appeal too. The polymers' physical makeup enables the development of mechanical devices with no gears, bearings, or other complex mechanisms responsible for large costs and complexity [5].

Active materials can convert electrical or chemical energy directly to mechanical energy through the response of the material. This capability is of great use in rapidly shrinking mechanical components due to the miniaturization of robots [7]. Realistically looking and behaving robots are believed possible, using artificial intelligence, effective artificial muscles, and biomimetic technologies [8]. Autonomous, human-looking robots can be developed to inspect structures with configurations that are not predetermined. A multifunctional automated crawling system developed at NASA/JPL, operates in field conditions and scans large areas using a wide range of NDE instruments [9].

There are many types of active polymers with different controllable properties, due to a variety of stimuli. They can produce permanent or reversible responses; they can be passive or active by embedment in polymers, making smart structures. The resilience and toughness of the host polymer can be useful in the development of smart structures that have shape control and self-sensing capabilities [2].

Depending on the type of actuation, the materials used are broadly classified as nonelectrically deformable polymers (actuated by nonelectric stimuli such as pH, light, temperature, *etc.*) and electroactive polymers (EAPs) (actuated by electric inputs). Different types of nonelectrically deformable polymers are chemically activated polymers, shape-memory polymers, inflatable structures, light-activated polymers, magnetically activated polymers, and thermally activated gels [2].

Polymers that change shape or size in response to electrical stimulus are called electroactive polymers (EAP) and are classified depending on the mechanism responsible for actuation as electronic EAPs (which are driven by electric field or coulomb forces) or ionic EAPs (which change shape by mobility or diffusion of ions and their conjugated substances). A list of leading electroactive polymers is shown in Table 1.1.

**Table 1.1.** List of leading EAP materials

| Electronic EAP | Ionic EAP |
|---|---|
| Dielectric EAP | Ionic polymer gels (IPG) |
| Electrostrictive graft elastomers | Ionic polymer metal composite (IPMC) |
| Electrostrictive paper | Conducting polymers (CP) |
| Electro-viscoelastic elastomers | Carbon nanotubes (CNT) |
| Ferroelectric polymers | |
| Liquid crystal elastomers (LCE) | |

The electronic EAPs such as electrostrictive, electrostatic, piezoelectric, and ferroelectric generally require high activation fields (>150V/μm) which are close to the breakdown level of the material. The property of these materials to hold the induced displacement, when a DC voltage is applied, makes them potential materials in robotic applications, and these materials can be operated in air without major constraints. The electronic EAPs also have high energy density as well as a rapid response time in the range of milliseconds. In general, these materials have a glass transition temperature inadequate for low temperature actuation applications.

In contrast, ionic EAP materials such as gels, ionic polymer-metal composites, conducting polymers, and carbon nanotubes require low driving voltages, nearly equal to 1–5V. One of the constraints of these materials is that they must be operated in a wet state or in solid electrolytes. Ionic EAPs predominantly produce bending actuation that induces relatively lower actuation forces than electronic EAPs. Often, operation in aqueous systems is plagued by the hydrolysis of water. Moreover, ionic EAPs have slow response characteristics compared to electronic EAPs. The amount of deformation of these materials is usually much more than electronic EAP materials, and the deformation mechanism bears more resemblance to a biological muscle deformation. The induced strain of both the electronic and ionic EAPs can be designed geometrically to bend, stretch, or contract [2].

Another way to classify actuators is based on actuator mechanisms. The various mechanisms through which EAPs produce actuation are polarization, mass/ion transportation, molecular shape change, and phase change. Dielectric elastomers and piezoelectric polymers produce actuation through polarization. Conducting polymers and gel polymers produce actuation basically through ion/mass transportation. Liquid crystal elastomers and shape-memory polymers produce actuation by phase change.

As can be observed, various stimuli can be used to actuate active polymers. Development of polymers that can respond to a noncontact mode of stimuli such as electrical, magnetic, and light can lead to the diversification of the applications of active polymers. Electrical stimulation is considered the most promising, owing to its availability and advances in control systems. There has been a surge in the amount of research being done on the development of electro-active polymers (EAPs), but other kinds of stimulation have their own niche applications.

Initially, the electrical stimulation of polymers produced relatively small strains, restricting their practical use. But nowadays, polymers showing large strains have been developed and show great potential and capabilities for the development of practical applications. Active polymers which respond to electric stimuli, electroactive polymers (EAPs), exhibit two-to-three orders of magnitude deformation, more than the striction-limited, rigid and fragile electroactive ceramics (EACs). EAPs can have higher response speed, lower density, and greater resilience than shape-memory alloys (SMAs). However, the scope of practical applications of EAPs is limited by low actuation force, low mechanical energy density, and low robustness. Progress toward actuators being used in robotic applications with performance comparable to biological systems will lead to great benefits [2].

In the following paragraphs, all types of active polymers are briefly described and thoroughly reviewed in cited references. Also, some of the most recent developments for certain polymers are presented. Some of the applications of active polymers are given as well.

## 1.2 Nonelectroactive Polymers

### 1.2.1 Chemically Activated Polymers

A polymer can change in dimension by interacting with chemicals, but it is a relatively slow process. For example, when a piece of rubber is dropped into oil, it slowly swells by interacting with the solvent [2].

The first artificial muscle was a pH actuated polymeric gel developed in 1950. Since then, a wide variety of polymer gel materials have been developed that can respond to stimuli such as pH, temperature, light, and solvent composition. The interaction with surroundings causes a change in shape or size of these polymers. Some of these polymers are sensitive to pH in aqueous environments.

Most of the earlier work on the gel muscles was done on pH actuation. Cross-linked polyacrylic acid gel is the most widely studied polymer for chemical actuation. This gel increases dimensionally when moved from an acid solution to a base solution and shows weak mechanical properties. To find stronger polymers, different materials were developed during the last 20 years.

Yoshida *et al.* [10] developed an oscillating, swelling-deswelling, pH-sensitive polymer gel system. Rhythmic swelling-deswelling oscillations were achieved by coupling temperature and pH-sensitive poly (*N*-isopropylacrylamide-co-acrylic acid-co-butylmethacrylate) gels with nonlinear oscillating chemical reactions. A pH-oscillating reaction was generated in a continuous-flow-stirred tank reactor, in which the pH of the system changed after a specific time interval. When polymer gels are coupled with reactions in a reactor, an oscillating response is produced.

One of the interesting materials in the this family is the polyacrylonitrile (PAN) gel fiber [11], which when oxidized and saponified shows behavior similar to that of polyacrylic acid gels. The strength of the PAN fibers is higher, and the response time is minimal. A change in length of 70% was observed in a few seconds when the system was moved from an acid to a base, which is very fast compared to polyacrylic acid gels (which could take days or weeks). A volume change of more than 800% was observed for PAN fibers [12]. Moreover, among the available polymer based actuator materials, PAN fiber is already produced commercially in large volumes and used in the production of textiles and as a precursor for making carbon fibers. Coupled with a simple activation process, the easy availability of PAN fiber makes it one of the most suitable materials for use in the development of practical applications. It was found that when fibers transform into gels, they have stronger mechanical properties and larger volume change, more closely resembling biological muscle than any other polymer gel actuators [11]. The diameter of commercially available PAN fiber is on the order of microns in its swollen state, so the response time is rapid as the response depends on the dimension (diameter) of the fibers. The response characteristics of the PAN fibers were found superior to other chemically activated polymer materials, but still not comparable to the response characteristics of skeletal muscles. To improve the response characteristics, sub-micron diameter PAN fibers were produced using a process called "electrospinning." Macroscopic observation of a PAN nanofiber mat made from electrospinning showed more than 600% deformation in a few seconds, but the mechanical properties of electrospun fiber-mat were found to be poorer than the commercial PAN fibers. Typically, the PAN fibers used in those of the textile industry are co-polymerized with a small amount of another polymer such as acrylamide, methyl acrylate, methyl methacrylate; therefore, there may be some differences in the mechanical properties of such modified PAN fibers. Efforts are underway to improve the mechanical properties and observe the deformational characteristics of the fibers on a microscale. The use of these PAN fibers has more potential in the development of the linear actuators and artificial muscles. For example, the force to weight ratio from experimentation in our lab showed that 0.2g of PAN fiber (5g in an activated state) can generate more than 150$gm_f$ (30–750 times of one weight) [13].

### 1.2.2 Shape-Memory Polymers

Shape-memory materials are stimuli-responsive materials that change shape through the application of external stimuli. The thermally-induced shape-memory effect is used widely. Thermally responsive shape-memory polymers change shape when heated above a certain temperature and can be processed into two shapes. One form, the permanent shape, is obtained through conventional processing techniques such as extrusion and injection molding. During this process, the material is heated above the highest thermal transition temperature ($T_{perm}$). The phase above $T_{perm}$ forms physical cross-links which enable the polymers to form permanent shapes. The second phase fixes the temporary phase, acting as a molecular switch. The switching segments can be fixed above the transition temperature ($T_{trans}$), either the glass transition temperature ($T_g$) or the melting temperature ($T'_m$). This transition temperature is usually less than $T_{perm}$. The material can be formed into a temporary shape by thermal processing or cold drawing and cooling below the transition temperature. When the material is heated above the $T_{trans}$, the physical cross-links in the switching phase are broken, forcing the material into a permanent shape known as recovery [14]. The operation of a shape-memory polymer is schematically depicted in Figure 1.1.

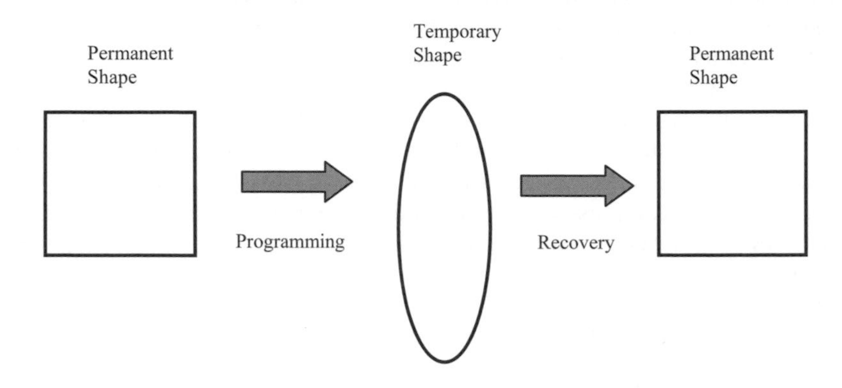


**Figure 1.1.** Cartoon showing one-way shape-memory effect produced by thermal activation. The permanent shape is transformed into a temporary shape through a programming process. The permanent shape is recovered when the sample is heated above the switching temperature.

As early as the 1930s, scientists discovered that certain metallic compounds exhibited the shape-memory effect when heated above a transition temperature. Since then, shape memory alloys (SMAs), such as the nickel-titanium alloy, have found uses in actuators and medical devices, such as orthodontic wires that self-adjust and stents for keeping blood vessels open. Despite their broad range of applications, SMAs are expensive and nondegradable, and in many cases, lack biocompatibility and compliance, allowing for a deformation of about 8% for Ni-Ti alloys [15].

Linear, phase-segregated multiblock copolymers, mostly polyurethanes, are the commonly used shape-memory polymers. Note that the shape-memory effect is not the property of one single polymer, but it is a combined effect of polymer structure and polymer morphology along with processing and programming technology. Programming refers to the process used to fix the temporary phase. The shape-memory effect can be observed in polymers with significantly different chemical compositions. A significant, new development in the design of shape-memory polymers is the discovery of families of polymers called *polymer systems*. The properties of these polymer systems can be tailored for specific applications by slightly varying their chemical composition [14]. The memory effect of shape-memory polymers is due to the stored mechanical energy obtained during reconfiguration and cooling of the material [16].

Shape-memory polymers (SMPs) are finding applications in varied fields from deploying objects in space to manufacturing dynamic tools [16]. The versatile characteristics of SMPs make them ideal for applications in dynamic configurable parts, deployable components, and inexpensive, reusable custom molds [16]. One type of SMP is the cold hibernated elastic memory (CHEM) structure that can be compressed into a small volume at a temperature higher than the glass transition temperature ($T_g$) and stored at temperatures below this $T_g$. When this material is heated again above $T_g$, the original volume of the structure is restored. Volume ratios of up to forty times have been obtained [2]. Structures having different sizes and shapes can be erected by the self-deployable characteristics of these CHEM materials due to their elastic recovery and shape-memory properties. One of the advantages of these materials is that they are a fraction of their original size when compressed and stored below $T_g$ and are lightweight. Commercial applications of these materials include building shelters, hangars, camping tents, rafts, and outdoor furniture. CHEM materials have good impact and radiation resistance as well as strong thermal and electrical insulation properties. One of the disadvantages of these materials is their packing needs: a pressure mechanism which may not be available readily in the outdoors, where they are most applicable [2].

Biodegradable and biocompatible SMPs are being developed which have tremendous potential in the development of minimal, invasive surgery technologies [14]. The permanent shapes of these fibers are programmed into a wound stitch, stretching to form thin fibers. This fiber is then heated above the transition temperature of the material inducing permanent deformation in the material sealing the wound. Biodegradable shape-memory polymers also show strong promise for implantable devices in biomedical applications [17].

### 1.2.3 Inflatable Structures

Pneumatic artificial muscles (PAMs, often called McKibben muscle) can be defined as contractile linear motion gas pressured engines. Their simple design is comprised of a core element that is a flexible reinforced closed membrane attached at both the ends to the fittings, acting as an inlet and an outlet. Mechanical power is transferred to the load through the fittings. When the membrane inflates due to gas pressure, it bulges outward radially, leading to axial contraction of the shell. This contraction exerts a pulling force on its load. The actuation provides unidirectional linear force and motion. PAMs can be operated underpressure or overpressure, but they are usually operated overpressure as more energy can be transferred. In PAMs, the force generated is related to the applied gas pressure, whereas the amount of actuation is related to the change in the volume. Therefore, the particular state of PAM is determined by the gas pressure and length [18]. The unique, physical configuration of these actuators gives them numerous variable-stiffness, spring like characteristics: nonlinear passive elasticity, physical flexibility, and light weight [19]. Like biological muscles, they are pull-only devices and should be used in antagonistic pairs to give better control of the actuation. Using an antagonistic pair provides control of the actuator stiffness allowing a continuum of positions and independent compliances. Like a human muscle, stiffness can be increased without change in the angle at the joint, giving an actuator control of both its stiffness and compliance [6].

PAMs, which are only one membrane, are extremely light compared to other actuators. Their power-to-weight ratio of 1 kW/kg was observed. They have easily adjustable compliance depending on the gas compressibility and varying force of displacement. PAMs can be directly mounted onto robot joints without any gears, eliminating inertia or backlash. They are easy to operate without such hazards as electric shock, fire, explosion and pollution.

The design of PAMs dates back as far as 1929, but, due to their complex design and poor reliability, they did not attract the attention of the research community. One of the most commonly used PAMs is the McKibben muscle (Figure 1.2, [20]) also called braided PAM (BPAM) due to its design and assembly. The muscle consists of a gas-tight bladder or tube with a double helically braided sleeve around it. The change in the braid angle varies the length, diameter, and volume of the sleeve. BPAMs have been widely used for orthotic applications because their length–load characteristics are similar to those biological muscles, but, due to the lack of availability of pneumatic power storage systems and poor valve technology, the interest in McKibben muscles has slowly faded in the scientific community. The Bridgestone Co. in Japan reintroduced the BPAMs for industrial robotic applications such as the soft arm, and Festo AG introduced an improved variant of PAM.

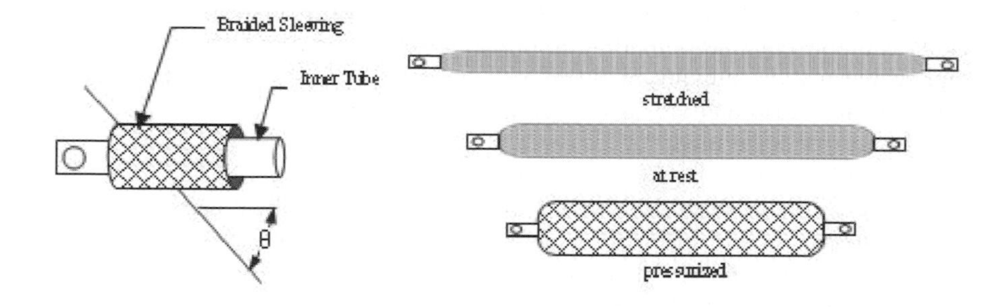


**Figure 1.2.** Braided muscle or McKibben muscle

Most of the PAMs used are in anthropomorphic robots, but various weak points exist in the design of braided muscle. They show considerable hysteresis due to the friction between the braid and shell, causing an adverse effect on the behavior of actuator, and a complex model is needed to determine the characteristics. PAMs generate low force and need an initial threshold pressure to generate actuation. They are plagued by low cycle life, but their generated force, threshold pressure and life cycle are dependent on material selection. The wires in the sleeve also snap from the ends during actuation, and they have limited actuation capacity (20 to 30%). A new design of PAM called netted Muscle (ROMAC) was designed to have better contraction and force characteristics with little friction and material deformation, but they have complex designs [18].

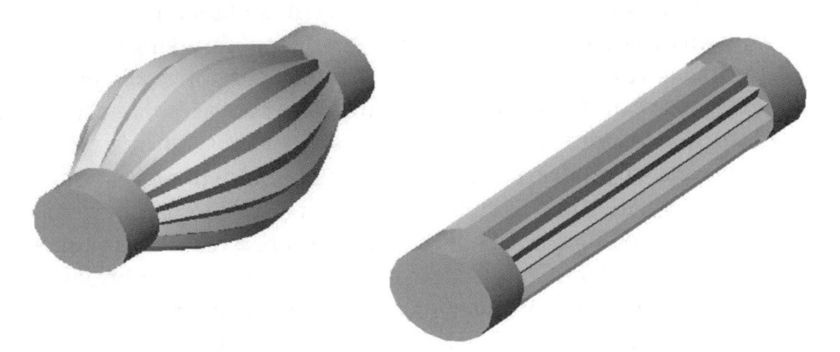

**Figure 1.3.** Schematic of a pleated, pneumatic artificial muscle in a stretched and inflated state

Another new PAM called pleated PAM (PPAM) (Figure 1.3) has a membrane rearrangement. The membrane is folded along its central axis to form an accordion bellows that unfurls during the inflation of the membrane. The membrane is made of a highly tensile, flexible material. Both ends of the membrane are tightly locked to the fittings. This design eliminates friction and hysteresis because the folded faces are laid out radially so the unfolding of the membrane needs no energy, giving a higher force output. PPAMs were found to be strong, operating with a large stroke and virtually no friction. They are very light in weight; a 60 g actuator pulls a 3500 N load and are easy to control when providing accurate positioning. PPAMs provide safe machine-man interaction. By using the right material, the material deformation can be eliminated while getting high tensile forces. Depending on the number of pleats, a uniform membrane loading can be obtained. As the number of pleats increases, a more uniform loading can be obtained. PPAMs need low threshold pressure to give high values of maximum pressure output. A maximum contraction of 45% was obtained that depended on the slenderness of the material [18].

A short actuation response time can be obtained to improve the flexibility of the actuator by employing high flow rate valves. This will occur through the development of a better closed-loop controller. These valves will be large and heavy and need high control energy which leads to a decrease in the energy efficiency of the whole system [19]. The diameter of the usable, transferable tubing is limited by the increase in gas viscosity, which increases the diameter. The flexibility is also compromised by large diameter tubing, and the efficiency of the system depends on the gas sources. Gas can be obtained from a reservoir or compressor motor or engine, or from a low-pressure reservoir with a heating chamber. The use of a compressor with a motor or engine will decrease the energy efficiency of the system and make it heavy and noisy. Using a heat chamber with a gas reservoir will enable higher efficiency as the heat energy is directly converted to mechanical energy [19].

It was found that the static characteristics of actuators are very similar to those of biological muscles, but actuators have a narrow, dynamic range. Actuators can be improved by employing lubricants to decrease the coulomb friction and viscous material is used to increase the viscous friction. One of the positive aspects of actuators is their high tension intensity compared to biological muscles. Their passive elastic characteristics can be improved using parallel and serial elastic elements. The pneumatic system used to drive the actuator needs more work to improve the efficiency of the whole system, and a lighter valve that can give a high flow rate needs to be designed. A light, quiet gas source with reasonable energy efficiency is needed, and to solve the tubing length and wrapping problem, better integration of tubing needs to be developed [19].

One of the main limitations of BPAMs for practical applications is short fatigue life (~10,000 cycles). Festo Corporation built a fluidic muscle to have a longer fatigue life by impregnating the fiber mesh into an expandable bladder [6]. The bladder, made from natural latex, was found to have 24 times more life than a synthetic silicone rubber bladder [2].

McKibben muscles have attractive properties for the development of mobile robots and prosthetic applications [21]. Most of the models used to predict the characteristics of McKibben muscles are concentrated on the effect of the braided sheath, but introducing the properties of the bladder into the design gave improved prediction of properties such as output force. A mathematical model is needed to understand the design parameters and improve desirable properties such as output force and input pressure, while minimizing undesirable properties such as fatigue properties. By coupling the effect of the properties of the braid and bladder, the performance prediction of the actuator was improved. Still, some discrepancies observed between the predictions of the model and the experimental results are believed to be due to mechanisms of elastic energy storage, the effects of friction between the bladder and braid, and friction between the fibers of the braid. The above effects are believed to be functions of the properties of braid and bladder, the actuation pressure, and the instantaneous actuator length [21].

A cockroach like robot with reasonable forward locomotion was built using only a feed-forward controller without any feedback circuit. The passive properties of BPAMs compensate for controller instabilities, acting as filters in response to perturbations, without the need for intervention of a controller. The speeds of BPAMs are higher when compared to biological muscles which are inherently slow because of neurological inputs [6].

### 1.2.4 Light Activated Polymers

The phenomenon of dimensional change in polyelectrolyte gels, due to chemically induced ionization, is explained by mechanochemistry. The deformation of polyelectrolyte gels produced by light-induced ionization was observed and labeled as the mechanophotochemicaleffect [22]. Observed irradiation with ultraviolet light caused the gel to swell by initiating an ionization reaction, developing an internal osmotic pressure. The gel collapsed when the light was removed and switched to its neutral state. The phase transition observed was slow due to the slow photochemical ionization and subsequent recombination of ions [23]. Phase transition due to visible light was observed later so harmful ultraviolet rays could then be eliminated when performing a phase transition.

Poly(p-*N,N*-dimethylamine)-*N*-gamma-*D*-glutamanilide) produces a dilation of 35% in each dimension, when exposed to light [22]. When irradiated for 10 minutes, poly(methylacrylate acid) gels buffered with *cis-trans* photoisomerizable (p-phenylazophenyl)trimethylammonium iodide dye produced a 10% elongation. The physical properties governing the deformation are (1) high polymeric amorphous or crystalline structures; (2) distinguishing features of porous, cross-linked gel matrices; and (3) suitable combinations of ionizeable groups. While (1) and (2) cannot be manipulated, the deformation properties of the gels can be controlled through (3). The deformation produced is independent of the stimuli used for ionization. The main demand on photoionization is that the charged species produced should have a sufficiently long life span to induce deformation; therefore, a suitable photoionization technique should be used.

A high-intensity light source is needed to produce meaningful concentrations of ions [22]. The observed transition was due to direct heating by the radiation, giving fast response. Gels were made from *N*-isopropylacrylamide with a light-sensitive chromophore and trisodium salt of copper chlorophyllin, and a 100-micrometer inner diameter capillary was used to form the gels. The phase transition experiments were carried out in a glass chamber where the temperature can be controlled within a ±0.1°C range. Argon laser radiation with a 488 nm wavelength was used, and the light intensity varied from 0–150 mW. The incident beam had a Gaussian diameter of ~7 mm and focused diameter of ~20 μm, using a lens with a 19 cm focal length. At a temperature of ~35°C, the gels gave a sharp, but continuous, volume change without any radiation. The transition temperature decreased as the intensity of light radiation increased. A more pronounced volume change was observed at a temperature of 33°C when a 60 mW light was applied, and discontinuous volume transition was observed with 120 mW of radiation. The light-sensitive gels collapsed when radiation in the visible wavelength was used (Figure 1.4 [22]). Shrinkage was observed throughout the whole temperature range, but the largest effect was observed at a transition region. A discontinuous transition was observed at an appropriate "bulk" equilibrium temperature, when the intensity was varied from 0–150 mW [22]. The light intensity at the transition state varied from gel to gel, believed to be due to the variation in the ratios of gel and beam diameters or bleaching conditions [23].

The effect of irradiation was observed to transform continuous transition to discontinuous transition and decrease the transition temperature. The chromophores incorporated in the gel absorb light energy and dissipate heat locally by causing radiationless transitions, increasing the local temperature of the polymer. The temperature increase in the gel, due to radiation, is proportional to the light intensity and chromophore concentration [23]. The rate of observed deformation was dependent on the intensity of the light source and was found to be due to dilation instead of phase transition induced by photoionization. A 5% cross-linked polymer was too stiff to produce photo-deformation, but deformation was observed with 1.5% cross-linking. Potential applications envisaged include printing, photocopying and actinometry [22]. It was observed that the phase transitions were due to the radiation forces instead of local heating, as observed previously. A direct influence on the balance of forces was caused when a gel was irradiated with a laser beam and became shrinkage in the gel. The shear relaxation process induced gel shrinkage of several 10s of microns [24].

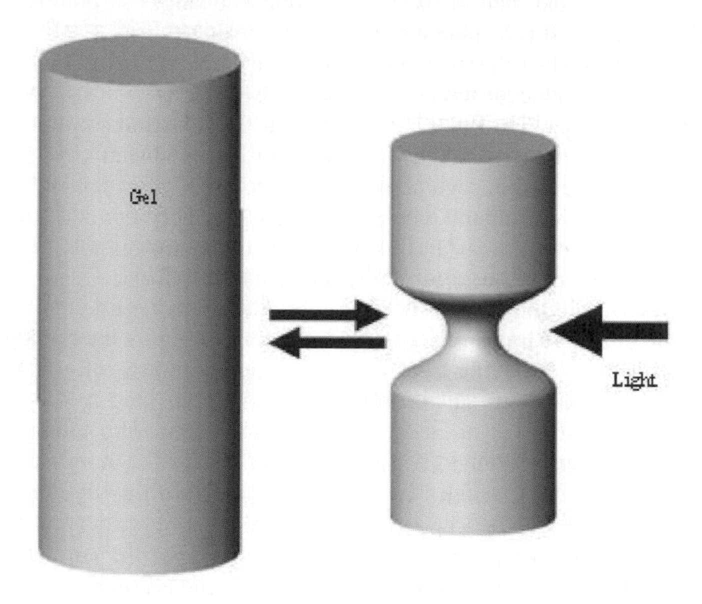


**Figure 1.4.** Cartoon showing the collapse of a light-activated gel under illumination

The combination of stimuli-responsive polymer gels and laser lights enables the development of a new gel-based system for actuation and sensing applications. It is known that radiation force immobilizes particles against Brownian motion and any convection [24]. These photoresponsive gels are used in such applications as artificial muscles, switches, and memory devices [23].

Azobenzene polymers and oligomers show surface relief features, when irradiated with polarized laser light. An atomic force microscope investigation of the amplitude mask irradiation of side-chain azobenzene polymers showed trenches and peaks, depending on the architecture of the polymer. Mass was transferred long distances, enabling the development of nanostructure replication technology. This technology, using polarized light, allows the storage of microscopic images as topographic features on produced polymer surfaces [25].

Extensive research is being conducted to discover other polymeric materials that change volume due to light exposure. These polymers are considered to be made of "jump molecules"—molecules that change in volume due to light exposure. Experiments have revealed that the volume change is not due to the heating of the water of hydration in the gel; instead, it is considered due to the contraction obtained by the attraction between the excited molecules in the illuminated region and surrounding molecules. Therefore, shrinkage is due to laser-induced phase transitions [2].

### 1.2.5 Magnetically Activated Polymers

Sensitive polymeric materials showing strain due to changes in the magnetic field are called magnetoelastic or magnetostrictive polymer materials, also often called ferrogels. The gradient of the magnetic field applied acts as the driving force [26]. A magnetic field induces forces on all kinds of materials; solid materials experience more forces than fluids. By combining fluidlike and solidlike properties in a material, the effect of magnetic force can be enhanced [3]. A magneto-controlled medium can be considered a specific type of filler-loaded swollen network. Ferrogels are a chemically cross-linked polymer network, swollen by a ferrofluid, which is a colloidal dispersion of monodomain, magnetic particles. In these gels, the magnetic particles are attached to the polymer chains by strong adhesive forces [26]. Under a uniform magnetic field, no net forces are observed on the gel, except the Einstein-de Haas effect which is caused by a change in the magnetic field vector. When these gels are subjected to a magnetic field gradient, the particles experience a net force toward the higher magnetic field. These particles carry the dispersing fluid and polymer network with them, producing a macroscopic deformation of the gel. Elongation, contraction, bending, and rotation can be obtained depending on the geometric arrangement of these materials. With their ability to create a wide range of smooth motions along with quick operation and precise controllability, these magnetic fields controlling soft and wet gels show good promise in the development of stimuli-responsive gels and actuators [26].

Electric and magnetic field-induced shape and movement was obtained in a polymer gel with a complex fluid as the swelling agent. Magnetic particles were incorporated into poly(*N*-isopropylacrylamide) and poly(vinyl alcohol) gel beads. The beads aligned as a chainlike structure in uniform magnetic field lines, and they aggregated in a nonuniform field due to magnetophoretic force. These magnetic gels give quick and controllable changes in shape, which can be exploited in applications mimicking muscular contraction [3]. The use of polymer gels as actuators creates a quick and reliable control system, and the use of electric or magnetic stimuli facilitates the development of these control systems.

A PVA gel, with magnetic nanoparticles, contracted in a nonuniform magnetic field (Figure 1.5 [26]), which is smaller than the field strength observed on the surface of common permanent magnets. By coordinating and controlling the magnetic field, muscle-like motion can be obtained, leading to the development of artificial muscles [3]. To better exploit these materials, the basic relationship between the magnetic and elastic properties of these materials should be investigated. The applied magnetic field on the gel can be better controlled using an electromagnet, where the current intensity gives the controllability. The relationship between deformation and current intensity needs to be determined for the efficient use of electromagnets [26].

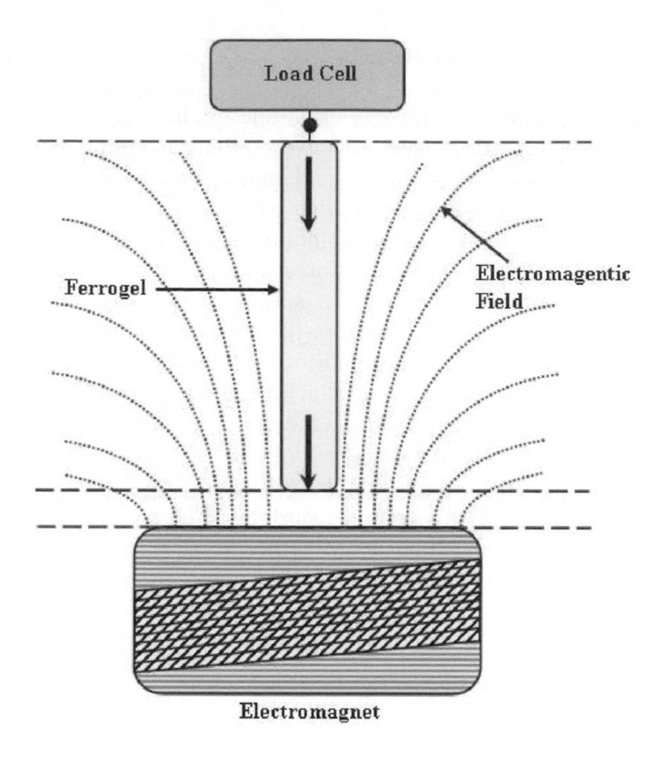


**Figure 1.5.** A schematic representation of the setup used to study the magnetoelastic properties of ferrogels

In a ferrogel, magnetic particles are under constant, random agitation when not under a magnetic field. Due to this random agitation, there is no net magnetic field in the material. It was observed that the magnetization of the ferrogel is directly proportional to the concentration of the magnetic particles and their saturation magnetization. In small fields, it was determined that the magnetization is linearly dependent on the field intensity, whereas in high fields, saturation magnetization was achieved [26].

For a ferrogel suspended along the axis of the electromagnet, the elongation induced by a nonuniform magnetic field depends on a steady current flow. A very small hysteresis was observed. It was determined that the modulus of the ferrogel is independent of the field strength and the field gradient. The relationship between elongation and current intensity found was a function of cross-linking densities as well. For small uni-axial strains, the elongation produced is directly proportional to the square of the current intensity [26]. The response time is only one-tenth of a second and observed to be independent of particle size. Ferrogels are generally incompressible and do not change in volume during activation [2]. Voltairas *et al.* [27] developed a theoretical model, in constitutive equations, to study large deformations in ferrogels when the hysteresis effect was not considered. This model can be used for quantitative interpretation of the magnetic field's dependent deformation of ferrogels for valve operations [27].

Through induction, magnetically heated, triggerable gels have been developed, where the heat generated from various loss mechanisms in the gel produces a thermal phase transition. The loss mechanisms include ohmic heating from eddy current losses, hysteresis losses, and mechanical (frictional) losses. Volume change was observed in these materials when a quasi-static (frequency of 240 kHz to 3 MHz) magnetic field was applied. When the field is removed, the gel returned to its initial shape, due to cooling of the material. Power electronic drives are being developed which will aid in the development of closed-loop servomechanisms for actuators. These materials show the potential in contact-less actuation and deformation wherever the magnetic field can reach, *e.g.* triggering gels under the skin [28, 29, 32].

MR rubber materials are being used in the development of adaptively tuned vibration absorbers, stiffness-tunable mounts and suspensions, and automotive bushings. These materials usually show continuously controllable and reversible rheological properties while under an applied magnetic field [30].

Magnetic polymers, with magnetic particles dispersed in a rubber matrix, have been used in magnetic tapes and magnetic gums for more than three decades [31].

### 1.2.6 Thermally Activated Gels

Thermally activated gels produce a volume change due to thermal phase transitions, usually within a temperature range of 20°C to 40°C. These polymers exhibit a contractile force of 100 kPa with a response time of 20–90 seconds [2]. Most of the studies on thermal phase transitions of gels were done on N-substituted polyacrylamide derivatives. Hirokawa and Tanaka (1984) first reported the volume phase transition of poly(*N*-isopropylacrylamide) (PNIPAAm) gel [69].

Poly(vinyl methyl ether) (PVME) is one of the most widely used thermo-responsive polymers. It undergoes phase transition at 38°C; at a temperature below the phase-transition temperature, PVME is completely soluble in water. The polymer precipitates with an increase in the temperature, and the polymer network is transformed from a hydrophilic to a hydrophobic structure. When a gel was employed, the transition produced a volume change. PVME can be cross-linked into a hydrogel by gamma-ray radiation. High-energy radiation is the one of the most widely used methods to make cross-linked polymer hydrogels. With an increase in the temperature, water is expelled from the gel network, causing it to shrink. The volume phase transition, induced by temperature change, can be exploited in the development of thermoresponsive soft actuators, thermo-responsive separation, *etc*. [33].

The deformation characteristics of a thermoresponsive hydrogel can be controlled by incorporating surfactants, or ionic groups, into a polymer network. The deformation properties of the hydrogel vary depending on the type and concentration of the surfactant or ionic groups. Quick, responsive thermo-responsive hydrogels are being developed using porous PVME gels, which swell and shrink much faster than homogeneous gels. A 1 cm cube of PVME porous gel showed a response time of 20–90 seconds, with a change in temperature from 10–40°C, where as a homogenous gel showed no response within the same time period. PVME porous gels show potential in the development of practical actuating devices due to this rapid temperature change [33].

Thermally sensitive polymer gels show great potential in the development of artificial muscles. Hot and cold water can be used for actuation, a favorable option compared to acid and base in chemically activated polymer gels. As the temperature increases, the swelling ratio of the PVME gel fiber decreases; this reaction increases as the temperature nears the transition point. A contractile force of 100 kPa was generated when the temperature was raised from 20 to 40°C [33].

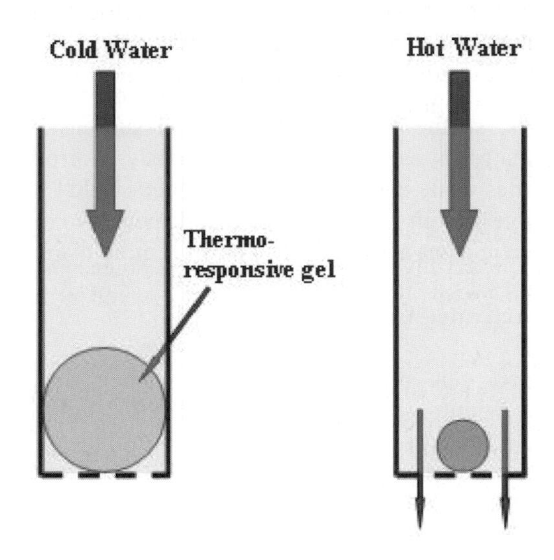


**Figure 1.6.** Automatic gel valve made of a thermoresponsive gel, which allows only hot water through the pipe

Thermoresponsive polymer gels are being studied for different applications. Modified NIPAAm gels are being developed for metered drug release by thermally controlling drug permeation. Gels can be used as a substrate for the immobilization of enzymes. In thermoresponsive gels, the activity of the immobilized enzyme was controlled by thermal cycling. Artificial finger and gel valve models were also developed using thermoresponsive polymer gels. The gel valve shrinks to allow only hot water while blocking the flow of cold water [33]. The solid-phase transition of a polymer was also used in the development of paraffin-based microactuators. Although large thermal expansion at the solid–liquid phase transition is a general property of long-chained polymers, the low transition temperature of paraffin was exploited in these actuators, using micromachining techniques which allow the production of many actuators on the same die. A deflection of 2.7 micrometers was obtained using a 200–400 micrometer radius device with a response time in the range from 30–50 milliseconds [34].

Thermally activated microscale valves are being developed for lab-in-a-chip applications. These valves will open and close due to a temperature-change induced phase transition (Figure 1.6). The valves also provide an advantage in production using lithographic techniques; noncontact actuation, which employs heating elements; or using heat from the fluid itself [35].

## 1.3 Electroactive Polymers

As stated earlier, since the last decade there has been a fast growing interest in electroactive polymers. The non-contact stimulation capability, coupled with the availability of better control systems that can use electrical energy, is driving the quest for the development of a wide range of active polymers. These polymers are popularly called electroactive polymers (EAPs), and an overview of various types of EAPs is given in the following sections.

### 1.3.1 Electronic EAPs

Based on the mechanism of actuation, EAPs are classified into electronic and ionic EAPs. Various characteristics of electronic EAPs have been discussed in previous paragraphs, but an overview of electronic EAPs is covered in this section.

*1.3.1.1 Ferroelectric Polymers*

Ferroelectric materials are analogous to ferromagnets, where the application of an electric field aligns polarized domains in the material. Permanent polarization exists even after the removal of the field, and the curie temperature in ferroelectric materials, similar to ferromagnetic materials, disrupts the permanent polarization through thermal energy [36].

Poly(vinylidene fluoride-trifluoroethylene) (P(VDF-TrFE)) is commonly used ferroelectric polymer. Local dipoles are created on the polymer backbone due to the high electronegativity of fluorine atoms. Polarized domains are generated by these local dipoles aligning in an electric field. The alignment is retained even after the removal of electric field, and the reversible, conformational changes produced by this realignment are used for actuation [36].

The polymers have a Young's modulus of nearly 1–10 GPa, which allows high mechanical energy density to be obtained. Up to 2% electrostatic strains were obtained with the application of a large electric field (~200 MV/m) which is nearly equal to the dielectric breakdown field of the material [2]. Up to a 10% strain was observed in ferroelectric polymers during the transition from the ferroelectric phase to the paraelectric phase, but the presence of hysteresis is a drawback. Hysteresis in ferroelectric materials is due to the energy barrier present when switching from one polarization direction to the other or when transforming from one phase to another [37]. A large field, in a direction opposite to the initial field, is required to reverse the polarization, dissipating substantial energy [36].

The energy barrier can be significantly reduced by decreasing the size of the coherent polarization regions to the nanoscale. This reduction is achieved by introducing defects in the polymer chains, which are created by electron radiation. Proper high electron irradiation eliminated the large hysteresis, and exceptionally large electrostatic strain was achieved. It is crucial to note that effective structures, induced by electron irradiation, cannot be recovered by applying a high electric field. For soft material, Maxwell stress can generate high strains. Ferroelectric polymers show better performance in strain and strain energy density compared to traditional piezoceramic and magnetostrictive materials [37]. Ferroelectric relaxors are practical, useful materials which show strong performance characteristics. When the Curie point in these materials is brought near to room temperature–the normal operating temperature–a nonpolar, paraelectric phase is present. This is achieved by introducing imperfections in the structure either by using radiation or incorporating a disruptive monomer along the chain [36]. These imperfections break the long-range correlation between the polar groups. Polarization is induced when an electric field is applied to these materials, but, due to the decrease in the energy barrier to the phase change, the hysteresis is reduced or eliminated [36]. The large molecular conformational changes (introduced) associated with the ferroelectric-to-paraelectric transition lead to macroscopic deformations that are used to generate actuation [36].

P(VDF-TrFE) contracts in a direction of the field and expands in the direction perpendicular to the field. The strain can be enlarged by prestraining, and moderate strains (up to 7%), with high stresses (reaching 45 MPa) have been achieved. High stiffness (70.4 GPa) was achieved but was dependent on the density of imperfections and a large work per cycle (approaching 1 $MJ.m^{-3}$) [36].

Ferroelectric polymers are easy to process, cheap, lightweight, and conform to complicated shapes and surfaces, but the low strain level and low strain energy limit the practical applications of these polymers [37]. Ferroelectric polymers can be easily patterned for integrated electronic applications. They adhere to wide variety of substrates, but they are vulnerable to chemical, thermal, and mechanical effects [38]. Ferroelectric EAPs can be operated in air, a vacuum, or water in a wide range of temperatures [2].

Limitations of ferroelectric polymers include fatigue of the electrodes, high electric fields, and high heat dissipation. Procurement of the fluorocarbons is also a problem due to environmental restrictions, and the *e*-beam irradiation process is expensive. The maximum strain of the polymers can be achieved only at an optimal loading condition that is dependent on the material used. This strain can decrease substantially above and below the optimal value [36].

The potential use of ferroelectric polymers can be extended by decreasing their operating potential. This can be achieved by using thin films (100 nm) or by increasing the dielectric constant. The film thickness is limited by the relative stiffness of the electrode material but can be overcome by using more compliant electrodes. The dielectric constant can be increased by adding high dielectric constant filler material. The operating temperature depends on the density of imperfections, which can be fine-tuned up or down to change the temperature range of operation. The typical range is between 20 and 80°C [36]. Instead of electrostatic energy, heat can also be used to activate ferroelectric polymers. Reversible actuation can be obtained when the materials are heated and cooled above and below their Curie points, which is just below room temperature [36].

#### *1.3.1.2 Dielectric Elastomers*

Dielectric elastomer actuators are made with an incompressible and highly deformable dielectric medium. When an electric field is applied across the parallel plates of a capacitor, the coulombic forces between the charges generate a stress, called the *Maxwell stress*, causing the electrodes to move closer. This movement squeezes the elastomer, causing an expansion in the lateral direction[39]. Dielectric elastomers are often called electrostatically stricted polymers (ESSP) actuators [2]. Figure 1.7 illustrates the operational mechanism of a dielectric elastomer with compliant electrodes. Dielectric elastomers show efficient coupling between electrical energy input and mechanical energy output [36]. Also, applying prestrain to dielectric elastomers can prevent the motion along an arbitrary direction and also introduce the motion to specific directions. It has also been observed that prestrain results in a higher breakdown potential of strains. These materials can be used as both actuators and sensors. With careful design, efficiencies as high as 30% can be obtained and be operated satisfactorily over large temperature ranges (e.g. silicone −100 to 250°C). Operation below the glass-transition temperature leads to the loss the of elastic characteristics of the material. Three commercially available materials are Dow Corning HS3 Silicone, Nusil CF 19-2186 Silicone, and 3M VHB 4910 acrylic. VHB is available in adhesive ribbons and silicones can be cast into thin films. The silicone surfaces are coated with conductive paint, grease, or powder to act as electrodes, and the typical voltages applied are in kilovolts (~10 kV) with currents in the range of less than several milliamperes [36]. Extensive theoretical and experimental studies have been done by de Rossi *et al.* [40] to characterize the effect of different electrodes and prestrain on the dielectric elastomers. The data presented help in the selection of the best electrode and prestrain values to obtain efficient response for different ranges of electric fields [40].

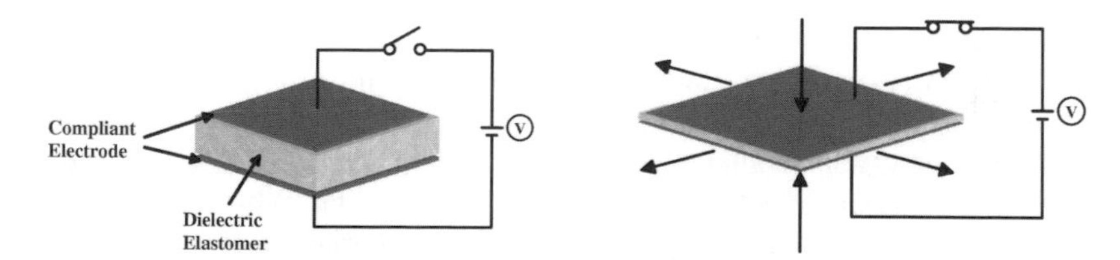


**Figure 1.7.** Operating principle of a dielectric elastomer

In general, the strain induced in a material is proportional to the square of the electric field and the dielectric constant. One of the ways to induce large strains is to increase the electric field, but the high electric fields involved in the actuation of dielectric elastomers can result in dielectric breakdown of the material. The strain can be increased using either a material with a high dielectric constant or films with low thicknesses. An electric breakdown field is defined as the maximum electric field that can be applied to dielectric elastomers without damaging them [41]. It was observed that the breakdown field increases with the prestrain of the elastomer. Dielectric elastomers require high electric fields for actuation (~100 V/μm), and it is a challenge to increase the breakdown strength of the elastomer at these fields. The small breakdown strength of air (2–3 V/μm) presents an additional challenge [36].

An actuator with three degrees of freedom (DOF) made of a dielectric elastomer, was developed recently. The structure has a wound helical spring with a dielectric elastomer sheet. The electrodes are patterned into four sections which can be connected to respective driving circuits. With this arrangement, the actuator can bend in two directions and also extend, giving it three degrees of freedom. Much larger deflections can be obtained from the above, and other envisaged applications include speakers (tweeters), pumps, and legged walking robots [36].

A newly designed lightweight, hyperredundant manipulator was developed which is driven by dielectric elastomers [41]; i.e. can produce precise and discrete motions without the need for sensing and feedback control. The manipulator showed great potential in the development of miniaturized actuators that have high DOFs; these binary robotic systems can have various applications from robotics to space applications. Dielectric elastomers are in the advanced stages of development for practical microrobots and musclelike applications, such as the biomimetic actuator developed by Choi *et al.* [42], which can provide compliance controllability [42].

The development of practical applications of dielectric elastomers requires the development of models for their design and control. The modeling of dielectric elastomers involves multiphysics, including electrostatic, mechanical, and material terms [43].

#### *1.3.1.3 Electrostrictive Graft Elastomers*

The electrostrictive graft elastomer is a new type of electroactive polymer developed in the NASA Langley Research Center in 1999 [44]. The graft elastomer consists of two components: flexible macromolecular backbone chains and crystallizable side chains attached to the backbone, called grafts (Figure 1.8(a)). The grafts on the backbone can crystallize to form physical cross-linking sites for a three-dimensional elastomer network and to generate electric field-responsive polar crystal domains (Figure 1.8(b)). The polar crystal domains are primary contributors to electrostromechanical functionality. When the materials is under an electrical field, the polar domains rotate to align in the field direction due to the driving force generated by the interaction between the net dipoles and the applied electric field. The rotation of grafts induces the reorientation of backbone chains, leading to deformational change and the polar domains randomize when the electric field is removed, leading to dimensional recovery. The dimensional change generated demonstrates quadradic dependence on the applied electric field as an electrostrictive material does [44].

From the experimental observations [44], it was noted that the negative strains were parallel to electric field and positive strains were perpendicular to the field. The same deformation was observed for a $180^{\circ}$ shift in the electric fields, and the direction of strains remained unchanged. The amount of strain is dictated by the electric field strength [44]. According to Wang *et al.* [45], the deformation of the graft elastomers can be described by considering two mechanisms: crystal unit rotation and reorientation of backbone chains. Crystal unit rotation draws the backbone chains toward themselves, causing an increase in the atomic density near the crystal units, that causes a negative strain. Local reorientation of backbone chains was considered to occur in three stages. In the first stage, a negative strain is generated in the direction parallel to the electric field and a positive strain perpendicular to the electric field. In the second stage, a positive strain is generated in both directions. In the third stage, negative strains will also be generated in both directions, due to the Maxwell stress effects [45].

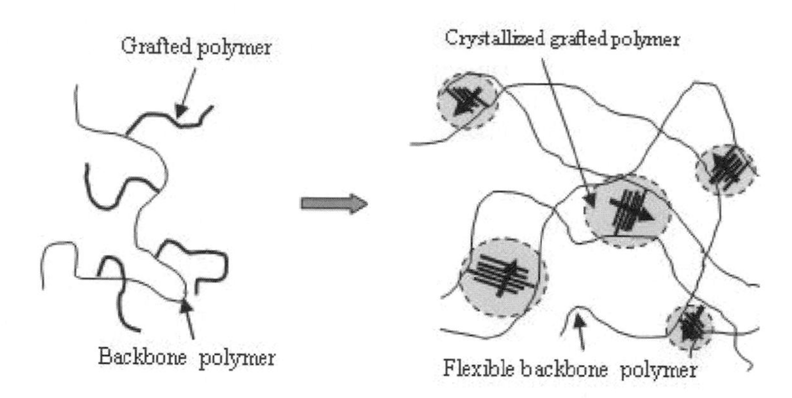


**Figure 1.8.** Schematic showing (a) molecular structure and (b) morphology of a grafted elastomer

One of the distinctive properties of graft polymers compared to other electrostrictive polymers is their high stiffness. Polyurethane has a modulus between 15 and 20 MPa, whereas modules of a graft elastomer are around 550 MPa, approximately thirty times more [44]. This property can be exploited in the development of an actuator that provides higher output power and mechanical energy density. Electrostrictive graft elastomers offer large electric-field-induced strains (4%) [44] and have several advantages such as good processability and electrical and mechanical toughness. Various bending actuators based on bilayers have been designed and fabricated. The sensitivity studies done by Wang *et al.* [45] showed that for a bilayer bending actuator, the curvature of the beam can be tailored by varying the thickness of the active layer. In this study, a 10% decrease in the thickness of active layer gave 30% more curvature in the beam.

An electrostrictive-piezoelectric multifunctional polymer blend was developed [44] that exhibits high piezoelectric strain and large electric-field induced strain responses. A material with the above combination can function both as an electrostrictive actuator and a piezoelectric sensor [72]. Various electrical, mechanical, and electromechanical properties of these elastomer-piezoelectric blend systems can be optimized by adjusting the composition, molecular design, and processing techniques [73].

#### *1.3.1.4 Electrostrictive Paper*

Paper, as an electrostrictive EAP (EAPap) actuator, was first demonstrated at Inha University, Korea [2]. The EAPap was made by bonding two silver-laminated papers with silver electrodes placed on the outside surfaces (Figure 1.9). A bending displacement was produced when an electrical field was applied to the electrodes. The performance of the actuator depends on the host paper, excitation voltage, frequency and type of adhesive used to bond thc papers. Fabrication of these lightweight actuators is quite simple [2].

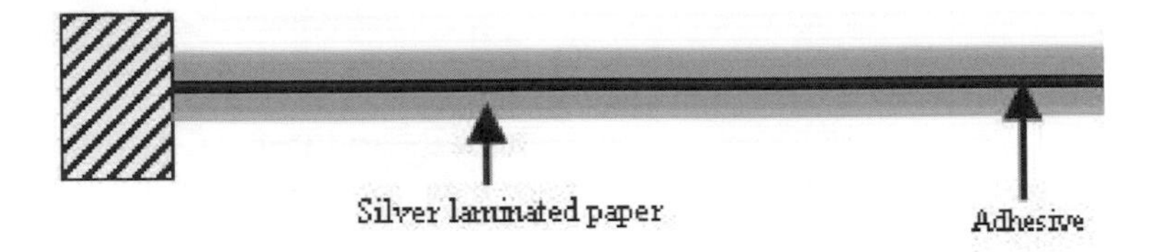


**Figure 1.9.** Schematic of the electrostrictive paper cantilever actuator

The successful development of a paper actuator for practical applications requires addressing various issues such as small displacement output, large excitation voltage, sensitivity to humidity, and performance degradation with time. In the initial studies, the electrostrictive effects were observed to be dependent on the adhesives used to make laminated layers. Different types of paper fibers such as softwood, hardwood, cellophane, and Korean traditional paper, all tested with various chemicals, were used to improve the bending performance of an EAPap actuator [46]. To eliminate the predominant effect of the electrodes, two different techniques were studied: the direct adhesion of aluminum foil and the gold-sputtering technique. It was determined, owing to the lower stiffness, that gold-sputtered electrodes gave better performance than aluminum foil electrodes. The paper with more cellulose, in an amorphous structure, gave a stronger response than the paper with crystalline cellulose. Cellophane gave a better response because of its amorphous cellulose with a low degree of polymerization. A combination of the piezoelectric effect and the ionic migration effect both associated with the dipole moment of the paper constituents is considered responsible for the strain observed in electrostrictive paper [46]. Although electrostriction may be an important mechanism of actuation, studies are needed to elucidate the fundamental physics of the actuation principle.

Various applications envisaged include active sound-absorbing materials, flexible speakers, and smart shape-control devices [2]. One of the unique applications being considered for the EAPap paper is an electronic acoustic tile, which broadcasts antinoise to cancel out sound or white noise in a room.

*1.3.1.5 Electroviscoelastic Elastomers*

Electroviscoelastic elastomers are the solid form of an electrorheological fluid (ER), which is a suspension of dielectric particles. When these ER fluids are subjected to an electric field, the induced dipole moments cause the particles to form chains in the directions of the field, forming complex anisotropic structures. During this process, the viscosity of the fluid increases greatly. An ER solid is obtained if the carrier in the ER fluid is polymerized. By careful selection, the carrier can be an elastomeric material and result in an electroviscoelastic elastomer. The ER elastomers have stable anisotropic arrangements of polarizable particles [2]. When an electric field is applied in the chain direction, these particles tend to move toward each other, creating stress, which causes deformation of the materials. Work can be obtained by opposing this deformation.

ER gels have unique advantages compared to ER fluids: no leakage, no sedimentation of particles, and ease of fabricating custom-made shapes and sizes. The key aspects of the structure of ER materials that are important for performance include the size and shape of particles, the dielectric properties, and the organization of the particles.

The numerical analysis of the ER response showed that it is strain dependent, and its response depends on the interparticle forces that increase as the particle spacing decreases [47]. It was observed that a polymer with low carrier density and mobility in an electric field would form an ideal matrix, and maximum ER responses can be observed when the particles align as a body-centered structure.

Metallic particles can be considered ideal due to their high polarizability, but their high density and conductivity precludes their use in ER fluids. This problem was eliminated in ER elastomers by maintaining enough gaps between particles in the elastomer to avoid a short circuit and by swelling the cured elastomer with curable silicone prepolymers [47]. The prepolymer acted as insulation between the particles to prevent shorts in the chain direction. The trapped particles in the swollen polymer act as isolated dipoles. The combined effect of these dipoles gives better performance.

Diluents were added to reduce viscosity and to increase swelling of the gel; the modulus of the gel doubled with an application of a 2 $kVmm^{-1}$ field, with only a 1% particle concentration. Potential applications include actuators, artificial muscles, smart skins and coating, displays, switches, valves, *etc.* [47].

#### *1.3.1.6 Liquid Crystal Elastomers*

Liquid crystal elastomers (LCE) can be activated by electrical energy applied through joule heating. LCEs are composite materials made of a monodomain nematic liquid crystal elastomer with conductive polymers distributed in the network structure [2]. In LCEs, actuation is produced through the stresses generated by the order change and alignment of liquid-crystalline side chains. These alignment changes are due to the phase changes induced by thermal or electrostatic energy [36]. Usually, flow induced in liquid crystals by stress fields prevents the build up of static forces. In liquid crystal elastomers, the liquid crystal molecules are bonded to cross-linked polymer backbones. This flexible polymer backbone allows the polymer chains to reorient themselves but prevents the flow of molecules leading to the build up of static forces that produce stresses and strains. These stresses are instead transferred, via the polymer backbone, to do mechanical work. This reorientation of the mesogens can be induced by temperature changes or by the application of an electric field [36]. The response times are typically less than a second, but the relaxation process is slower, in the range of 10 seconds. Cooling is needed to expand the material to its original dimensions [2].

In a thermally driven system, the rate-limiting factor is the heat transfer time constant that is governed by the thermal diffusivity of the base material. It is expected that the rate is proportional to the inverse of the square of thickness of the flat films. A time constant of 0.25 and 0.5 seconds was observed for a 100 μm thick flat film with heat transferred on one side, where as a time constant $< 0.125$

seconds was observed when heat was transferred from both sides. A time constant in microseconds may be obtained by using thin films (~1 μm) [36]. Efforts are underway to decrease the nematic isotropic transition temperature to below sub-ambient to obtain an effective elastic response [2].

The mesogen units in ferroelectric liquid crystals have intrinsic polarization or an anisotropic dielectric constant. These LCEs contract or expand in an electric field due to the reorientation of the mesogen units which induce bulk stresses and strains in the backbone. A better response can be obtained in electrostatically driven films because the electric field can be applied very quickly through the bulk of the material [36].

Factors influencing the response time include specific mesogens, the structure of the polymer backbone, and the degree of cross-linking. Faster responses can be obtained with smaller mesogens that have less cross-linked matrices [36]. Fast response time (~10ms) and large strains (~45%) were obtained from electrostatically driven and thermally driven LCEs, respectively. The applied electric fields are lower (1.5 to 25 MV/m) compared to those used in ferroelectrics and dielectric elastomers (approx 100 MV/m). Faster responses can also be obtained by irradiative heating of a thermally driven LCE film, but the cooling is still by conduction which is relatively slow in thick samples [36]. A reduction in response time was also demonstrated by employing a photoabsorption mechanism without the least effect on mechanical properties. The LCE films are coated with a thin layer of carbon on the surfaces to absorb more light energy and increase the absorption of heat by radiation [48]. Generally, the LCE samples consisted of 50–200 μm thick monodomain films, where the measured contraction stress, strain, and frequency gave values of 210 kPa, 45% and subsecond relaxation time, respectively. Optimization of the mechanical properties of these materials can be achieved through the effective selection of the liquid crystalline phase, the density of cross-linking, the flexibility of the polymer backbone, the coupling between the backbone and the liquid crystal group, and external stimuli [2].

LCEs are in the early stages of development, but it is known that due to the low stiffness and tensile strength of these materials, a relatively small change in load can induce large strains. Laser heating can be used to overcome the limitations caused by heat transfer during thermal actuation but at the cost of reduced efficiency [36]. LCEs are being studied to develop membranes that can separate right-handed and left-handed forms of drugs in the pharmaceutical industry [49]. Along with applications in artificial muscles and actuators, the elastomers can also be developed to be mechanically tunable optical elements. A swimming motion was obtained when a light was shone on a dye-doped LCE sample floating on water [50].

LCEs exhibit piezoelectric behavior: when the liquid crystals are in a smectic phase, the mesogens arrange themselves in a distinctive layered structure. The mesogens within a layer form a smectic C phase by tilting to one side at a constant tilt angle, and if the mesogen has a chiral center near the core, the chirality breaks symmetry in the unit cells, creating a degree of polarity without an external field. This material exhibits ferroelectric behavior in response to spontaneous polarization. The production of piezoelectric sensors and actuators from LCEs is easy because they do not require poling [2].

### 1.3.2 Ionic-EAPs

The following paragraphs give a brief overview of the various "ionic" electroactive polymers being developed.

*1.3.2.1 Ionic Polymer Gels*

As stated in the chemically activated polymer section, pH activated polymers such as PAN hold tremendous promise in actuator technology. However, pH changes using chemical solutions typically cause deformation of gels and are somewhat inconvenient. The formation of salts is inevitable due to the chemical reaction between the acid solution and the basic solution. The salts created may be attached to the polymer surface and block contact between the polymer chain and the protonated hydrogen ion environment, affecting the response time. The chemical actuation of synthetic gels is undesirable except in some underwater applications, and the development of the electrically driven system makes it a potential material in robotic applications. Control of the electrically driven system is easy, and the electrochemistry of the system comes into to play when electrical actuation is considered.

Electrical actuation of polymeric gels was first studied by Tanaka *et al.* [51] on polyacrylic acid gels. The gel changed shape and size when placed between electrodes that were surrounded by an aqueous solution. The gel shrank near the anode when it touched the electrode, but when the gel was not touching an electrode, it swelled near the anode. The phenomenon can be reversed by reversing the polarity of the electric field. The electrolysis of water is considered to affects the pH of the solution near the electrodes and can be superimposed on the electrical response of the gel actuators. Due to electrolysis, the region near the cathode (a negative electrode) will become more basic, as $OH^-$ is released. Similarly, the region near the anode will become more acidic. Due to this phenomenon, when the gel system is placed near the anode, it will contract. The bending of an acrylic acid-acrylamide copolymer gel was observed by Shiga and Kurauchi [52] when the gel was placed between the electric field in aqueous solution.

Note that the response of gels is not a material property but rather is dependent on the dynamics of the electrolytic reactions of the system, the geometry of the system, the composition of the electrolyte and electrodes, and the previous history of the gel. A detailed study of the relationship between the electric field and the volume change mechanism of polymer gels is needed to determine the possibility of building a practical mechanical system with an electrically induced polyelectrolyte actuator and to better control an electrochemically driven gel actuator that is free from undesired chemical reactions. Poor mechanical properties of ionic gels are a major constraint in the development of practical applications.

Electrical activation of PAN fiber bundles has been studied by Schreyer *et al.* [53]. When the PAN muscle is near the anode side, the muscle shrinks, and when in the cathode vicinity, the PAN fiber is elongated. The elongation and contraction is done simply by changing the polarity. Furthermore, using a high-conductivity material–such as graphite or platinum coating–the PAN fiber itself can be used as an electrode. The advantage of this method is that the control motion of the PAN actuator is quite simple; the disadvantage is that the conductive material on the

PAN surface is delaminated after three to four cycles. The actuation is completely stopped after the conductive material is wiped out [13].

For electrochemically driven PAN filaments, it was observed that the diameter changes in PAN filaments are quite similar to the pH activated ones, but with a longer response time (approximately ten minutes). Single strands of fibers produced an approximate force of 10 $gm_f$ for both aforementioned activation methods and have similar standard deviation ranges. Force generation reached a steady state within a few seconds of the chemical activation system. On the other hand, it took approximately 10 minutes for the force generation to reach a steady state in the electrochemically-driven system [13]. Efforts are currently underway to improve the response characteristics utilizing submicron-diameter electrospun PAN fibers. Actuation of ionic polymer gels is slow due to the required diffusion of ions, and the large displacement produced by electrodes deposited on the gel surface causing damage. Better performance can be obtained by using thin layers and robust electroding techniques [2]. Figure 1.10 shows an electochemically driven PAN crane system [70].

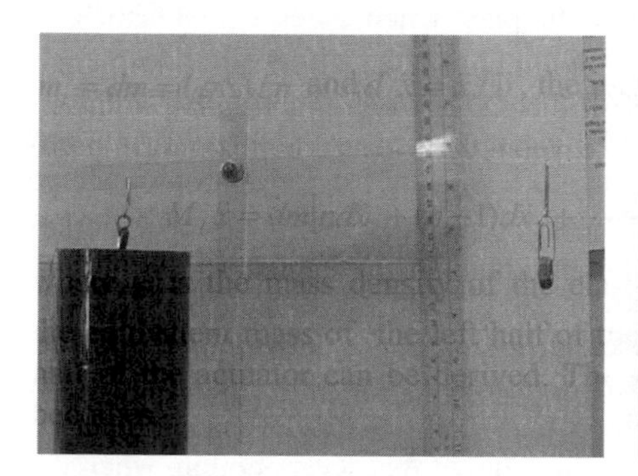
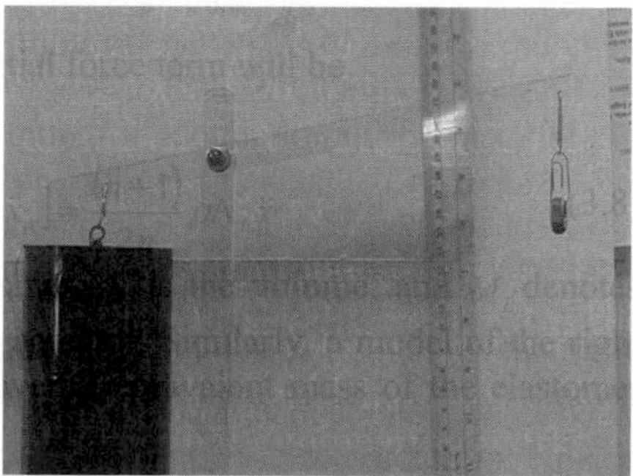

**Figure 1.10.** Working of PAN crane with electrochemically driven actuation system (left) before and (right) after 20minutes [70]

### *1.3.2.2 Ionic Polymer-Metal Composite (IPMC)*

Ionic polymer-metal composites have been studied extensively in the past 15 years. Oguro *et al.* [54] initially determined that the composite of a polyelectrolyte membrane-electrode, which is a perfluorinated, sulfonate membrane (Nafion® 117) chemically coated with platinum electrodes on both sides of the membrane, deforms and bends when a low voltage (~1–5V) is applied across the electrodes in an aqueous solution [55]. A change in the ion concentration, induced by an electric field, attracts water and causes deflection toward one of the metal electrodes. Swelling occurs on one side and shrinkage on the other side due to the nonuniform distribution of water in the polymer electrolyte network [36].

An ion exchange membrane (IEM) forms the bulk of the material of the IPMC. IEMs are permeable to cations but impermeable to anions because of the unique ionic nature of the fixed perfluorinated polymer backbone. Different membranes can be used to make IPMCs, but the most commonly used are Nafion™ from DuPont, USA and Flemion from Asahi, Japan. The primary applications of these polymers are in fuel cells used in hydrolysis. Nafion can exchange $H^+$ with other cations. The top and bottom electrode surfaces are formed when the metal ions are reduced by exchanging $H^+$ ions with metal ions. These two metallic surfaces have good conductivity and can be manipulated by the chemical process. IPMC's cross section resembles a sandwich with electrode layers outside and a polymer matrix in the center [55].

An ionic polymer consists of a fixed network with negative charges, balanced by mobile positive ions. The polymer network consists of pockets of solvents, and two, thin boundary layers are generated by the application of anelectric field. A cation-poor layer forms on the anode side while a cation-rich layer forms on the cathode side. Due to the accumulation of cations on the cathode side, water molecules move to this side and cause hydrophilic expansion. The stresses in the polymer matrix cause bending toward the anode. With time, the back diffusion of water molecules causes a slow relaxation toward the cathode. The degree of actuation obtained is a function of the type of polymer used, the type of counter-ion, the amount of water, the quality of metallization, and the thickness and surface area of the polymer membrane [36]. When a voltage higher than the electrolysis voltage of water is applied, blistering and damage to the electrodes was observed, causing degraded performance of the material [55].

The IPMC matrix is made of a hydrophobic polymer backbone and hydrophilic anionic sidechains and forms clusters of concentrated anions which neutralize cations and water within the polymer network. These ionic polymers, with a standard thickness, are commercially available. The solution-recasting technique developed by Kim *et al.* [56, 71] enabled control over the thickness of the films. Films with thickness in the range of 30 μm to 2 mm were produced. Ionic polymers were transformed into IPMCs by depositing metal on both sides. Metal particles (3–10 nm) were loaded on both sides penetrating the polymers up to 10–20 μm. These metal particles balance the charging at boundary layers. Metal particles are chemically loaded by soaking in $Pt(NH_3)_4HCl$ and then reducing through $LiBH_4$ or $NaBH_4$. Because this process is expensive, an inexpensive loading process was developed by Shahinpoor and Kim (2002) [56], where metal particles are physically loaded with electrode layer deposition. IPMCs produced in this way show properties comparable to chemically loaded IPMCs. This procedure also gives better flexibility in the selection of the electrode material: a wide variety of metals such as platinum, palladium, silver, gold, carbon, and graphite can be deposited by this procedure. Platinum is a widely used electrode material due to its high corrosion resistance and its higher deflection and work densities. The metal electrode decreases the surface resistance and increases current densities, giving faster actuation. IPMCs are being developed for wide range of applications and for hydrodynamic propulsion. Various swimming and flapping applications have been also been researched [36].

#### *1.3.2.3 Conducting Polymers*

Conducting polymers are electronically conducting organic materials. Actuation is produced in these materials when the electronically changing oxidation state–usually positive charges–leads to the flux of ions into or out of the polymer backbone, causing deformation. Solvent flux may also occur when there is a difference in ion composition. The insertion and removal of ions between polymer chains is considered the primary factor for dimensional change, whereas conformational change and solvent flux are considered secondary factors [36]. The basic structure of a conducting polymer actuator is a sandwich of two polymer strips with electrolyte between them (Figure 1.11). The polymer strips act as electrodes in electrochemical cells. When a potential is applied to the electrodes, oxidation occurs at the anode and reduction at the cathode. To balance the charges, ions are transferred into and out of the polymer and electrolyte. Swelling occurs when there is an addition of ions, and contraction is present when ions are removed. The sandwich bends when one electrode swells and the other shrinks, as shown in Figure 1.11 [2,57]. The chain orientation in the polymer network affects the rate of doping and redoping. The greater the orientation, the lower the rate of doping, which affects the response time of the actuators. But the higher orientation provides more achievable strength and modulus [57]. Linear actuators can easily be made by separating the electrodes from each other. Strains in this case are measured to be between 2 and 20 %.

The response of a conducting polymer actuator depends on the molecular diffusion when the electrodes are thick. When they are thin, the electrodes are limited by the RC time due to electric double-layer and electrolyte resistance effects. These issues can be addressed by using very thin electrodes with small interelectrode separation and filled with electrolytes that have high conductivity. By using this method, the conducting polymer actuator is a suitable material for the development of micromechanical actuator elements [57]. The most widely used conducting polymers include polypyrrole and polyaniline; thin films of these materials are usually produced by electrodeposition or chemical synthesis. The properties of these conducting polymers depend on the solvent and salt used during the electrodeposition synthesis. The cycle life of these polymers can be increased to hundreds of thousands using ionic liquid electrolytes. Forces up to tens of newtons are being obtained along with displacements of several millimeters. A displacement of 100 mm was obtained through mechanical amplification. Currents in the range of several hundreds of milliamperes have been used with voltages up to 10 V. In steady state, minimal current is needed to provide a catch state, where the deformation state can be held with minimal energy. This deformation can be exploited in robotic applications [36].

Compared to piezoelectric materials, conducting polymers are predicted to have higher work densities per cycle, slightly lower force generation, lower power densities, and require lower operational voltages. Conducting polymers suffer from disadvantages such as low cycle life and energy conversion efficiency, and they also need electrodes with large surface areas to achieve high actuation rates. Various applications for conducting polymer actuators being considered by researchers include actuators for micromachining and micromanipulation, microflaps for aircraft wings, micropumps, and valves for "labs on a chip"; actuators for adaptive optics and steer-able catheters; and artificial muscles for robotic and prosthetic devices [57]. Conducting polymer actuators need low actuation voltage, which is a special advantage for medical actuator applications such as catheters or for microactuators [57]. Conducting polymers are considered a suitable material as a matrix for enzymes in biosensors, which is believed to enhance speed, sensitivity, and versatility [58].

Conducting polymers have several properties–high tensile strength (>100MPa), large stresses (34 MPa), stiffness (~1 GPa modulus), and low actuation voltage (~2V)–that make them attractive actuator materials. However, in general, like other ionic EAPs, these materials have low electromechanical coupling (<1%), leading to low efficiency. Their efficiency can be improved by recovering a significant amount of energy. The low electromechanical coupling and voltages may require high currents for operation, a major constraint in the development of large autonomous applications. Practical applicability should also address encapsulation issues. Moderate strains (typically ~2–9%) can be obtained by mechanical amplification, and a maximum strain rate of 12%/s was observed. The strain rate is limited by the internal resistance of polymers and the electrolytes and ionic diffusion rate in the polymer backbone [36].

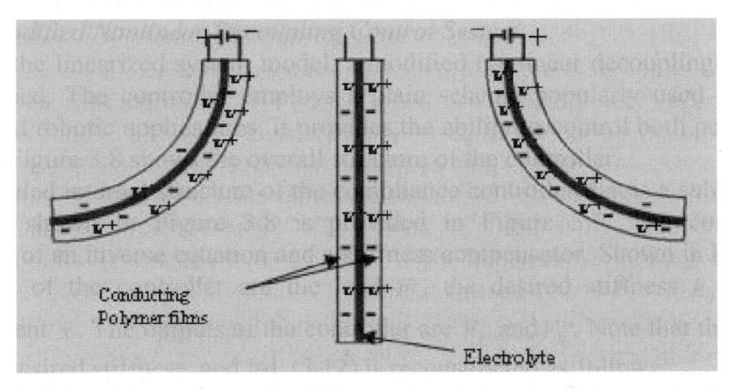


**Figure 1.11.** Schematic representation of three states during the electromechanical cycle of a rocking-chair type of bimorph-conducting polymer actuator. Both electrodes have the same concentration of dopant ($K^+$) when the cantilever is undistorted, and electrochemical transfer of dopants between electrodes causes bending either to the right or to the left.

The response rate of a polypyrrole (PPy) conducting polymer is being improved by using polymers doped with bis(trifluoromethanesulfonyl)imide (TFSI) and lithium bis(trifluoromethanesulfonyl)imide/propylene carbonate (LiTFSI/PC) as an electrolyte instead of water or propylene carbonate (PC). A response of 10.8% $s^{-1}$ was observed, where the conventional PPy polymers gave a peak response rate of 0.1% $s^{-1}$, and a maximum strain of 23.6% was observed [59]. A linear actuator was developed using a PPy-metal coil composite which gave a strain of 11.6%. The composites provide tremendous flexibility in the design of actuators with a wide range of displacement and force capabilities, as they can be connected and bundled to suit the requirement. The composites can also be encapsulated and used in air [60].

Encapsulated polypyrrole actuators have been developed using a gel, doped with salt, as the electrolyte. The gel electrolyte was made of agar or polymethylmethacrylate (PMMA) and gave good actuation responses [61]. Composite conducting polymers, reinforced with textile fibers such as polyester-PPy or nylon-PPy, have good electrical properties and good structural properties, similar to textile fibers. These composites are believed to have great potential, with stealth and camouflage capabilities, in the design of aircraft fuselages. The composites can also be used in the design of continuous transport belts in coal mines, where static dissipation is of great importance. They are also being studied in the design of solar cells and displays [62].

*1.3.2.4 Carbon Nanotube Actuators*

Carbon nanotubes (CNTs) emerged as a formal EAP in 1999, bringing their exceptional mechanical and electrical properties to the realm of actuator technology [2]. Typically, single-walled carbon nanotubes (SWCNTs) have a minimum diameter of 1.2 nm but can be larger. Carbon nanotubes form bundles, due to van der Waal forces, are used in actuator studies, and have a typical diameter of 10 nm. The actuation of carbon nanotubes depends on charging the surface; therefore, multiwalled carbon nanotubes (MWCNTs) are considered inefficient because of their less accessible surface area [36]. Actuation was observed in nanotubes that were suspended in an electrolyte. The change in bond length, due to the injection of large charges into nanotubes, is considered responsible for the observed deformation. In a carbon nanotube, the electron flow path is provided by a network of conjugated bonds connecting the carbon atoms. The electrolytes form an electric double layer around nanotubes, creating an ionic imbalance between nanotubes and electrolytes (Figure 1.12 [63]). The C-C bond length also increases because of the repulsion between positively charged carbon atoms formed by electron removal. These dimensional changes are translated into a macroscopic deformation in a network of entangled nanotubes. When this network is used, the bond length changes translate into macroscopic deformation [2]. Coulombic forces dominate from low to moderate levels of charging, giving a parabolic relationship between the strain and the applied potential. At higher potentials, the relationship is lost because ions and solvent in the solution start exchanging electrons with nanotubes, discharging the double layer. The loss of the double layer limits the amount of maximum strain that can be obtained from CNT actuators; approximately 0.1% to 1% strains were observed. The problem of low strain is overshadowed by the huge work densities (~200 $MJ/m^3$) that can be obtained owing to the high elastic modulus (640 GPa) and enormous tensile strength (>>1 GPa) of these materials [36].

Present studies are being done on nanotube sheets or papers that are composed of bundle of nanotubes joined by mechanical entanglement and van der Walls forces. Actuators are fabricated by attaching strips of a CNT sheet on both sides of double-sided scotch tape (a mechanism similar to the one shown in Figure 10). Voltage applied in electrolyte bending was observed; the direction of the bending reversed with a change in the direction of potential applied [63]. Due to the random orientation and weak van der Walls forces between the nanotubes in CNT paper, the properties of these fibers and sheets were many orders in magnitude less than a single nanotube. Advances in carbon nanotube spinning techniques are providing ways to make a variety of macroscopic objects for different applications. It is believed that the unique mechanical and electrical properties of CNT strands will extend to larger scales. E-textiles are being developed composed of distributed layers of actuator and sensor segments, each segment consisting of woven CNT-based filament yarn sandwiched between two electrodes. These nanotube-based fabrics are believed to provide technology for a variety of macroscopic textile applications such as lightweight sensor systems, membrane structures with actuation and shape-changing capabilities, and power generation [64].

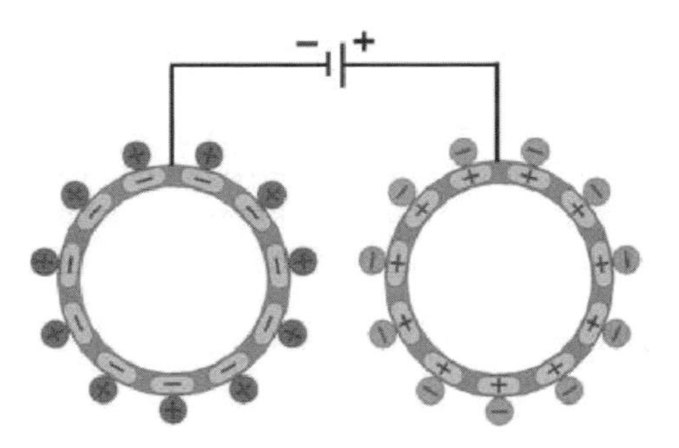

**Figure 1.12.** Schematic illustration of a charge injection in a nanotube-based electromechanical actuator [63]. Redrawn with pwermission from *Science* 284:1340. Copyright 1999 AAAS.

Nanotube actuators need significantly fewer volts compared to 100s of volts needed for piezoelectric actuators. Nanotube actuators have been operated at 350°C. If the mechanical properties of single nanotube can be translated to nanotube sheets, a strain of 1% will provide order-of-magnitude advantages compared to commercial actuators in work per cycle and stress generation capabilities. A maximum isometric actuator stress of 26 MPa was reported to work for a SWCNT actuator. The achievable strain is independent of an applied load, so the work done during constant load contraction increases linearly until the failure of the material occurs. High-stress applications are limited because of the creep effect. The potential applications of these nanotube actuators depend on the ability to improve the properties of nanotube sheets. These sheets are created by increasing the alignment and binding of the nanotubes. The strain rate of nanotube actuators is low compared to piezoelectrics, *etc.*, and depends on ion diffusion [65]. The actuation strain rate and amount of strain for carbon nanotube sheet actuators have been improved by employing the resistance compensation technique. In the resistance compensation technique, a higher input voltage is applied to compensate for the ohmic drop that occurs across the electrolyte [66]. The strain rate of a CNT actuator depends on the rate of charge injection. Due to the availability of a huge internal surface area, enormous capacitance exists in carbon nanotubes. To obtain high strain rates, one of the critical requirements is to decrease the internal resistance of the cell. Another factor to be considered is the speed of ion transport in carbon nanotube fibers and papers [36]. It appears that the rate drops rapidly in large devices after several seconds. The rate drops further in composite carbon nanotube fibers due to the slow rate of ion transport in a polymeric binder. To develop large nanotube actuators, electrode spacing, diffusion distances, and conduction paths should be minimized by microstructuring. Due to the limited strains obtained, mechanical amplification is needed for practical applications, but their limitations include the high cost of materials, low efficiency, and poor bulk mechanical properties compared to single nanotubes [36]. At high positive potentials, a pneumatic mechanism was observed to provide giant actuation up to 300% of the thickness in the direction of carbon nanotube sheets [67].

Carbon nanotube-electroactive polymer composites are also exploited for their superior mechanical and electrical properties. They exhibit a deformation mechanism similar to polyaniline but with improved mechanical properties, allowing higher strains with higher stresses and mechanical energy densities [68]. A large electromechanical response was also reported for a carbon nanotube-nematic liquid crystal elastomer. Carbon nanotubes were aligned along the nematic director during preparation, creating a very large dielectric anisotropy. A uniaxial stress of ~1kPa at a constant field of ~1MV/m was reported showing a potential in the development of electrically driven actuators [7].

## 1.4. Concluding Remarks

It can be seen from the above reported research and the scale of the academic interest in active polymer materials, that they have the potential to become an indispensable part of future technological developments. With each polymer having its own niche applications, they are bound to be the materials of future. With growing emphasis on interdisciplinary research, different active materials can be combined to develop tailor-made, multifunctional properties, where single materials can act as sensors, actuators, structural elements, *etc.*

To date, the robotics community has adopted only two major active polymer technologies: dielectric elastomers and ionic polymer-metal composites because the maturity of these two technologies is inevitable. However, other technologies are also quite promising and leaves one the great potentials to use them in robotic applications. Two other technologies that the robotics community is currently considering are conducting polymers and electrostrictive graft elastomers. In later chapters, we will focus on four major active polymer technologies: dielectric elastomers (Chapters 2 and 3), electrostrictive graft elastomers (Chapter 4), conducting polymers (Chapter 5), and ionic polymer-metal composites (Chapters 6–10). We all expect that the robotics community will adopt other promising active polymer materials as their maturity and availability improve.

## 1.5. References

[1] K. Gurunathan, A.V. Murugan, R. Marimuthu, U.P. Mulik, and D.P. Amalnerkar (1999) Electrochemically synthesized conducting polymeric materials for applications towards technology in electronics, optoelectronics and energy storage devices. Materials Chemistry and Physics, 61:173–191.

[2] Y. Bar-Cohen (2001) Electroactive Polymer (EAP) Actuators as Artificial Muscles (Reality, Potential, and Challenges). SPIE Press, Bellingham, Washington, USA.

[3] M. Zrínyi (2000) Intelligent polymer gels controlled by magnetic fields. Colloid & Polymer Science, 278(2):98–103.

[4] M. Ayre (2004) Biomimicry – A Review. European Space Agency, Work Package Report.
[5] Y. Bar-Cohen (2003) Actuation of biologically inspired intelligent robotics using artificial muscles. Industrial Robot: An International Journal, 30(4):331–337.
[6] D.A. Kingsley, R.D. Quinn, and R.E. Ritzmann (2003) A cockroach inspired robot with artificial muscles. International Symposium on Adaptive Motion of Animals and Machines (AMAM), Kyoto, Japan.
[7] S. Courty, J. Mine, A. R. Tajbakhsh, and E. M. Terentjev (2003) Nematic elastomers with aligned carbon nanotubes: New electromechanical actuators. Europhysics Letters, 64(5): 654–660.
[8] Y. Bar-Cohen and C. Breazeal (2003) Biologically inspired intelligent robotics. Proceedings of SPIE International Symposium on Smart Structures and Materials, EAPAD
[9] Y. Bar-Cohen (2004) Biologically inspired robots as artificial inspectors – science fiction and engineering reality. Proceedings of 16th WCNDT – World Conference on NDT.
[10] R. Yoshida, T. Yamaguchi, and H. Ichijo (1996) Novel oscillating swelling-deswelling dynamic behavior of pH-sensitive polymer gels. Materials Science and Engineering, C(4):107–113.
[11] S. Umemoto, N. Okui, and T. Sakai (1991) Contraction behavior of poly(acrylonitrile) gel fibers. Polymer Gels, 257–270.
[12] K. Salehpoor, M. Shahinpoor, and M. Mojarrad (1996) Electrically controllable artificial PAN muscles. SPIE 1996, 2716:116–124.
[13] K. Choe (2004) Polyacrylonitrile as an Actuator Material: Properties, Characterizations and Applications, MS thesis, University of Nevada, Reno.
[14] A. Lendlein and S. Kelch (2002) Shape-memory polymers. Angewandte Chemie International Edition, 41: 2034–2057.
[15] http://www.azom.com/details.asp?ArticleID=1542
[16] http://www.crgrp.net/shapememorypolymer/smp.html
[17] A. Lendlein and R. Langer (2002) Biodegradable, elastic shape-memory polymers for potential biomedical applications. Science, 296:1673–1676.
[18] F. Daerden and D. Lefeber (2001) The concept and design of pleated pneumatic artificial muscles. International Journal of Fluid Power, 2(3):41–50.
[19] C-P. Chou and B. Hannaford (1996) Measurement and modeling of McKibben pneumatic artificial muscles. IEEE Transactions on Robotics and Automation, 12:90–102.
[20] F. Daerden and D. Lefeber (2002) Pneumatic artificial muscles: actuators for robotics and automation. European Journal of Mechanical and Environmental Engineering, 47(1):10–21.
[21] G.K. Klute and B. Hannaford (2000) Accounting for elastic energy storage in McKibben artificial muscle actuators. ASME Journal of Dynamic Systems, Measurement, and Control, 122(2):386–388.
[22] A. Aviram (1978) Mechanophotochemistry. Macromolecules, 11(6):1275–1280.
[23] A. Suzuki and T. Tanaka (1990) Phase transition in polymer gels induced by visible light. Nature, 346:345–347.
[24] S. Juodkazis, N. Mukai, R. Wakaki, A. Yamaguchi, S. Matsuo, and H. Misawa (2000) Reversible phase transitions in polymer gels induced by radiation forces. Nature, 408:78–181.
[25] N.C.R. Holme, L. Nikolova, S. Hvilsted, P.H. Rasmussen, R.H. Berg, and P.S. Ramanujam (1999) Optically induced surface relief phenomena in azobenzene polymers. Applied Physics Letters, 74(4):519–521.

[26] M. Zrínyi, L. Barsi, and A. Büki (1996) Deformation of ferrogels induced by nonuniform magnetic fields. Journal of Chemical Physics, 104(21):8750–8756.

[27] P.A. Voltairas, D.I. Fotiadis, and C.V. Massalas (2003) Modeling of hyperelasticity of magnetic field sensitive gels. Journal of Applied Physics, 93(6):3652–3656.

[28] D.K. Jackson, S. B. Leeb, A.H. Mitwalli, P. Narvaez, D. Fusco, and E.C. Lupton Jr (1997) Power electronic drives for magnetically triggered gels. IEEE Transactions on Industrial Electronics, 44(2):217–225.

[29] N. Kato, S. Yamanobe, Y. Sakai, and F. Takahashi (2001) Magnetically activated swelling for thermosensitive gel composed of interpenetrating polymer network constructed with poly(acrylamide) and poly(acrylic acid). Analytical Sciences, 17, supplement:i1125–i1128.

[30] M. Lokander (2004) Performance of Magnetorheological Rubber Materials. Thesis, KTH Fibre and Polymer Technology.

[31] M. Kamachi (2002) Magnetic polymers. Journal of Macromolecular Science Part C-Polymer Reviews, C42(4):541–561.

[32] P.A. Voltairas, D. I. Fotiadis, and L.K. Michalis (2002) Hydrodynamics of magnetic drug targeting. Journal of Biomechanics, 35:813–821.

[33] H. Ichijo, O. Hirasa, R. Kishi, M. Oowada, K. Sahara, E. Kokufuta, and S. Kohno (1995) Thermo-responsive gels. Radiation Physics and Chemistry, 46(2):185–190.

[34] E. T. Carlen, and C. H. Mastrangelo (1999) Simple, high actuation power, thermally activated paraffin microactuator. Transducers '99 Conference, Sendai, Japan, June 7–10.

[35] C. Folk, C-M. Ho, X. Chen, and F. Wudl (2003) Hydrogel microvalves with short response time. 226th American Chemical Society National Meeting, New York.

[36] J.D.W. Madden, A. N. Vandesteeg, P.A. Anquetil, P.G.A. Madden, A. Takshi, R.Z. Pytel, S.R. Lafontaine, P.A. Wieringa, and I.W. Hunter (2004) Artificial muscle technology: Physical principles and naval prospects. IEEE Journal of Oceanic Engineering, 20(3):706–728.

[37] Q.M. Zhang, V. Bharti, and X. Zhao (1998) Giant electrostriction and relaxor ferroelectric behavior in electron-irradiated poly(vinylidene fluoride-trifluoroethylene) copolymer. Science, 280:2101–2104.

[38] S. Ducharme, S. P. Palto, L. M. Blinov, and V. M. Fridkin (2000) Physics of two-dimensional ferroelectric polymers. Proceedings of the Workshop on First-Principles Calculations for Ferroelectrics, Feb 13–20, Aspen, CO, USA.

[39] G. Kofod (2001) Dielectric Elastomer Actuators. Dissertation, The Technical University of Denmark.

[40] F. Carpi, P. Chiarelli, A. Mazzoldi, and D. de Rossi (2003) Electromechanical characterization of dielectric elastomer planar actuators: Comparative evaluation of different electrode materials and different counterloads. Sensors and Actuators A, 107:85–95.

[41] A. Wingert, M. Lichter, S. Dubowsky, and M. Hafez (2002) Hyper-redundant robot manipulators actuated by optimized binary dielectric polymers. Proceedings of SPIE International Symposium on Smart Structures and Materials, EAPAD

[42] H.R. Choi, K. M. Jung, S.M. Ryew, J.-D. Nam, J.W. Jeon, J.C. Koo, and K. Tanie (2005) Biomimetic soft actuator: Design, modeling, control, and application, IEEE/ASME Transactions on Mechatronics, 10(5): 581-586.

[43] C. Hackl, H-Y Tang, R.D. Lorenz, L-S. Turng, and D. Schroder (2004) A multi-physics model of planar electro-active polymer actuators. Industry Applications Conference, 3:2125–2130

[44] J. Su, J.S. Harrison, and T. St. Clair (2000) Novel polymeric elastomers for actuation. Proceedings of IEEE International Symposium on Application of Ferroelectrics, 2:811–819.

[45] Y. Wang, C. Sun, E. Zhou, and J. Su (2004) Deformation mechanisms of electrostrictive graft elastomer. Smart Materials and Structures, 13:1407–1413.
[46] J. Kim and Y.B. Seo (2002) Electro-active paper actuators. Smart Materials and Structures, 11:355–360.
[47] Y. An and M.T. Shaw (2003) Actuating properties of soft gels with ordered iron particles: Basis for a shear actuator. Smart Materials and Structures, 12:157–163.
[48] D.K. Shenoy, D.L. Thomse III, A. Srinivasan, P. Keller, and B.R. Ratna (2002) Carbon coated liquid crystal elastomer film for artificial muscle applications. Sensors and Actuators A, 96:184–188.
[49] http://www.azom.com/news.asp?newsID=1220
[50] M. Camacho-Lopez, H. Finkelmann, P. Palffy-Muhoray, and M. Shelley (2004) Fast liquid-crystal elastomer swims into the dark. Nature Materials, 3:307–310.
[51] T. Tanaka, I. Nishio, S-T. Sun, and S. Ueno-Nishio (1982) Collapse of gels in an electric field. Science, 218:467–469.
[52] T. Shiga and T. Kurauchi (1990) Deformation of polyelectrolyte gels under the influence of electric field. Journal of Applied Polymer Science, 39:2305–2320.
[53] H. B. Schreyer, G. Nouvelle, K. J. Kim, and M. Shahinpoor (2000) Electrical activation of artificial muscles containing polyacrylonitrile gel fibers. Biomacromolecules, 1:642–647.
[54] K. Oguro, Y. Kawami, and H. Takenaka (1992) Bull. Government Industrial Research Institute Osaka, 43, 21.
[55] K.J. Kim, and M. Shahinpoor (2002) Development of three dimensional ionic polymer-metal composites as artificial muscles, Polymer, 43(3):797–802.
[56] M. Shahinpoor and K.J. Kim (2002) A novel physically-loaded and interlocked electrode developed for ionic polymer-metal composites (IPMCs), Sensors and Actuator: A. Physical, 96:125–132.
[57] R.H. Baughman (1996) Conducting polymer artificial muscles. Synthetic Metals, 78:339–353.
[58] M. Gerard, A. Chaubey, and B.D. Malhotra (2002) Application of conducting polymers to biosensors. Biosensors & Bioelectronics, 17:345–359.
[59] S. Hara, T. Zama, W. Takashima, and K. Kaneto (2005) Free-standing polypyrrole actuators with response rate of 10.8%$s^{-1}$. Synthetic Metals, 149:199–201.
[60] S. Hara, T. Zama, W. Takashima, and K. Kaneto (2004) Polypyrrole-metal coil composite actuators as artificial muscle fibres. Synthetic Metals, 146:47–55.
[61] J.D. Madden, R. A. Cush, T.S. Kanigan, C.J. Brenan, and I. W. Hunter (1999) Encapsulated polypyrrole actuators. Synthetic Metals, 105:61-64.
[62] A. Bhattacharya, and A. De (1996) Conducting composites of polypyrrole and polyaniline: A review. Progress in Solid State Chemistry, 24:141–181.
[63] R.H. Baughman, C. Cui, A.A. Zakhidov, Z. Iqbal, J.N. Barisci, G.M. Spinks, G.G. Wallace, A. Mazzoldi, D. De Rossi, A.G. Rinzler, O. Jaschinski, S. Roth, and M. Kertesz (1999) Carbon nanotube actuators. Science, 284:1340–1344.
[64] N. Jalili, B.C. Goswami, A. Rao, and D. Dawson (2004) Functional fabric with embedded nanotube actuators/sensors. National Textile Center Research Briefs – Materials Competency (NTC Project: M03-CL07s).
[65] R.H. Baughman, A.A. Zakhidov, and W.A. de Heer (2002) Carbon nanotubes–the route toward applications. Science, 297:787-792.
[66] J.N. Barisci, G.M. Spinks, G.G. Wallace, J.D. Madden, and R.H. Baughman (2003) Increased actuation rate of electromechanical carbon nanotube actuators using potential pulses with resistance compensation. Smart Materials and Structures, 12:549–555.

[67] G.M. Spinks, G.G. Wallace, L.S. Fifield, L.R. Dalton, A. Mazzoldi, D. De Rossi, I.I. Khayrullin, and R.H. Baughman (2002) Pneumatic carbon nanotube actuators. Advanced Materials, 14(23):1728–1732.

[68] M. Tahhan, V-T Truong, G.M. Spinks, and G.G. Wallace (2003) Carbon nanotube and polyaniline composite actuators. Smart Materials and Structures, 12:626–632.

[69] Y. Hirokawa and T. Tanaka, (1984) Volume phase transitions in a non-ionic gel. Journal of Chemical Physics, 81:6379–6380.

[70] K. Choi, K.J. Kim, D. Kim, C. Manford, and S. Heo (2006) Performance characteristics of electro-chemically driven polyacrylonitrile fiber bundle actuators. Journal of Intelligent Material Systems and Structures (in print).

[71] K.J. Kim and M. Shahinpoor (2002) Development of three dimensional ionic polymer-metal composites as artificial muscles. Polymer, 43(3):797–802.

[72] J. Su, Z. Ounaies, J.S. Harrison, Y. Bar-Cohen, and S. Leary (2000) Electromechanically active polymer blends for actuation. Proceedings of SPIE-Smart Structures and Materials, 3987:140–148.

[73] J. Su, K. Hales, and T.B. Xu (2003) Composition and annealing effects on the response of electrostrictive graft elastomers. Proceedings of SPIE Smart Structures and Materials, 5051:191–199.

2

# Dielectric Elastomers for Artificial Muscles

J.-D. Nam[1], H.R. Choi[2], J.C. Koo[2], Y.K. Lee[3], K.J. Kim[4]

[1] Department of Polymer Science and Engineering, Sungkyunkwan University, 300 Chunchun-dong, Jangan-gu, Suwon, Kyunggi-do 440-746, South Korea, dnam@skku.edu
[2] School of Mechanical Engineering, Sungkyunkwan University, 300 Chunchun-dong, Jangan-gu, Suwon, Kyunggi-do 440-746, South Korea
[3] School of Chemical Engineering, Sungkyunkwan University, 300 Chunchun-dong, Jangan- gu, Suwon, Kyunggi-do 440-746, South Korea
[4] Active Materials and Processing Laboratory, Mechanical Engineering Department (MS312), University of Nevada, Reno, NV 89557, USA

## 2.1 Introduction

Natural muscles have self-repair capability providing billions of work cycles with more than 20% of contractions, contraction speed of 50% per second, stresses of ~0.35 MPa, and adjustable strength and stiffness [1]. Artificial muscles have been sought for artificial hearts, artificial limbs, humanoid robots, and air vehicles. Various artificial muscles have been investigated for large strain, high response rate, and high output power at low strain using their own material characteristics [2]. Among the candidates for artificial muscles, the dielectric elastomer has typical characteristics of light weight, flexibility, low cost, easy fabrication, *etc.*, which make it attractive in many applications. Applications of dielectric elastomer include artificial muscles and also mobile robots, micro-pumps, micro-valves, disk drives, flat panel speakers, intelligent endoscope, *etc.* [3–7].

Dielectric elastomer actuators have been known for their unique properties of large elongation strain of 120–380%, large stresses of 3.2 MPa, high specific elastic energy density of 3.4 J/g, high speed of response in $10^{-3}$ s, and high peak strain rate of 34,000%/sec [1,2,4,8]. They transform electric energy directly into mechanical work and produce large strains. Their actuators are composed primarily of a thin passive elastomer film with two compliant electrodes on the surfaces, exhibiting a typical capacitor configuration. As with most rubbery materials, the elastomer used in actuator application is incompressible (Poisson's ratio = 0.5) and viscoelastic, which consequently exhibits time- or frequency-dependent characteristics that could be represented by stress relaxation, creep, and dynamic-mechanical phenomena under stressed and deformed states [9–10]. When the electrical voltage is applied to the electrodes, an electrostatic force is generated between the electrodes. The force is compressive, and thus the elastomer film expands in the in-plane direction.

As an advantage of dielectric elastomer actuators, the performance of elastomer actuators can be tailored by choosing different types of elastomers, changing the cross-linking chemistry of polymer chains, adding functional entities, and improving fabrication techniques with ease and versatility in most cases. The deformation of elastomers complies with the theories of rubber elasticity and nonlinear viscoelasticity. When an electrical field is applied, the elastomer deformaton is influenced primarily by the intrinsic properties of moduli and dielectric constants of elastomers in a coupled manner. In addition, maximum actuation capabilities are often restricted by the dielectric strength (or breakdown voltage) of elastomer films. Although the low stiffness of elastomers may increase strain, maximum actuator stroke, and work per cycle, it should be considered that the maximum stress generation decreases with decreased moduli. Accordingly, the property-processing-structure relationship of elastomers especially under the electrical field and large deformation should be understood on the basis of the fundamental principles of deforming elastomers and practical experience in actuator fabrication.

## 2.2 General Aspects of Elastomer Deformation

Elastomers can be stretched several hundred percent; yet on being released, they contract back to their original dimensions at high speeds. By contrast, metals, ceramics, or other polymers (linear or highly cross-linked polymers) can be stretched reversibly for only about 1%. Above this level, they undergo permanent deformation in an irreversible way and ultimately break. This large and reversible elastic deformation makes elastomers unique in actuator applications. The fundamental dielectric and mechanical properties of most common elastomers are summarized in Table 2.1.

Elastomers are lightly-crosslinked polymers. Without cross-linking, the polymer chains have no chemical bonds between chains, and thus the polymer may flow upon heating over the glass transition temperature. If the polymer is densely cross-linked, the chains cannot flow upon heating, and a large deformation cannot be expected upon stretching. Elastomers are between these two states of molecular conformation. The primary chains of elastomers are cross-linked at some points along the main polymer chains. For example, commercial rubber bands or tires have molecular weights of the order of $10^5$ g/mol and are cross-linked every 5–10x$10^3$ g/mol, which gives 10–20 cross-links per primary polymer molecule. The average molecular weight between cross-links is often defined as $M_c$ to the express degree of cross-linkage.

A raw elastomer is a high molecular weight liquid with low strength. Although its chains are entangled, they readily disentangle upon stressing and finally fracture in viscous flow. Vulcanization or curing is the process where the chains of the raw elastomer are chemically linked together to form a network, subsequently transforming the elastomeric liquid to an elastic solid. The most widely used vulcanizing agent is sulfur that is commonly used for diene elastomers such as butadiene rubber (BR), styrene-butadiene rubber (SBR), acrylonitrile-butadiene rubber (NBR), and butyl rubber (IIR). Another type of curing agent is peroxides, which are used for saturated elastomers such as ethylene propylene rubber, chlorinated polyethylene (CSM), and silicone elastomers. The mechanical behavior of an elastomer depends strongly on cross-link density. When an uncross-linked elastomer is stressed, chains may readily slide past one another and disentangle. As cross-linking is increased further, the gel point is eventually reached, where a complete three-dimensional network is formed, by definition. A gel cannot be fractured without breaking chemical bonds. Therefore, the strength is higher at the gel point, but it does not increase indefinitely with more cross-linking. The schematic of elastomer properties is shown as a function of cross-link density in Figure 2.1. The elastomer properties, especially the modulus, are significantly changed by the cross-link density in most elastomer systems, and thus the actuator performance can be adjusted by controlling the degree of elastomer vulcanization (degree of cross-link density). The cross-link density can be adjusted by the kinetic variables of vulcanization reactions such as sulfur (or peroxide) content, reaction time, reaction temperature, catalyst (or accelerator), *etc*. Note that the elastomer vulcanization process is not a thermodynamic process but a kinetically-controlled process in most cases.

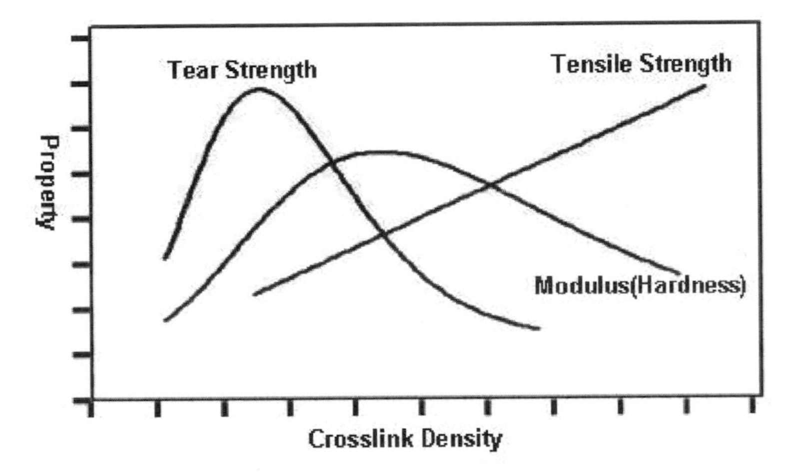


**Figure 2.1.** Elastomer properties schematically plotted as a function of cross-link density

Another significant phenomena in elastomers is the hysteresis loop of stress-strain curves. As seen in Figure 2.2, the stress under loading and unloading is different in the pathway. Furthermore, the unloading curve usually does not return to the origin. As the elastomer is allowed to rest in a stress-free state, the strain will reach the origin. It should also be mentioned that the shape of the hysteresis loop changes with loading-unloading cycles, especially in the early stage of cycles, eventually reaching an identical hysteresis loop. The hysteresis phenomena of elastomers should be considered in the development of actuators for long-term durability.

**Table 2.1.** Dielectric and mechanical properties of elastomers [11, 12]

| | Dielectric constant at 1kHz | Dielectric loss factor at 1kHz | Young's modulus [x$10^6$ Pa] | Eng. stress [MPa] | Break stress [MPa] | Ultimate strain [%] |
|---|---|---|---|---|---|---|
| Polyisoprene, natural rubber(IR) | 2.68 | 0.002–0.04 | 1.3 | 15.4 | 30.7 | 470 |
| Poly(chloroprene)(CR) | 6.5–8.1 | 0.03/0.86 | 1.6 | 20.3 | 22.9 | 350 |
| Poly(butadiene)(BR) | – | – | 1.3 | 8.4 | 18.6 | 610 |
| Poly(isobutene–co–isoprene)butyl rubber | 2.42 | 0.0054 | 1 | – | 17.23 | – |
| Poly(butadiene–co–acrylonitrile)(NBR, 30% acrylonitrie constant) | 5.5 ($10^6$ Hz) | 35 ($10^6$ Hz) | | 16.2 | 22.1 | 440 |
| Poly(butadiene–co–styrene) (SBR, 25% styrene constant) | 2.66 | 0.0009 | 1.6 | 17.9 | 22.1 | 440 |
| Poly(isobutyl–co–isoprene rubber)(IIR) | 2.1–2.4 | 0.003 | – | 5.5 | 15.7 | 650 |
| Chlorosulfonated polyethylene(CSM) | 7–10 | 0.03–0.07 | – | – | 24.13 | – |
| Ethylene–propylene rubber(EPR) | 3.17–3.34 | 0.0066–0.0079 | – | – | 20.68 | – |
| Ethylene–propylene diene monomer (EPDM) | 3.0–3.5 | 0.0004 at 60 Hz | 2 | 7.6 | 18.1 | 420 |
| Urethane | 5–8 | 0.015–0.09 | – | – | 20–55 | – |
| Silicone | 3.0–3.5 | 0.001–0.010 | – | – | 2–10 | 80–500 |

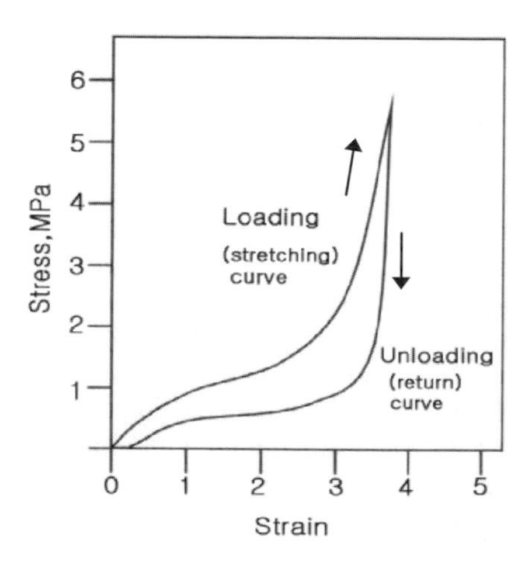


**Figure 2.2.** Stress-strain curve of elastomers under loading and unloading a exhibiting hysteresis loop

## 2.3 Elastic Deformation of Elastomer Actuator under Electric Fields

The electrostatic energy (*U*) stored in an elastomer film with thickness *z* and surface area *A* can be written as

$$U = \frac{Q^2}{2C} = \frac{Q^2 z}{2\varepsilon_o \varepsilon_r A} \tag{2.1}$$

where Q, C, $\varepsilon_o$ , and $\varepsilon_r$ are the electrical charge, capacitance, free-space permittivity ($8.85 \times 10^{-12}$ F/m), and relative permittivity, respectively. The capacitance is defined as $C = \varepsilon_o \varepsilon_r A / z$. From the above equation, the change in electrostatic energy can be related tohe differential changes in thickness (*dz*) and area (*dA*) with a constraint that the total volume is constant (*Az* = constant). Then the electrostatic pressure generated by the actuator can be derived as [5]

$$P = \varepsilon_o \varepsilon_r E^2 = \varepsilon_o \varepsilon_r \left(\frac{V}{z}\right)^2 \tag{2.2}$$

where *E* and *V* are the applied electric field and voltage, respectively. The electrostatic pressure in Eq. (2.2) is twofold larger than the pressure in a parallel-plate capacitor due to that fact that the energy would change with the changes in both the thickness and area of actuator systems.

Actuator performance has been derived by combining Eq. (2.2) and a constitutive equation of elastomers. The simplest and the most common equation of state may combine Hooke's law with Young's modulus (*Y*), which relates the stress (or electrostatic pressure) to thickness strain ($s_z$) as

$$P = -Y s_z \tag{2.3}$$

where $z = z_o(1 + s_z)$ and $z_o$ is the initial thickness of the elastomer film. Using the same constraint that the volume of the elastomer is conserved, $(1+s_z)(1+s_x)(1+s_y) = 1$ and $s_x = s_y$, the in-plane strain ($s_x$ or $s_y$) can be derived from Eqs. (2.2) and (2.3). For example, when the strain is small (*e.g.*, less than 20%), which may be not the case in practical actuator application, *z* in Eq. (2.2) can be simply replaced by $z_o$, and the resulting equation becomes

$$s_z = -\frac{\varepsilon_o \varepsilon_r}{Y}\left(\frac{V}{z_o}\right)^2 \tag{2.4}$$

or the in-plane strain can be expressed because $s_x = -0.5 s_z$,

$$s_x = -\frac{\varepsilon_o \varepsilon_r}{2Y}\left(\frac{V}{z_o}\right)^2 \tag{2.5}$$

This equation often appears in published literature. When the strain is large, however, the strain should be derived by combining Eqs. (2.2) and (2.3) in a quadratic equation:

$$s_z = -\frac{2}{3} + \frac{1}{3}\left[f(s_o) + \frac{1}{f(s_o)}\right] \tag{2.6}$$

where $f(s_o) = \left[1 + 13.5 s_o + \sqrt{27 s_o (6.75 s_o - 1)}\right]^{1/3}$ and $s_o = -\frac{\varepsilon_o \varepsilon_r}{Y}\left(\frac{V}{z_o}\right)^2$

The through-thickness strain $s_z$ can be converted to in-plane strain by solving the quadratic equation for the strain constraint as

$$s_x = (1 + s_z)^{-0.5} - 1 \tag{2.7}$$

For low strain materials, the elastic strain energy density ($u_e$) of actuator materials has been estimated as [4]

$$u_e = \frac{1}{2} P s_z = \frac{1}{2} Y s_z^2 \tag{2.8}$$

However, for high strain materials, the in-plane area over which the compression is applied changes markedly as the material is compressed, and thus the elastic strain energy density can be obtained by integrating the compressive stress times the varying planar area over the displacement, resulting in the following relation [4]:

$$u_e = \frac{1}{2} P \ln(1 + s_z) \tag{2.9}$$

However, it should be mentioned that Eqs. (2.4) and (2.6) are based on Hooke's equation of state, which may not be applicable to all elastomer systems. Elastomers usually have nonlinear and viscoelastic behavior in stress-strain relations, and thus the performance of the elastomer actuator should be analyzed by using a more realistic equation of state based on the fundamental theory and modeling methodology of rubber elasticity.

## 2.4 Rubber Elasticity and Equation of State of Elastomers

The relationships among macroscopic deformation, microscopic chain extension, and entropy reduction have been derived by many researchers providing quantitative relations between chain extension and entropy reduction [9,11,13,14]. The fundamental principle is that the repulsive stress of an elastomer arises from the reduction of the entropy of elastomers rather than through changes in enthalpy. As a result, the basic equation between stress and deformation is given as (for unidirectional compression and expansion)

$$P = -nRT\left(\lambda - \frac{1}{\lambda^2}\right) \tag{2.10}$$

where $R$ is gas constant, $T$ is temperature, and $\lambda$ is the ratio of length (L) to the original length ($L_o$), *i.e.*, $\lambda = L/L_o$, and thus it is related to the thickness strain in Eq. (2.3) as $\lambda = 1 + s_z$. The quantity $n$ represents the number of active network chain segments per unit volume [15, 16]. It should be mentioned that the extension (or contraction) ratio is usually used in the description of deformation in elastomer systems instead of strain because the extent of deformation is relatively large. The quantity $n$ represents the number of active network chain segments per unit volume, which is equal to $\rho / M_c$, where $\rho$ and $M_c$ are the density and the molecular weight between cross-links. As can be seen in Eq. (2.8), the equation is nonlinear and consequently the Hookean relation does not hold. However, Eq. (2.10) is often valid for relatively small extensions in most elastomers. The actual behavior of cross-linked elastomers in unidirectional extension is well described by the empirical equation of Mooney-Rivilin [17]:

$$P = -\left(C_1 + \frac{C_2}{\lambda}\right)\left(\lambda - \frac{1}{\lambda^2}\right) \tag{2.11}$$

According to the above equation, a plot of $P/\left(\lambda - 1/\lambda^2\right)$ versus $1/\lambda$ should be linear especially at low elongation, where $C_1$ and $C_2$ are obtained from the slope and intercept of the plot. The value of $C_1 + C_2$ is nearly equal to the shear modulus (or $Y/3$). Table 2.2 summarizes typical values of $C_1$ and $C_2$ of several elastomers.

## 2.5 Nonlinear Viscoelasticity of Elastomers in Creep and Stress Relaxation

For the application of elastomer actuators, the prestrain condition (~50–100%) is a substantial factor in actuator design and application. The prestrained condition of elastomers can be suited to nonlinear viscoelastic creep or stress relaxation characteristics in a large deformation. Various theories and models have been developed to analyze the nonlinear viscoelastic behavior of polymers in a large strain [9,10,13]. Here, we will introduce several methods applicable to the development and analysis of elastomer actuators under prestrained or actuating conditions.

**Table 2.2.** Constants of the Mooeny-Rivlin equation [14]

| Elastomer | $C_1$ | $C_2$ | $(C_1+C_2)$ | $C_2/C_1+C_2$ |
|---|---|---|---|---|
| Natural | 2.0(0.9–3.8) | 1.5(0.9–2) | 3.5 | 0.4(0.25–0.6) |
| Butyl rubber | 2.6(2.1–3.2) | 1.5(1.4–1.6) | 4.1 | 0.4(0.3–0.5) |
| Styrene–butadiene rubber | 1.8(0.8–2.8) | 1.1(1.0–1.2) | 2.9 | 0.4(0.3–0.5) |
| Ethane–propene rubber | 2.6(2.1–3.1) | 2.5(2.2–2.9) | 5.1 | 0.5(0.43–0.55) |
| Polyacrylate rubber | 1.2(0.6–1.6) | 2.8(0.9–4.8) | 3 | 0.5(0.3–0.8) |
| Silicone rubber | 0.75(0.3–1.2) | 0.75(0.3–1.1) | 1.5 | 0.4(0.25–0.5) |
| Polyurethane | 3(2.4–3.4) | 2(1.8–2.2) | 5 | 0.4(0.38–0.43) |

To describe the creep behavior of elastomers with a large deformation (~100%), separable stress and time functions have been proposed. The model proposed by Pao and Marin is based on the assumption that the total creep strain is composed of an elastic strain, transient recoverable viscoelastic strain, and a permanent non-recoverable strain [18]:

$$s(t) = P/Y + KP^n\left(1 - s^{-qt}\right) + BP^n t \tag{2.12}$$

where $K$, $n$, $q$ and $B$ are constants for the material.

Findley *et al.* have fitted the creep behavior of many polymers to the following analytical relation in a form of power-law equation [19]:

$$s(t) = s_o + mt^n \tag{2.13}$$

where $s_o$ and $m$ are functions of stress for a given material and $n$ is a material constant. A more general relation for the single-step loading tests can be written as [9]

$$s(P,t) = s_o \sinh\frac{P}{P_o} + mt^n \sinh\frac{P}{P_m} \tag{2.14}$$

where $m$, $P_o$, and $P_m$ are constants for a material.

A similar relationship has been proposed by Van Holde [20]:

$$s(t) = s_o + mt^{1/3} \sin\alpha P \tag{2.15}$$

where $\alpha$ is a constant.

When a constant strain is applied to the elastomer actuators as a prestrain, the stress changes as a function of time and prestrain values. The nonlinear stress relaxation behavior with a large strain has taken a rheological approach. For example, the model proposed by Martin *et al.* is as follows [9];

$$P = Y\frac{s}{\lambda^2}\exp A\left(\lambda - \frac{1}{\lambda}\right) \tag{2.16}$$

where $\lambda$ is the extension ratio and $A$ is a constant.

## 2.6 Tunable Properties of Dielectric Constant and Modulus of Elastomers

According to Eq. (2.5), actuator performance is directly influenced by the stiffness and dielectric constant of an elastomer. In terms of actuator strain, lower values of the modulus and higher values of the dielectric constant are desirable. However, in terms of stress, the lower modulus values are not always desirable because the maximum stress attainable from an actuator increases with the modulus of elastomers. Accordingly, the dielectric constant and modulus should be optimized to give the desired performance of dielectric elastomer actuators.

The dielectric constant of elastomers can be enhanced simply by incorporating high dielectric materials. For example, copper phthalocyanine oligomere (CPO) has been blended with silicones to increase dielectric constants [2,21,22]. In this approach, the dielectric constant increased from 3.3 to 11.8 from the pristine silicone to a silicone blend with 40 wt% of CPO, which corresponds to a 250% increment in the dielectric constant. As discussed in Eqs. (2.4) and (2.5), the increased dielectric constant provides increased strain of the elastomer actuator [3]. However, it should be mentioned that the dielectric strength (or breakdown voltage) of an elastomer film is one of the most substantial factors in actuator applications. Incorporating heterogeneous entities or ionic chemicals usually decreases the dielectric strength, and subsequently the actuator performance should be limited by the applicable electric fields. Although the strain of the CPO/silicone blend system is increased, the maximum attainable strain is decreased by a deteriorated breakdown voltage [3].

It has also been reported that the gallery height of montmorillonite (MMT), a layered inorganic clay nanoplatelet system forms distributed effective nanocapacitors when incorporated in polymers [23]. The resulting dielectric constant and ionic conductivity have increased by 20–4300% and two to three orders of magnitude, respectively, in phenolic resin/MMT nanocomposite systems without a significant decrease in breakdown voltage. A similar nanocapacitance effect has been observed in polyurethane elastomer/MMT nanocomposite systems exhibiting an increment in dielectric constant from 2.6 to 5.6 [24]. It should be mentioned that the morphology of the layered nanoplatelets determines the dielectric constant as well as the modulus of nanocomposite systems. When nanoplatelets are homogeneously dispersed (or exfoliated) in a polymer, the modulus is significantly increased with a slight increment in the dielectric constant. On the other hand, when the polymer is intercalated between nanoplatelets, the modulus is not much increased but the dielectric constant is increased [24].

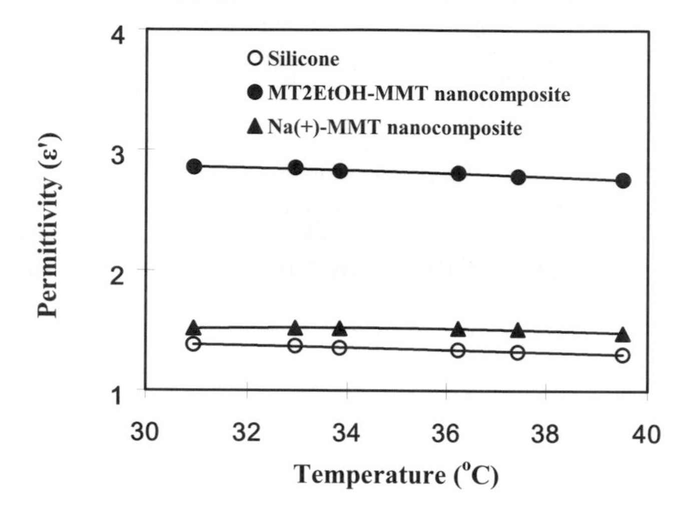


**Figure 2.3.** Dielectric constants of pristine silicone elastomer compared with its nanocomposite systems containing Na+ ion and MT2EtOH as intercalants in MMT

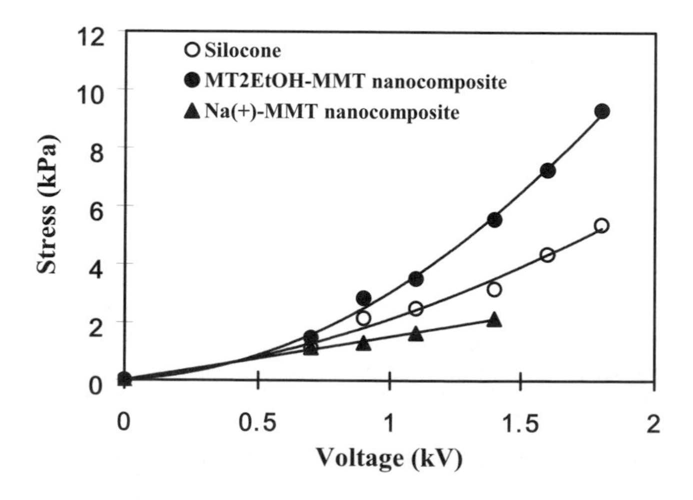


**Figure 2.4.** Comparison of electric-field-induced stress for pristine silicone and two nanocomposite systems measured up to breakdown voltages

In Figure 2.3, for example, the dielectric constant of a silicone-based polymer is compared with two nanocomposite systems containing two different types of commercial montmorillonite systems (Southern Clay Production): Na+/MMT and 92.6 meq/100 g ($d_{001}$=11.7Å), and methyl tallow bis-2-hydroxyethylammonium (MT2EtOH)/MMT 90 meq/100 g ($d_{001}$=18.5Å), where tallow is predominantly composed of octadecyl chains with small amounts of low homologues (~65% of $C_{18}$, ~30% of $C_{16}$ and ~5% of $C_{14}$). The layer thickness of an MMT sheet is around 1 nm and the lateral dimensions vary from several nanometers to micrometers. The Na+/MMT system gives an exfoliated or isotropically dispersed state of nanoplatelets, and the MT2EtOH/MMT system gives an intercalated structure maintaining the layered structure of MMT platelets. As with the urethane elastomer system, the intercalated MMT/silicone nanocomposite provides a higher dielectric constant than the exfoliated system seemingly due to the nanocapacitance effect. The well-known fact that the exfoliated nanocomposites give the highest modulus value is also demonstrated in these silicone nanocomposite elastomer systems; 55 kPa for silicone, 88 kPa for a $Na^+$-MMT nanocomposite, and 72 kPa for a MT2EtOH-MMT nanocomposite.

For these three systems of silicone-based materials, the generated stress is compared in Figure 2.4. As can be seen, the intercalated MT2EtOH-MMT actuator provides the highest stress values generated in the whole range of the electric field up to the breakdown voltage. It is due to the increased dielectric constant induced by the layered nanocapacitance effect. It should be pointed out that the breakdown voltage of the nanocomposite system is not decreased by the incorporation of nanoplatelets, which can hardly be achieved in other chemicals or fillers.

## 2.7 References


[1] R.H. Baughman, Science 308, 63 (2005).

[2] Y. Bar-Cohen, Electroactive Polymer (EAP) Actuators as Artificial Muscles-Reality, Potential and Challenges, Vol. PM136, SPIE-Society of Photo-optical Instrumentation Engineers, Bellingham, WA 2004.

[3] X. Zhang, C. Löwe, M. Wissler, B. Jähne, and G. Kovacs, Advanced Engineering Materials, 7(5), 361 (2005).

[4] R. Pelrine, R. Kornbluh, Q. Pei, J. Joseph, Science, 287, 836 (2000).

[5] R. Pelrine, R. Kornbluh, J.P. Jeseph, Sensors and Actuators A64, 77 (1999).

[6] H.R. Choi, K.M. Jung, J.C. Koo, J.D. Nam, Y.K. Lee, and M.S. Cho, Key Engineering Materials, 297-300, 622 (2005).

[7] J.C. Koo, H.R. Choi, M.Y. Jung, K.M. Jung, J.D. Nam, Y.K. Lee, Key Engineering Materials, 297-300, 665 (2005).

[8] J.D. Madden, IEEE Journal of Oceanic Engineering, 29, 706 (2004).

[9] I.M. Ward, Mechanical Properties of Solid Polymers, John Wiley & Sons, New York, (1985).

[10] J. D. Ferry, Viscoelastic Properties of Polymers, John Wiley & Sons, New York, (1980).

[11] A.N. Gent (ed.), Engineering with Rubber, Hanser, New York (1992).

[12] J. Brandrup, E.H. Immergut, E. A. Grulke (eds.), Polymer Handbook, 4th Ed., John Wiley & Sons, New York (1999).

[13] L. Nielsen and R.F. Landel, Mechanical Properties of Polymers and Composites, Marcel Dekker, New York (1994).

[14] D.W. Van Krevelen, Properties of Polymers, Elsevier, New York (1990).

[15] P.J. Flory, Polymer, 20, 1317 (1979).
[16] L.R.G. Treloar, The Physics of Rubber Elasticity, 3rd ed., Clarendon Press, Oxford, (1975).
[17] M. Mooney, Journal of Applied Physics, 11, 582 (1940); M. Mooney, Journal of Applied Physics, 19, 434 (1948); R.S. Rivlin, Transactions of the Royal Society (London), A240, 459, 491, 509 (1948); R.S. Rivlin, Transactions of the Royal Society (London), A241, 379 (1948).
[18] T.H. Pao and J. Martin, Journal of Applied Mechanics, 19, 478 (1952); Journal of Applied Mechanics, 20, 245 (1953).
[19] W.N. Findley and G. Khosla, Journal of Applied Physics, 26, 821 (1955).
[20] R, Van Holde, J. Polym. Sci., 24, 417 (1957).
[21] Q.M. Zhang, H. Li, M. Poh, F. Xia, Z.-Y. Cheng, H. Xu, C. Huang, Nature, 419, 284 (2002).
[22] B. Achar, G. Fohlen, J. Parker, Journal of Polymer Science Part A: Polymer Chemistry, 20, 1785 (1982).
[23] E.P.M. Williams, J.C. Seferis, C.L.Wittman, G.A. Parker, J.H. Lee, J.D. Nam, Journal of Polymer Science Part A: Polymer Physics, 42, 1-4 (2004).
[24] J.-D. Nam, S. D. Hwang, H. R. Choi, J. H. Lee, K. J. Kim, and S. Heo, Smart Materials and Structures, 14, 87-90 (2004).

# 3

# Robotic Applications of Artificial Muscle Actuators

H.R. Choi[1], K.M. Jung[2], J.C. Koo[2], J.D. Nam[3]

[1] School of Mechanical Engineering, College of Engineering,
Sungkyunkwan University, Suwon 440-746, Korea
hrchoi@me.skku.ac.kr
[2] School of Mechanical Engineering, College of Engineering,
Sungkyunkwan University, Suwon 440-746, Korea
jungkmok@me.skku.ac.kr
[3] School of Mechanical Engineering, College of Engineering,
Sungkyunkwan University, Suwon 440-746, Korea
jckoo@me.skku.ac.kr
[4] Department of Polymer Science and Engineering, College of Engineering
Sungkyunkwan University, Suwon 440-746, Korea
jdnam@skku.edu

## 3.1 Introduction

For the last few decades, the roles of robots have been widely expanded from handling of routine manufacturing processes to hosting various entertainment applications. With the evolution of the information technology that creates ubiquitous communication environments in human life, the expectation of advances in robotic technology has been more intensified. Obviously, development of new robot applications has outpaced the improvement of mechanical and electrical functionality of robot hardware. One of the most languid activities in the hardware development in robotics might exist in the field of sensors and actuators. For instance, the efficacy of control algorithms or information handling of the current cutting-edge robots is often constrained by actuator performance, sensing capabilities, mechanical locomotion, or power sources.

According to the development trends of robots, their functionality is recently concentrated on mimicking human movements or animal functions. Introduction of new kinds of actuators, so-called *soft* actuators, might be of key interest in the new robot technology development. However, the physical properties of traditional transducers such as electromagnetic motors, voice coil motors, are truly different from that of animal muscles so that operation of a robot equipped with actuators should be confined and efficient only in a structured environment. As a result, providing some level of flexibility to a robot skeleton and also to actuators will be the critical development path of the next generation of robots.

Energy transduction considered from the thermodynamic point of view does not provide ample research opportunities mainly because the macroscopic observation of energy flow has been well characterized. Especially energy flow in mechanical-electrical domain energy transformation that most likely relies on electromagnetic phenomena has been scrutinized for some decades. Consequently, most of the engineering applications, where the mechanical-electrical energy transformation is needed, employ electromagnetic transducers. Material development for energy transduction is, however, in its infancy. Recognizing the available number of transducer types that could be adopted in mechanical-electrical domain energy transformation, the development of an innovative new energy transformation material is well motivated.

Despite the tremendous engineering research opportunities in the development of *soft* actuators for robotic applications, this field of study has been in a lukewarm stage for years. The advent of EAPs (electroactive polymers) recently constitutes an enormous impact on lingering development activity. There are a couple of reasons why the materials deserve keen attention from the robotic engineering field. First, they could provide rectilinear motions without any assistance from a complicated power train. Recognizing that a complicated power transfer mechanism creates bulky robots and it hampers accomplishing delicate missions, total elimination or partial reduction of the power train mechanism benefits expansion of robot application where precise operation is required. Besides, reducing the number of power transmission stages, of course, improves energy efficiency. Second, the inherent flexibility of polymeric materials offers many engineering possibilities for creating biomimetic machines. Acknowledging the fact that animals are naturally soft, more precisely their actuation devices are soft, development of the soft energy transducer should be one of the most important prerequisites for biomimetic robot operations [1, 2].

Although actuators made with the polymers seem to provide many advantages over the traditional electromagnetic actuators, there are still some controversies over the feasibility of actuators. Stability and durability issues of the material, when it is manufactured as an actuator, are the principal concerns of the contention. However, considering the development progress that remains at a primitive stage and the lack of refined manufacturing technology, the debate might not be the limitation or failure of the technology but the consequences for improvement. There are many different polymeric materials available for development, and the research on the polymer transducers is truly multidisciplinary. Especially when a multidisciplinary technology is in its infancy, typically no dominant solution or consensus can be easily made to concentrate resources and accelerate research activities. Despite the ongoing arguments, the prominent beneficiary of polymer actuator technology might be the field of robotics. There are a number of books and articles available on polymer sensors and actuators. Most of them deliver quite a broad range of information about energy transformation from the basic principles to advanced physical phenomena. However, this chapter departs from the norm. In this chapter, robotic devices made with the polymer, especially dielectric elastomer, are introduced. All devices are successfully controlled with the standard feedback or feedforward algorithms. Focusing on robotic applications by coupling polymeric physics and robotic devices, this chapter would be a valuable asset and also an arsenal for readers to explain emerging robotic actuator technology.

There are various types of EAPs available for actuator development, and they are categorized in two groups by energy transduction characteristics. Ionic and nonionic are the groups of materials. Dielectric elastomers are a sort of non-ionic material. In addition, they are normally incompressible but highly resilient so that a high strain level can be achieved. Silicone and acrylic materials are typical examples of dielectric elastomers. An actuator made with dielectric material is basically a two-plate capacitor. A laminated dielectric elastomer sheet is coated with compliant electrodes on both sides. When an electrical field is applied across the electrodes, positive and negative charges are accumulated near the electrode. This generates coulombic attraction forces between the electrodes, and it results in mechanical pressure, called Maxwell stress. Compressive mechanical pressure moves the electrodes closer to each other. Consequently, the incompressible elastomer expands in lateral directions and yields displacement [3].

Although a wide range of active researches has been undertaken for the improvement of many different EAPs, emphases on dielectric elastomer actuators are made in this chapter because the materials are currently more popularly applied to industrial applications than the other EAPs. Two distinct implementations of the material to robotic actuation are introduced. The construction and functionality of the polymer actuators are closely dependent on the initial stretching of the material, so the design concepts and fabrication of polymer actuators with prestrain are provided and followed by those without prestrain.

## 3.2 Prestrained Dielectic Elastomer Actuator

### 3.2.1 Fundamentals of Prestrained Actuator

The basic operation of the dielectric elastomer actuator is simply that the polymer intrinsically deforms either in expanding or in contracting when electrical voltage is applied to the electrodes. Basic principle actuation mechanisms are well explained in many publications [4-6]. Although numerous authors recently have presented various polymer actuators, few have demonstrated practical feasibility of designs and controllable actuator systems that can be implemented with a reasonable amount of control action. In this section, an antagonistically configured dielectric elastomer actuator is presented. Given the material and the geometrical constraints which should be well accounted in controllable actuation, it successfully delivers controlled bidirectional rectilinear mechanical motions by changing its compliance and exerting force.

The principal operation is similar to the electromechanical transduction of a parallel two-plate capacitor, as shown in Figure 3.1. When an electric potential is applied across the polymer film coated with compliant electrodes on both sides, the material is compressed in thickness and expands in the lateral direction. By virtue of this contraction, mechanical actuation force is generated. This physics couples mechanical and electrical energy domains so that energy transduction happens. The effective mechanical pressure, called Maxwell stress along the thickness direction by electrical input, is given by

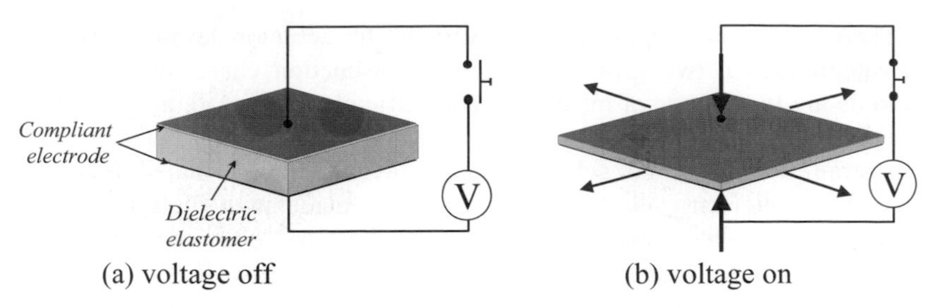


**Figure 3.1.** Basic actuation mode of a dielectric elastomer actuator

$$\sigma_e = -\varepsilon_o \varepsilon_r E^2 \tag{3.1}$$

where $E$ is an applied electric field, and $\varepsilon_o$ and $\varepsilon_r$ are the electric permittivity of free space and the relative permittivity, respectively.

Although the way to acquire mechanical actuation from the basic dielectric elastomer is straightforward, there is still a significant limitation for the basic actuation to be used for a practical application mainly due to its low exerting forces. It is simply because the actuation is provided by a soft and thin polymer film. In addition, the thin polymer film used for the basic operation can be easily ruptured by even small normal forces or buckled by lateral forces. Moreover, the operation is hardly controlled so that it is practically called a simple movement rather than an actuation. One of the possible solutions to overcome the critical limitation of the basic actuation mechanism is prestrained actuator design. This concept was originally proposed by researchers in SRI and they demonstrated a 100% length change of a dielectric polymer sheet by applying pretension in the actuation direction [5].

### 3.2.2 Antagonistic Configuration of a Prestrained Actuator

Noting the fundamental limitations of the basic actuation mechanism, a novel design of an antagonistically configured actuation mechanism, called ANTLA (antagonistically driven linear actuator) is introduced. A schematic illustration of the mechanism is provided in Figure 3.2. With the prestrain effect of the configuration, a polymer sheet may produce a relatively larger displacement [4], although a recent study proves that it is not necessary for acquiring a large strain [6]. Having the prestrained polymer sheet connected to both frame and output terminal, a combination of push-pull forces produces larger actuation displacements.

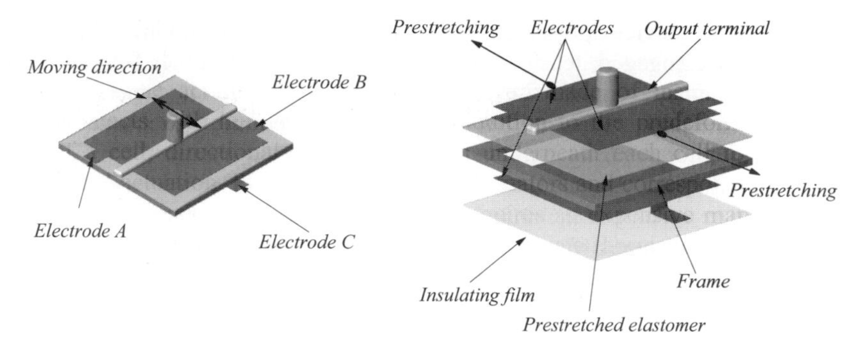


**Figure 3.2.** Fabrication of antagonistically driven actuator

In this design, the actuator is actually composed of a single prestretched elastomer film affixed to a rigid square frame, which provides uniform pretension in the direction of actuation. For the antagonistically driven mechanism that requires two independent polymer sections, a polymer sheet is stretched and a compliant electrode paste is placed on the top and bottom surfaces. Finally, the top compliant electrode is partitioned in two sections. They are called electrode A and B of the top surface, and the common electrode C of the bottom in Figure 3.2. In addition, a mechanical output terminal is attached at the boundary of the partitioned electrodes. Although it is fabricated from a single elastomer film, it works as the antagonistically driven actuator with partitioned electrodes so that it can provide bidirectional "push-pull" actuation.

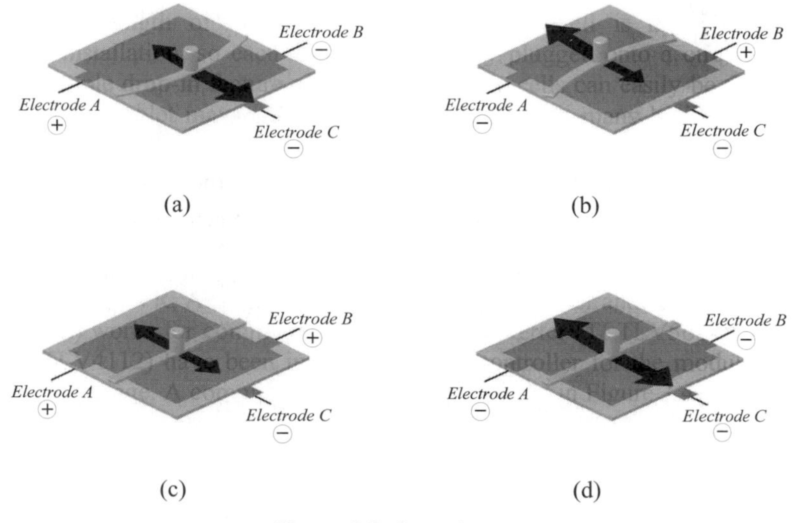


**Figure 3.3.** Operation

In detail, the device presented works as follows. Assuming that uniform pretensions are engaged, tensions on both sides of the elastomer film are initially balanced. When the electrical input is applied to one of the elastomers, it expands, and the force equilibrium is broken. The output terminal, therefore, moves to rebalance the broken elastic equilibrium. For example, as shown in Figure 3.3a, if a positive input voltage is given on electrode A, a negative one on B, and a negative one on C, then the output terminal moves toward electrode B because the elastomer on electrode A expands due to the input voltage on A. Similarly, if a positive voltage input is given on electrode B while a negative input is applied to A and C, then the output terminal moves toward electrode A. Besides the basic actuation, the design presented can provide an additional feature that is normally difficult to acquire from existing traditional actuators. The compliance of the actuator can be actively modulated by controlling input voltages. For instance, as shown in Figure 3.3c, if positive input voltages are given on electrodes A and B while applying negative voltage to C, the actuator becomes more compliant. On the contrary, it becomes stiffer when all applied voltages are the same. The actuator presented delivers four different actuation states; forward, backward, compliant and stiff, which are characteristics of human muscles.

### 3.2.3 Modeling and Analysis

*3.2.3.1 Static Model*

In this section, a static model of the actuator presented is discussed. The modeling process starts from longitudinally dividing the actuator into two sections, as shown in Figure 3.4.

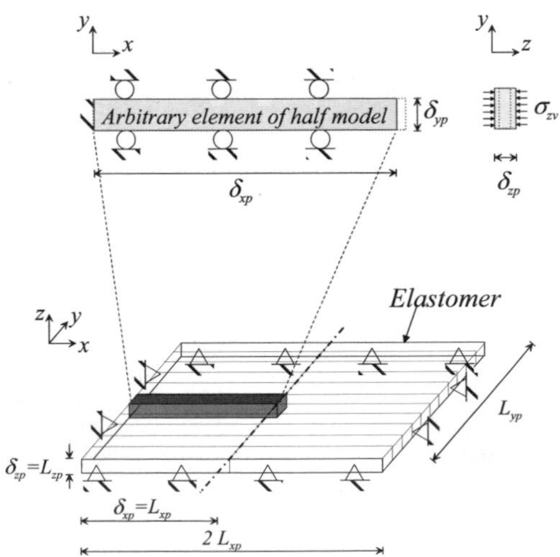


**Figure 3.4.** Discretization of a polymer sheet

It is then horizontally discretized into a number of elements with infinitesimal width. Then a force balance on an element can be derived, as shown in Figure 3.5. Initially, both the left and right sections of the dielectric elastomer sheet are in elastic force balance.

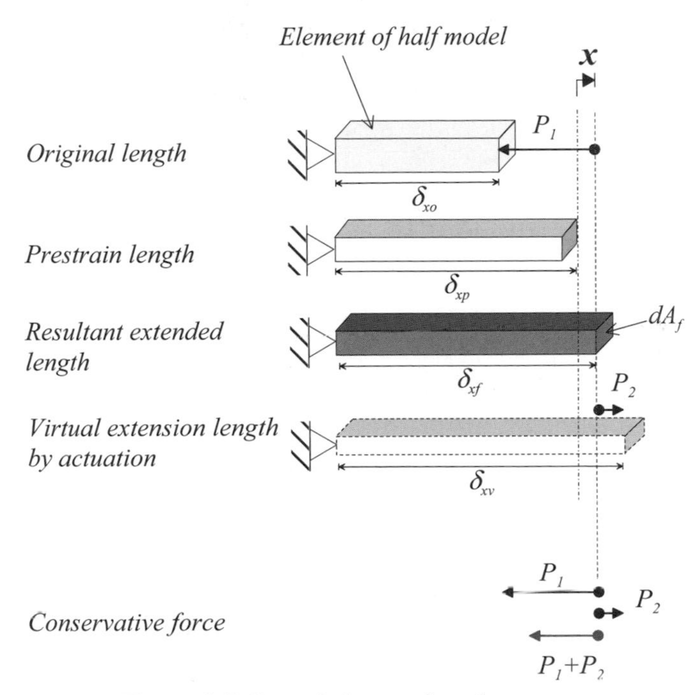


**Figure 3.5.** Force balance of an element

When electric voltage is applied to one of the sections of the dielectric elastomer, the force equilibrium is rebalanced due to the induced Maxwell stress, noted as $P_1$ in the figure. The forces $P_1$ and $P_2$ can be derived as

$$P_1 = -dA_f Y_x^o \varepsilon_{xp} + \frac{x}{\delta_{xo}} \stackrel{\Delta}{=} f_1(x)$$

$$P_2 = dA_f Y_x \left(\frac{\delta_{xf}}{\delta_{xo}}\right) \left[ \frac{Y_z}{Y_z - \varepsilon_r \varepsilon_o \left( \delta_{zo} / \delta_{zf} \right) \left( V / \delta_{zf} \right)^2} - 1 \right] \stackrel{\Delta}{=} f_2(x, V) \tag{3.2}$$

where the super or subscripts, $o$, $p$, and $f$ denote the original unstretched, prestretched, and actuated states, respectively. $Y_x$ denotes the $x$-directional effective elastic modulus, ${}^o\varepsilon_{xp}$ stands for $x$-directional strain caused by prestretching, $dA_f$ is the cross-sectional area of an element, and $\varepsilon_o$ and $\varepsilon_r$ denote the permittivity of free space and the relative permittivity of the elastomer, respectively. $f_1(x)$ represents an elastic restoration force caused by prestretching, whereas $f_2(x,V)$ represents an electrostatic force as a function of the displacement $x$ and the applied voltage $V$. Consequently, the resultant force on the strip is obtained by the summation of the restoration force caused by prestretching and the induced electrostatic force that can be given as

$$P = -Y_x \delta_{yp} \delta_{zf} \left( {}^{o}\varepsilon_{xp} + \frac{x}{\delta_{xo}} - \frac{\delta_{xf}}{\delta_{xo}} \right) \left[ \frac{Y_z}{Y_z - \varepsilon_r \varepsilon_o \left( \delta_{zo} / \delta_{zf} \right) \left( V / \delta_{zf} \right)^2} + 1 \right] \tag{3.3}$$

where $\delta_{yp}$ is the width of the infinitesimal strip of the actuator, as shown in Figure 3.5. The equations obtained from the half strip with infinitesimal width can be easily extended to the full model by integrating the forces from numerous strips with infinitesimal width. In the full model, the final displacement is determined at the equilibrium point between the force of the left half model and that of the right half one, as shown in Figure 3.5. Assuming that the positive direction is toward the right side, in an arbitrary displacement $x$, the total output force can be derived as

$$F = P_R - P_L \overset{\Delta}{=} g_k \cdot K(x) - g_e \cdot E(x, V_L, V_R) \tag{3.4}$$

where the forces on the output terminal by the left elastomer and the right one are represented by $P_L$ and $P_R$, $K(x)$ and $E(x,V_L,V_R)$ represent the force by the prestrain and electrostatic effect, and $V_L$ and $V_R$ are input voltages on the left and right sides of the elastomer in Figure 3.4, respectively. $K(x)$ and $E(x,V_L,V_R)$ are defined as

$$K(x) \overset{\Delta}{=} Y_x L_{yp} \left[ {}^{o}\varepsilon_{xp} \left( \delta_{Lzf} - \delta_{Rzf} \right) + \frac{x\left(1 + {}^{o}\varepsilon_{xp}\right)}{\delta_{xp}} \left( \delta_{Lzf} + \delta_{Rzf} \right) \right] \tag{3.5}$$

and

$$E(x, V_L, V_R) \overset{\Delta}{=}$$

$$Y_x L_{yp} \left\{ \delta_{Lzf} \frac{\delta_{Lxf}}{\delta_{Lxo}} \left[ \frac{Y_Z}{Y_Z - \frac{\delta_{Lzo}}{\delta_{Lzf}} \varepsilon_r \varepsilon_o \left( \frac{V_L}{\delta_{Lzf}} \right)^2} - 1 \right] - \delta_{Rzf} \frac{\delta_{Rxf}}{\delta_{Rxo}} \left[ \frac{Y_Z}{Y_Z - \frac{\delta_{Rzo}}{\delta_{Rzf}} \varepsilon_r \varepsilon_o \left( \frac{V_R}{\delta_{Rzf}} \right)^2} - 1 \right] \right\} \quad (3.6)$$

where $\delta_{Ljf}$ and $\delta_{Rjf}$ are the $j$ directional final length of the left side and the right side. $\delta_{Ljo}$ and $\delta_{Rjo}$ are the $j$ directional original length of the left and the right sides such as $j = x, z$, respectively, and $g_k$ and $g_e$ are called *effective restoration* and *electrostatic coefficients*, respectively. These coefficients are dependent on geometries such as frame size and thickness of the output terminals. They are determined experimentally in general although they have unit value in the ideal case. Eqs. (3.5) and (3.6) provide the static relations between the displacement and input voltages.

*3.2.3.2 Dynamic Model*

To derive a dynamic model the actuator presented is simplified as a lumped model , as shown in Figure 3.6. Nevertheless, the mathematical model still has a complicated form because the model has some nonlinear aspects like viscous damping. Based on the lumped model, the dynamic equation of the actuator can be expressed as

$$F(t) = M\,\ddot{x} + B(\dot{x}) + g_k K(x) - g_e E(x, V_L, V_R) \quad (3.7)$$

where $M\ddot{x}$ is the inertial force, $B(\dot{x})$ represents the damping force, and $F(t)$ denotes external forces. $K(x)$ and $E(x, V_L, V_R)$ are obtained from Eqs. (3.5) and (3.6), and the other terms, $M\,\ddot{x}$ and $B(\dot{x})$, are to be derived next. $M$ is the summation of the mass of the load, structural parts, and elastomer film of the actuator, and it varies during operation.

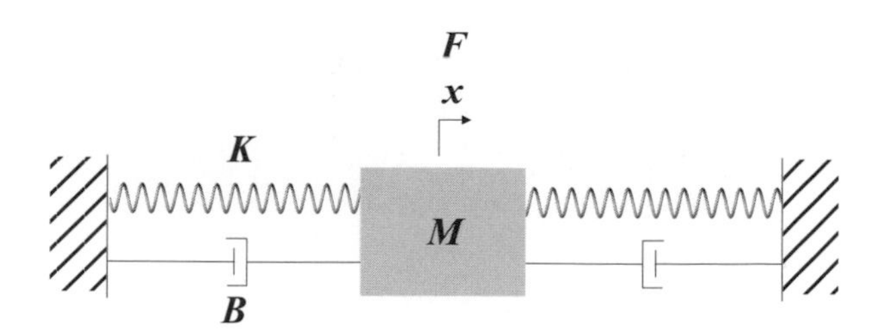


**Figure 3.6.** Dynamic model of the actuator

In the present formulation, however, only the equivalent mass of the elastomer is considered, and the other terms such as the mass of the output terminal may be included in the model by increasing the equivalent mass. To get the equivalent mass of the elastomer, the actuator is considered to be composed of n elements that are evenly divided, as depicted in Figure 3.7. Because each $q_{\text{th}}$ element has its own displacement $x_q = \Sigma_{i=1}^{q} dx_i$, and acceleration $\ddot{x}_q = \Sigma_{i=1}^{q} \frac{d^2 x_i}{dt^2}$, assuming that $m_i = dm = (\rho\Lambda)/n$ and $d\ \ddot{x} = \ddot{x}/n$, the inertial force term will be

$$M_L \ddot{x} = dm[nd\ddot{x}_1 + (n-1)d\ddot{x}_2 + \cdots + \ddot{x}_n] = \frac{(n+1)}{2n} \rho\Lambda_L \ddot{x} \tag{3.8}$$

where $\rho$ is the mass density of the elastomer, $\Lambda_L$ is the volume, and $M_L$ denotes the equivalent mass of the left half of the actuator. Similarly, a model of the right half of the actuator can be derived. The overall equivalent mass of the elastomer becomes

$$M = M_L + M_R = \frac{1}{2} \cdot \rho \cdot \Lambda \tag{3.9}$$

where $\Lambda$ denotes the total volume of the elastomer and $M_R$ means the equivalent mass of the right half of the elastomer. Because elastomers are viscoelastic materials, they have complicated energy dissipating mechanisms. Therefore, it is not easy to take all the effects into consideration for modeling. Instead, overall effects had better be included in the model by introducing the concept of equivalent damping, which may not be constant but in the form of an equation.

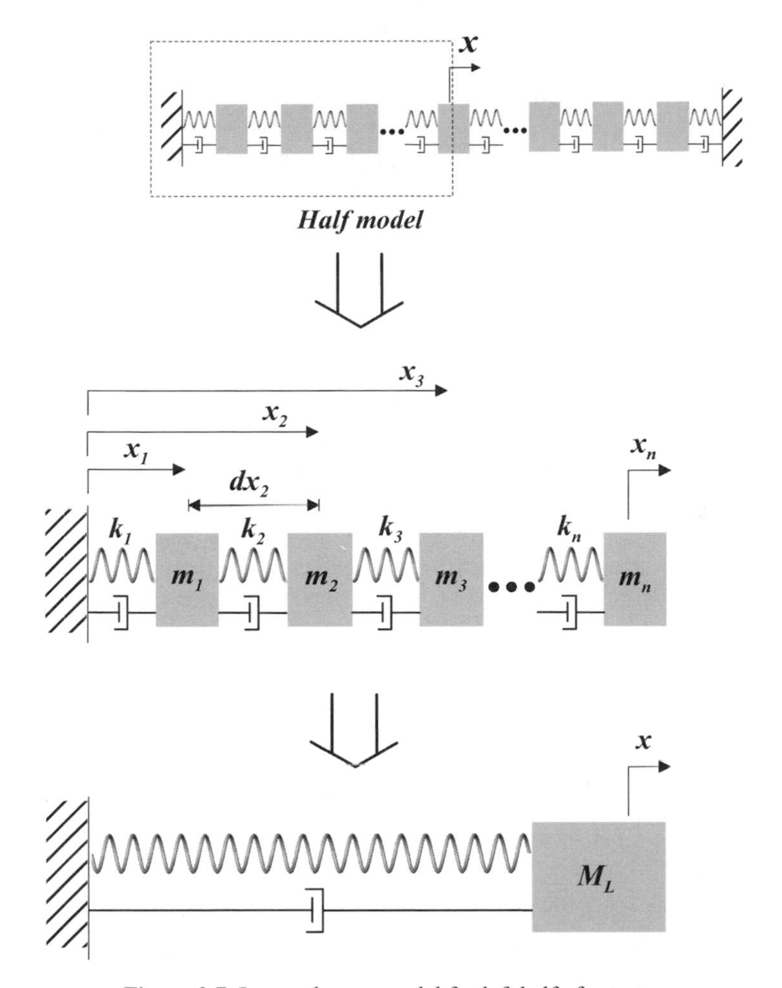


**Figure 3.7.** Lumped mass model for left half of actuator

For example, the equivalent viscous damping $B(\dot{x})$ of VHB4905 can be represented as

$$B(\dot{x}) = \frac{h}{\omega^{0.25}} \dot{x}^{1/2} \tag{3.10}$$

where $h$ denotes the hysteretic damping coefficient of VHB4905 and $\omega$ represents the driving frequency. The $h$ depends on the material, and it is 0.3 for VHB4905 with 200% prestrain. In the dynamic equation of Eq. (3.7), all terms are to be derived with Eqs. (3.5), (3.6), (3.9), and (3.10). This model will be useful for developing a control method for the actuator presented.

### 3.2.4 Control of Actuator

In this section, a method for control of both displacement and compliance of the antagonistically configured actuator is explained. For reasonable handling of the nonlinear characteristics of the actuator, a linearization of the actuator system is performed, and a modified nonlinear decoupling control method is applied to the linearized system.

*3.2.4.1 Linearized System Model*

From Eqs. (3.4) - (3.6), the stiffness of the actuator is calculated as

$$\kappa = \frac{\partial F}{\partial x} = \frac{\partial}{\partial x}\left[g_k K(x) - g_e E(x, V_L, V_R)\right] \tag{3.11}$$

where $\kappa$ denotes the stiffness. To control the compliance of the actuator, it is necessary to find the inverse function of Eq. (3.11). It is not easy to derive a closed form solution of the inverse because Eq. (3.11) has some complicated nonlinear terms. Therefore, a linearization about the equilibrium position is needed to elaborate the inverse solution and the control law. By employing Taylor's series expansion (limited to the first order), a linearized model for the overall system is obtained as follows:

$$F = \left( -\frac{\partial P_L}{\partial x}\Big|_{x=0} + \frac{\partial P_R}{\partial x}\Big|_{x=0} \right) x + \left( -P_L\big|_{x=0} + P_R\big|_{x=0} \right) \stackrel{\Delta}{=} \kappa x + \beta \tag{3.12}$$

where $\kappa$ and $\beta$ mean the stiffness and the output force at the equilibrium point of the whole system. The left and right halves of the actuator are symmetrical with respect to the $x-y$ plane, so linearized equations can be derived such as

$$\kappa = A_1 - A_2 \left[ \frac{V_R^2}{\left(A_3 - A_4 V_R^2\right)^2} + \frac{V_L^2}{\left(A_3 - A_4 V_L^2\right)^2} \right] \tag{3.13}$$

and

$$\beta = A_5 \left( \frac{1}{A_3 - A_4 V_R^2} - \frac{1}{A_3 - A_4 V_L^2} \right) \tag{3.14}$$

where

$$A_1 = g_\kappa \frac{2Y_x \delta_{zp} L_{yp}}{\delta_{xp}}$$
$$A_2 = g_e \frac{3Y_x Y_z \delta_{zp}^4 L_{yp} \delta_{z0} \varepsilon_o \varepsilon_r}{\delta_{x0}}$$
$$A_3 = Y_z \delta_{zp}^3 \qquad (3.15)$$
$$A_4 = \delta_{z0} \varepsilon_o \varepsilon_r$$
$$A_5 = g_e \frac{Y_x Y_z \delta_{zp}^4 L_{yp} \delta_{xp}}{\delta_{x0}}$$

*3.2.4.2 Modified Nonlinear Decoupling Control System*

By using the linearized system model, a modified nonlinear decoupling controller is developed. The controller employs a plain scheme popularly used in various control and robotic applications. It provides the ability to control both position and stiffness. Figure 3.8 shows the overall structure of the controller.

A detailed internal structure of the compliance controller that is a subpart of the controller shown in Figure 3.8 is provided in Figure 3.9. The controller is composed of an inverse equation and a stiffness compensator. Shown in Figure 3.9, the inputs of the controller are the load $F$, the desired stiffness $k_d$, and the displacement $x$. The outputs of the controller are $V_L$ and $V_R$. Note that the stiffness $k_d$ is the desired stiffness, and Eq. (3.12) is reconstructed as follows:

$$\kappa = k_d$$
$$\beta = F - k_d x \qquad (3.16)$$

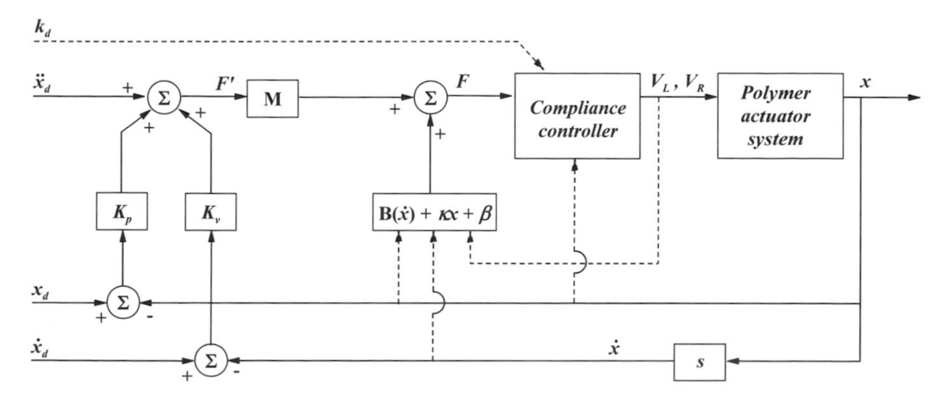


**Figure 3.8.** Structure of modified nonlinear decoupling controller

Therefore, the inverse solution for the stiffness input can be obtained by solving the two equations in Eq (3.16).

The input voltages are calculated by

$$V_L = \sqrt{\frac{C_1 + C_2}{C_3 - C_4}}$$

$$V_R = \sqrt{\frac{C_1 + C_2}{C_3 + C_4}} \tag{3.17}$$

where

$$C_1 = A_3 A_4^2 A_5^2 (k_d - A_1) + A_2 A_3^2 A_4 (F - k_d x)^2 - A_2 A_4 A_5^2$$

$$C_2 = \left[-2 A_2 A_3 A_4^3 A_5^4 (k_d - A_1) - A_2^2 A_3^2 A_4^2 A_5^2 (F - k_d x)^2 + A_2^2 A_4^2 A_5^4\right]^{\frac{1}{2}}$$

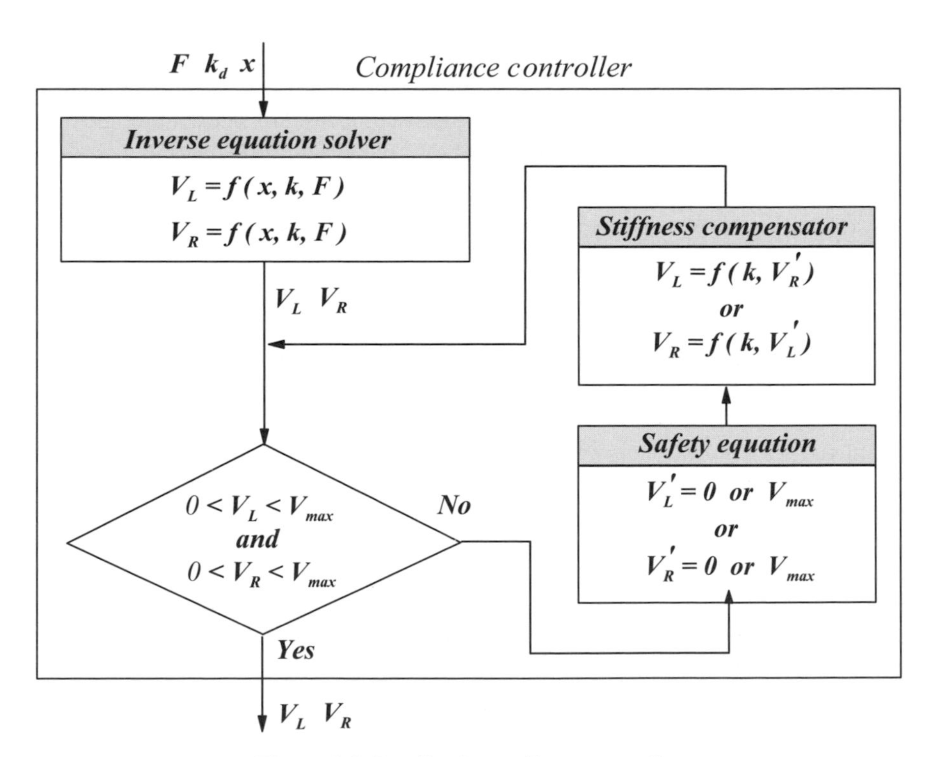


**Figure 3.9.** Details of compliance controller

$$C_3 = A_4^3 A_5^2 (k_d - A_1) + A_2 A_3 A_4^2 (F - k_d x)^2$$

$$C_4 = A_2 A_4^2 A_5 (F - k_d x) \tag{3.18}$$

As shown in Figure 3.9, the inverse outputs $V_L$ and $V_R$ obtained must be determined within the limit of the actuating voltages. Consequently, the stiffness compensator can be derived from Eq. (3.13) as follows:

$$V_L = \sqrt{\frac{-A_2 + A_3 G_R + \sqrt{A_2^2 - 2A_2 A_3 G_R}}{A_4 G_R}}$$

$$V_R = \sqrt{\frac{-A_2 + A_3 G_L + \sqrt{A_2^2 - 2A_2 A_3 G_L}}{A_4 G_L}} \tag{3.19}$$

where

$$G_L = 2A_4\left[k_d - A_1 + A_2 \frac{V_L'^2}{(A_3 - A_4 V_L'^2)^2}\right]$$

$$G_R = 2A_4\left[k_d - A_1 + A_2 \frac{V_R'^2}{(A_3 - A_4 V_R'^2)^2}\right] \tag{3.20}$$

Thus, Eq (3.17) can be termed an inverse stiffness solution.

### 3.2.5 Inchworm Microrobot Using Prestrained Actuator

The design of ANTLA can be directly applied to biomimetic microrobots. Some examples are shown in this section.

Annelid animals such as earthworms or maw worms, etc., contain metameric structures composed of numerous ringlike segments. In their movements, annelid animals use longitudinal muscles as well as circular ones and carry out locomotion by contracting these muscles alternatively without looping their metameric structures. To reproduce this muscle combination, three ANTLAs embedded in the metameric structure of the robot provide actuation forces, while they play the role of the frame as well. It not only enables efficient actuation but also makes it possible to manufacture a robot by a totally new fabrication method that enables mass production of a robot through processes such as injection molding or stamping [10,11]. In addition, there are possibilities to easily mimic the natural and delicate motions such as animal skin motions, wrinkling, and eyebrow movement without using many actuators. The design of the segment addressed in this section illustrates a realization of embedding actuators in the robot without using complicated mechanisms or their substitutes.

*3.2.5.1 Design of a Segment*

The metameric structure of annelid animals features a number of ringlike segments. The segment can be regarded as an independent actuator capable of exerting multiple DOFs of motion. As shown in Figure 3.10, the segment is composed of two parts, a lower body and the upper body. The outer diameter of the segment is 30 mm and the length is 18 mm. The lower body is composed of a plastic frame and three ANTLAs. The ANTLAs are fixed equidistantly along the circumferential direction on the frame. The actuators are attached to the frame of the segment body made with plastics, so there is a large cavity in the central part of the segment that can be used for embedding devices and equipment that a robot has to carry for its main missions. The lower body plays a role as a connector that contains a mechanical joint and a structure for attaching another adjacent segment.

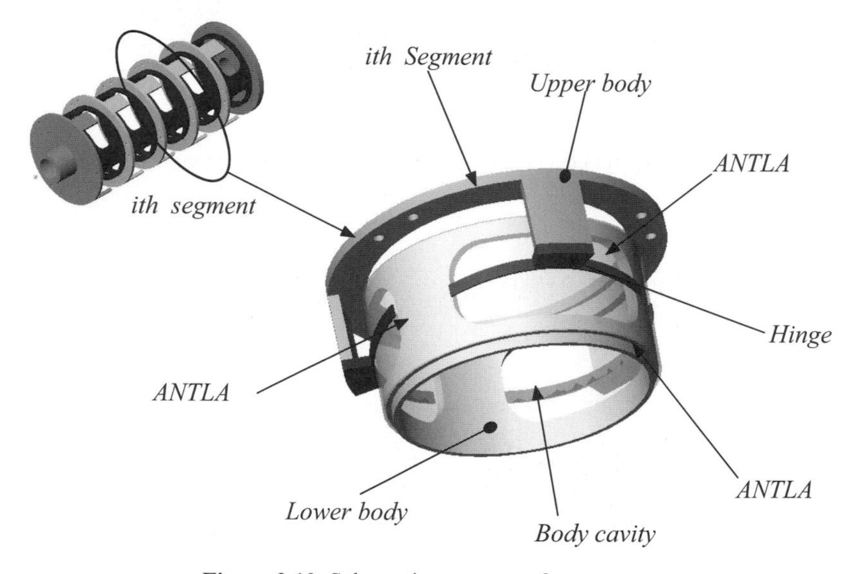


**Figure 3.10.** Schematic structure of a segment

Each ANTLA generates a single degree-of-freedom linear motion, so a segment composed of three of them produces three-degrees-of freedom motion. Actuation of the three actuators produces motion of the upper body relative to the lower body, thus motions like translation, panning, and tilting are possible. A segment can be modeled as a parallel mechanism with three linear actuators. The push-pull motion of the ANTLA is represented as a couple of active prismatic joints so are connectors between the upper and the lower body simplified as spherical joints.

In Figure 3.11, the motion of a segment is illustrated. The moving direction of the segment is designated with an arrow. As illustrated in Figures 3.11c and b, the segment moves up or down when all of the actuators (ANTLA) go up or down simultaneously. On the contrary, the segment turns if one ANTLA moves upward and the others go downward, as shown in Figure 3.11d. Consequently, the segment can generate three DOF motions.

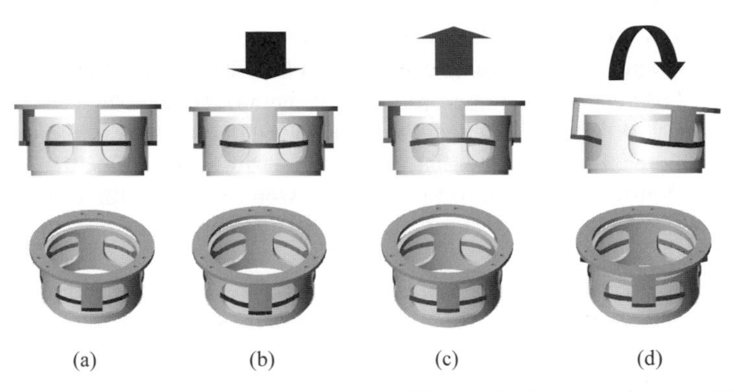


**Figure 3.11.** Actuating sequences of segment illustrated: (a) neutral state (b) moving upward (c) moving downward (d) turning by asymmetrical actuation

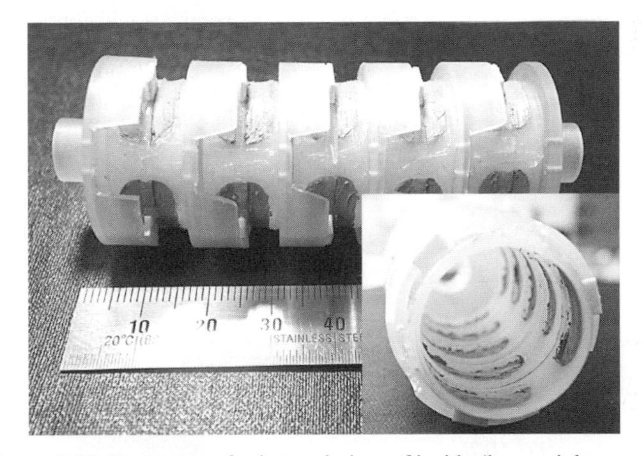


**Figure 3.12.** Prototype of robot and view of inside (lower right corner)

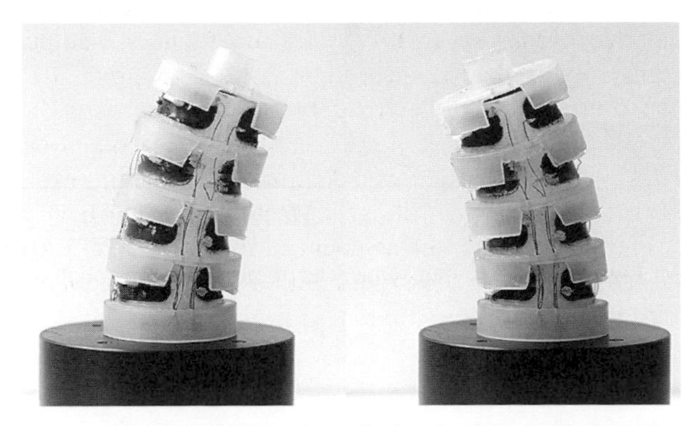

**Figure 3.13.** Bending motion of robot in right and left turning

*3.2.5.2 A Microrobot*

By connecting several segments in series, an inchworm microrobot can be created, as shown in Figure 3.12. The microrobot can generate multiple DOF motions with five segments (or can be extended to a hyperredundant one if the number of segments is increased). The robot has plenty of hollow space inside that provides a morphological structure similar to an annelid as well as a space for accommodating electronic parts, power supply, electrical wiring, and control systems for the robot operation.

An accumulated turning angle of each segment generates a bending motion of the robot, as shown in Figure 3.13. By a proper combination of sequential motion of each segment, the microrobot generates translational motion, as illustrated in Figure 3.14. The shaded region in the figure represents an active segment. First, the front segment of the actuator expands while the others remain still. Then, the next actuator expands while the front segment contracts. By a consecutive motion of the aforementioned actuation pattern from the front to the tail, the robot moves forward. A demonstration of the robot motion for translational movement in a tube is shown in Figure 3.15.

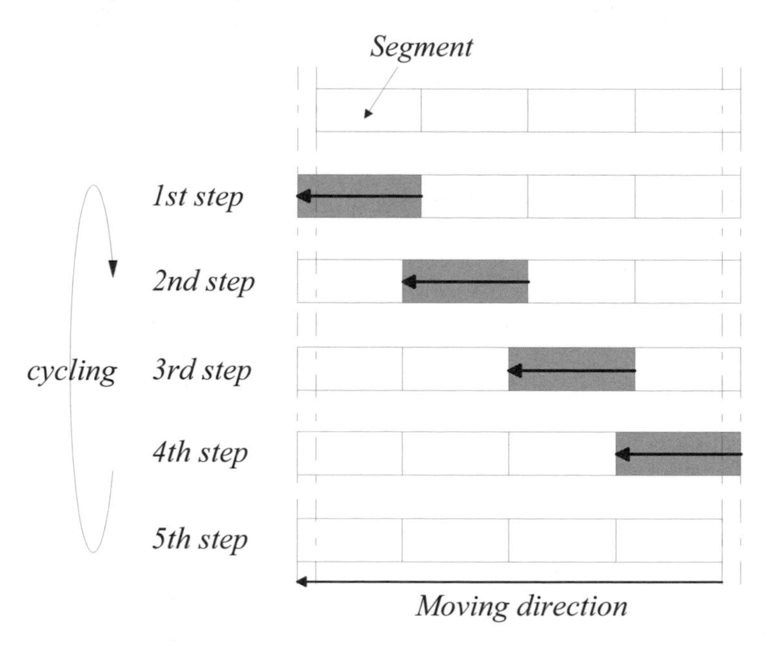


**Figure 3.14.** Moving sequences of an inchworm robot

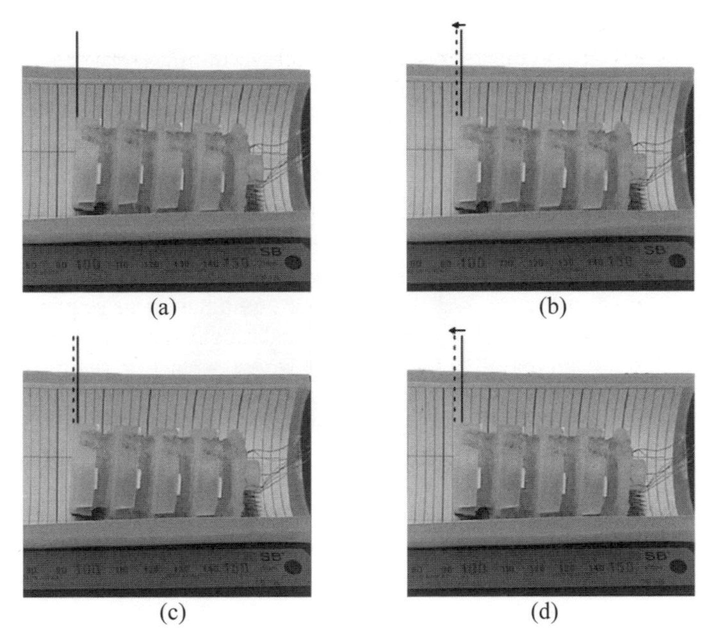

**Figure 3.15.** Translational motion of a microrobot (solid line: initial, dotted line: moved position)

### 3.2.6 Multi-DOF Linear Motor Using a Prestrained Actuator

In the previous sections, the basic working principles, benefits, and drawbacks of an antagonistically configured dielectric elastomer actuator have been delineated. Given the material and the geometrical constraints which should be well accounted for by controllable actuation of the material, the prestrained ANTLA successfully provides controlled bidirectional linear mechanical motions by changing its compliance and exerting force. Guaranteed controllability of motion with a simple control law is, of course, one of the most important design requirements of any kind of actuator. Especially, when they are applied to human-muscle like motion generators, simple architecture and easy control should be ensured.

One of the significant drawbacks of the simple ANTLA design is fragility of the film by direct external forces normal to its surface. Recognizing the disadvantages of the simple ANTLA design, a new actuator concept using dual films is introduced so that it successfully bears heavy transverse forces. This design provides a rugged linear actuation mechanism as well as expandability to multi-degree-of-freedom actuation. Furthermore, the new design is not limited to dielectric elastomer but can be directly applied to other types of polymeric actuators.

*3.2.6.1 Single DOF Linear Motor*

The actuator is constructed with two films made from a dielectric elastomer. A schematic of the design concept is provided in Figure 3.16.

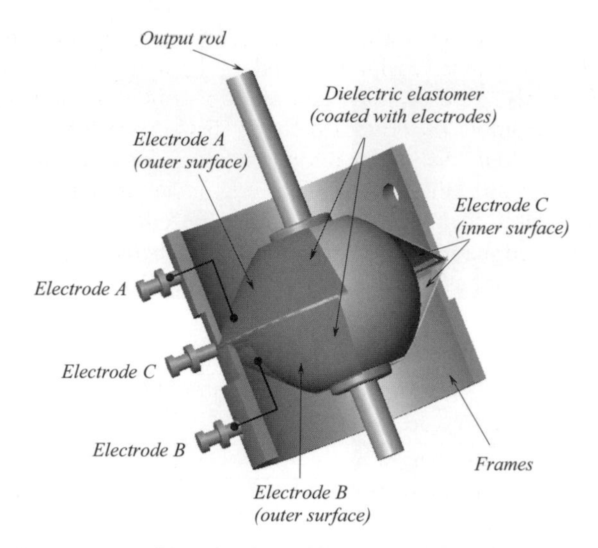


**Figure 3.16.** Schematic view of linear motor actuator concept

Two sheets of elastomer film are used for the design. Each film is mounted on a circular frame that works as a ground electrode. It is designated electrode C in the figure. The two mounted films are prestrained by sandwiching a pretensioning rod. Once the assembly is done, the rod is positioned and remains at the strain force equilibrium of the two pretensioned elastomer units unless electrical input is applied. If the elastic force equilibrium is broken for any reason, the rod will move either upward or downward and stops at a new equilibrium position. This is the fundamental working mechanism of the actuator.

A more detailed explanation of the actuator working mechanism is provided in Figure 3.17. When an electric potential is applied to electrodes A and C across the top film, the dielectric elastomer film expands, and it results in breaking the initial strain force balance.

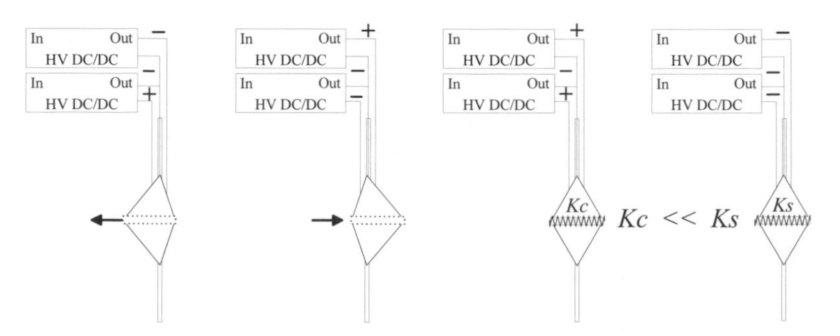


**Figure 3.17.** Principal actuation mode

Hence, the rod will move upward and stop at a new equilibrium position. Because both polymer films are constrained by each other through the rod, the proposed actuator motion should be controllable only by adjusting the stiffness of the films.

The actuator is to return to its original state as soon as the input is removed, and the actuation can be, of course, easily reversed by applying electrical input to the bottom film. Similarly, a push-pull actuation is possible by alternating actuations of the two. Some example patterns of actuator driving are listed in Table 1. This paradigm provides four distinct working modes, forward, backward, highly compliant, and highly stiff, those are similar to human muscle fibers.

Having controllability of the linear actuator introduced, a single-DOF linear actuator is constructed, named a *polymer motor*. A schematic cross-sectional view of a fully packaged motor assembly is shown in Figure 3.18.

**Table 3.1.** Linear motor driving paradigm

| State | Electrode (ABC) | | | | | | |
|---|---|---|---|---|---|---|---|
| Stiff state | - | - | - | or | + | + | + |
| More compliant | + | + | - | or | - | - | + |
| Action toward A | + | - | - | or | - | + | + |
| Action toward B | - | - | + | or | + | + | - |

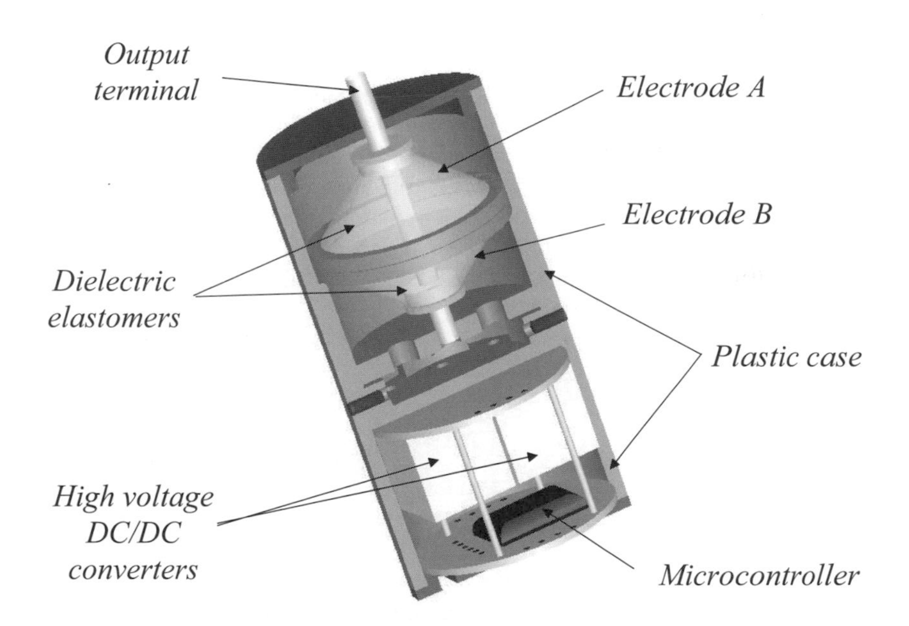


**Figure 3.18.** Cross-sectional view of polymer motor

The thrust force and physical size of the actuator are potentially scalable so that it could be applied for various actuator operations ranging from microrobotics to consumer electronics. Moreover, it generates a thrust force by a true linear mechanism without any mechanical transformers, which guarantees higher energy efficiency, quiet operation, and easy control. A prototype of the polymer motor is shown in Figure 3.19. All of the essential components such as the dielectric elastomer actuator, the driving circuits, the microcontroller, and an RS-232C serial interface are integrated in a unit.

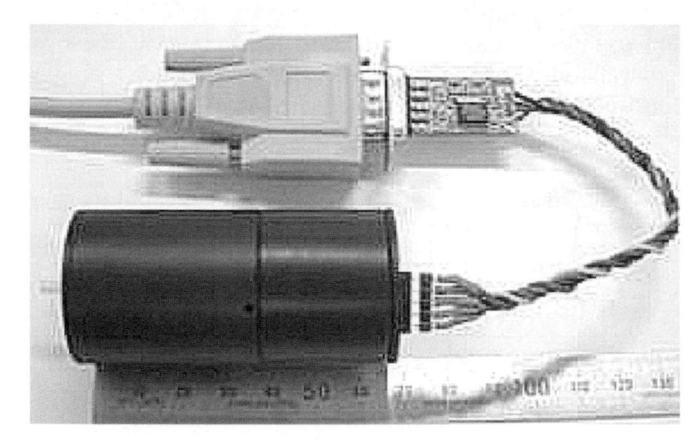

**Figure 3.19.** Prototype of polymer motor

*3.2.6.2 Multi-DOF Polymer Motor*

The concept presented of the single-DOF polymer motor can be easily extended to a multi-DOF motion actuator using a partitioned electrode coating scheme that is illustrated in Figure 3.20. The actuator consists of eight polymer film sections, four sections on each side. It could be manufactured by simply partitioning the elastomer surface and applying the carbon coating separately on each partitioned area during the electrode coating process. Each polymer section is divided with separate electrodes, so each quadrant must be controlled independently.

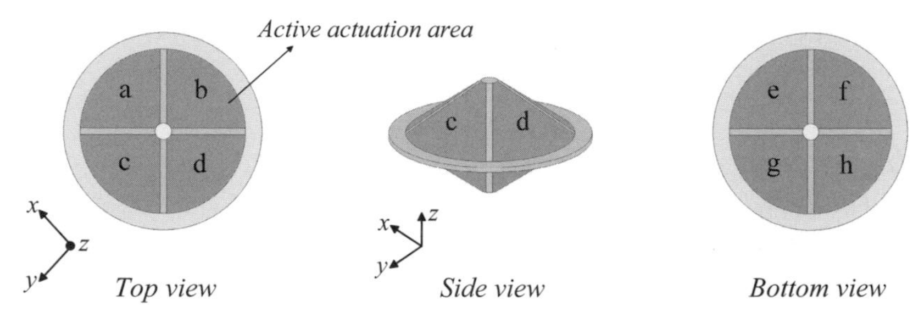


**Figure 3.20.** Schematic of a five DOF polymer motor design

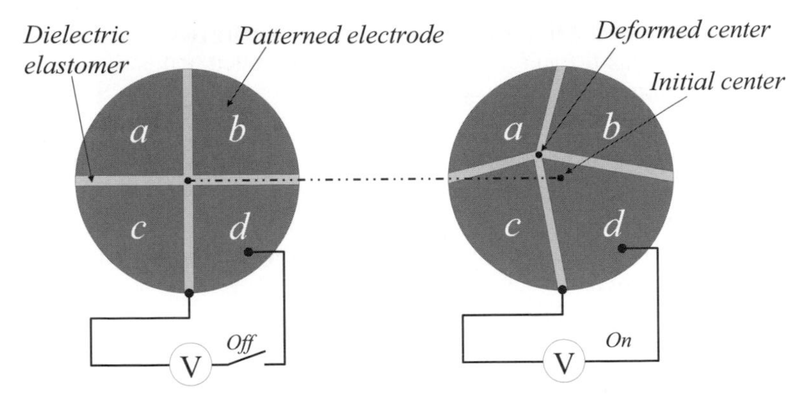


**Figure 3.21.** Generation of multi-DOF motions

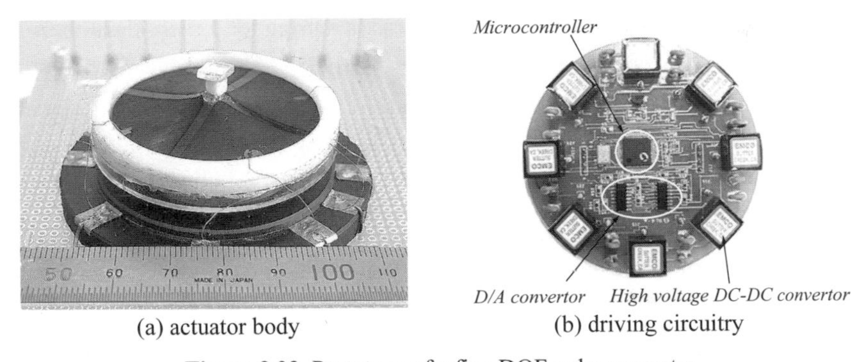


(a) actuator body (b) driving circuitry

**Figure 3.22.** Prototype of a five DOF polymer motor

A proper combination of individual motions of each section might provide continuous multi-DOF actuation. For example, when sections *d* and *h* are actuated, the output terminal moves in the positive *x* direction. Figure 3.21 shows the generic idea for creating translational motions. If sections *c* and *f* are turned on in a non-symmetrical input pattern, the output terminal will be tilted with respect to the positive *x* axis. Then, if the control action succeeds in providing electrical input to sections *d* and *e*, the terminal will rotate about the positive *y* axis.

Further continuous control action with proper adjustment of the input voltage during the transition of the rotating axis helps to keep the terminal in smooth rotation. A prototype of the multi-DOF polymer motor is shown in Figure 3.22 with its control circuit that could be combined in an actuator assembly.

## 3.3 Nonprestrained Dielectric Elastomer Actuator

### 3.3.1 Fundamentals of Nonprestrained Actuator

Having some amount of prestrain on dielectric elastomer actuators improves the deformation and helps in generating macromotions. However, there are a few counteractions due to the prestrain. One of the critical drawbacks is stress relaxation and hysteresis, although the extents depend on the materials, and it is still the key concern over the actuator.

Recognizing the disadvantages of prestrained actuator designs, in the following sections, a different actuator design concept without using pretension is introduced that maintains the initial performance without the rapid aging process that happens by relaxation. The stress relaxation of the prestrained actuator is illustrated in Figure 3.23. The working procedure of the actuator depicted in Figure 3.23 is (a) initial state - the restoration force of stretched actuator $F_L$ is balanced by the restoration force of elastic body $F_R$ such as a spring or other elastomer; (b) actuation state - a longitudinal expansion force $F_m$ , which leads from the electrostatic force $F_{Maxwwell}$ that is engaged on the actuator in the vertical direction by applying voltage, is added to the restoration force; (c) new equilibrium state - the two materials, dielectric elastomer actuator and elastic body, are deformed to rearrange the equilibrium condition of the restoration forces.

Assuming uniform pretensions are engaged along the prestrain, the tensions on both elastomer films are initially balanced in that direction. Once the electrical input is applied that induces Maxwell stress, the balance is broken. Then, the mid point is moved by the unbalanced force, and it stops at a new balanced position. This process generates a large force and displacement.

As mentioned, however, the actuator design has a couple of drawbacks. First of all, the most significant problem is caused by the viscoelastic characteristic of polymer materials. A viscoelastic material shows a combination of elastic and viscous behavior so that an applied stress results in an instantaneous elastic strain followed by a viscous time-dependent strain.

The time-dependent behavior of the material under a quasi-static state can be categorized into three types of phenomena: creep, stress relaxation, and constant rate stressing. Figures 3.24 and 3.25 show the viscoelastic behavior of VHB4905 (3M) that is one of the most representative dielectric elastomers.

Most prestrained actuator designs suffer from viscoelastic characteristics. Hence the stress induced by the stretching of the elastomer gradually decreases and the output force of the actuator may decrease because the maximum actuation force is dependent on the pretension of the elastomer. In addition, the actuation of the pre-strained elastomer normally requires another elastic counterpart such as a spring or elastomer, as shown for ANTLA.

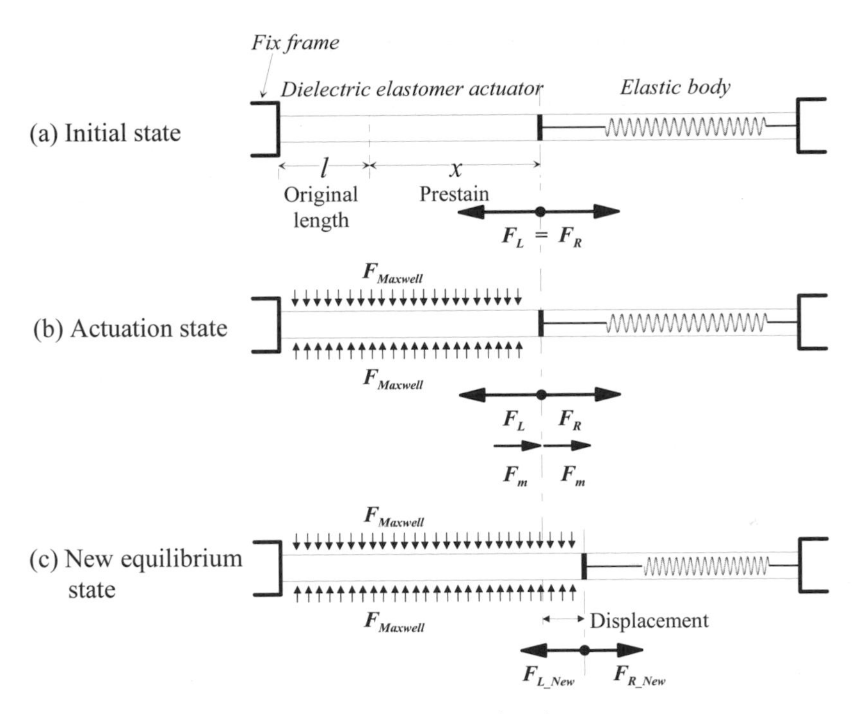


**Figure 3.23.** Stress relaxation

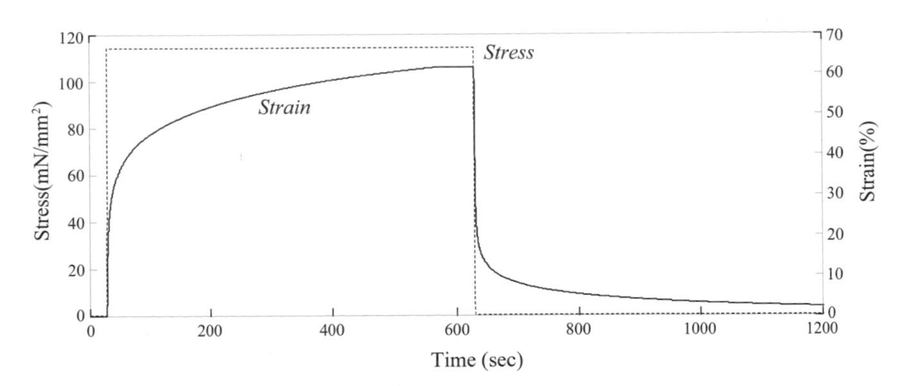


**Figure 3.24.** Creep at constant stress

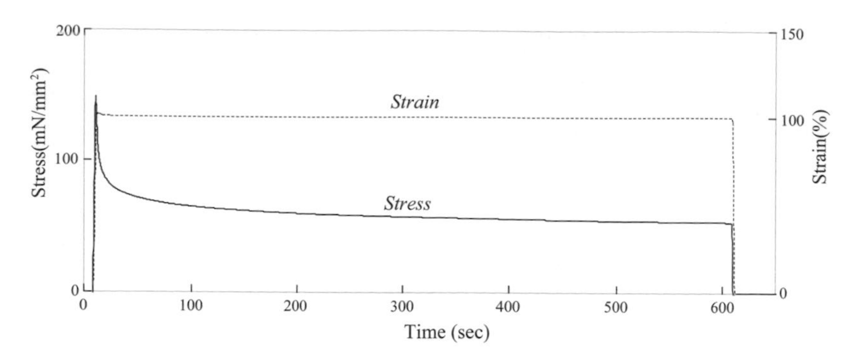


**Figure 3.25.** Stress relaxation at constant strain

To avoid the time-dependent behavior of the dielectric elastomer actuator, the pretension should be removed and only a pure compressive force induced by the Maxwell stress should be used for actuation. For the first step of the nonprestrained actuator design, the amount of deformation of the dielectric elastomer caused by the Maxwell stress must be calculated. The governing equation should be modified for the vertical strain $\delta_z$ according to the compression stress $\sigma_z$.

$$t = (1+\delta_z)t_0 \tag{3.21}$$

$$\frac{\sigma_z}{Y} = \delta_z = -\frac{1}{Y}\varepsilon_o\varepsilon_r\left(\frac{V}{(1+\delta_z)t_0}\right)^2 = -\frac{1}{Y}\varepsilon_o\varepsilon_r\left(\frac{V}{t_0}\right)^2\frac{1}{(1+\delta_z)^2} \tag{3.22}$$

$$\delta_z^3 + \delta_z^2 + \delta_z = -\frac{1}{Y}\varepsilon_0\varepsilon_r\left(\frac{V}{t_0}\right)^2 \tag{3.23}$$

where $Y$ denotes the elastic modulus, $\delta_z$ is the strain in the vertical direction, and $t_0$ is the initial thickness.

Figure 3.26 shows the vertical strain $\delta_z$ curve versus voltage for silicone KE441(ShinEtsu) whose material properties are shown in Table 3.2. As shown in Figure 3.26, the estimated amount of compressive strain is about 1-3.5 %, although that is dependent on the material properties and the applied input voltage. Most of dielectric elastomers are incompressible, so if the actuator is assumed to be a thin circular disk, the strain is derived as

$$(1+\delta_x)(1+\delta_y)(1+\delta_z) = (1+\delta_r)^2(1+\delta_z) = 1 \tag{3.24}$$

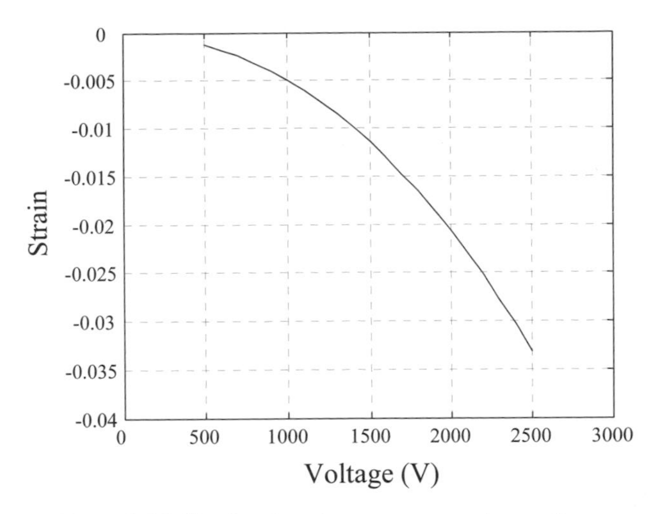


**Figure 3.26.** Simulated strain curve versus given voltage

**Table 3.2.** Material properties of KE441 silicone

|  |  |
|---|---|
| Elastic modulus (Mpa) | 2 |
| Breakdown voltage (kV/mm) | 20 |
| Relative permittivity | 2.8 |

where

$$\delta_r = 1/\sqrt{1+\delta_z} - 1 \tag{3.25}$$

Approximation of Eq. (3.25) yields

$$\delta_r \approx -(1/2)\delta_z \tag{3.26}$$

Eq. (3.26) means that the usable strain is only half of the vertical strain. For that reason, either a material with a higher dielectric constant or very high input voltage is required for a better actuator performance. However, neither seems to be very practical because the polymeric materials commercially available have limited dielectric characteristics and the electrical circuit devices handling high voltage are also limited. Therefore, a new actuating method has to be developed for the nonprestrained actuator.

The basic operating concept of the nonprestrained dielectric actuator is illustrated in Figure 3.27. As shown in Figure 3.27a, a thin dielectric elastomer sheet is confined by rigid boundaries. Once a compressive force is applied to the sheet, it must expand. That induces buckling situation in the sheet and the sheet has to become either convex or concave. This idea makes an efficient actuation without prestrain. The relation between the curvature $r$, the angle $\theta$, and the strain $\delta_a$ can be derived as follows:

$$b = a(1+\delta_a) = r\theta \tag{3.27}$$

$$r = \sin(\theta/2) = a/2 \tag{3.28}$$

$$\frac{\theta}{\sin(\theta/2)} = 2(1+\delta_a) \tag{3.29}$$

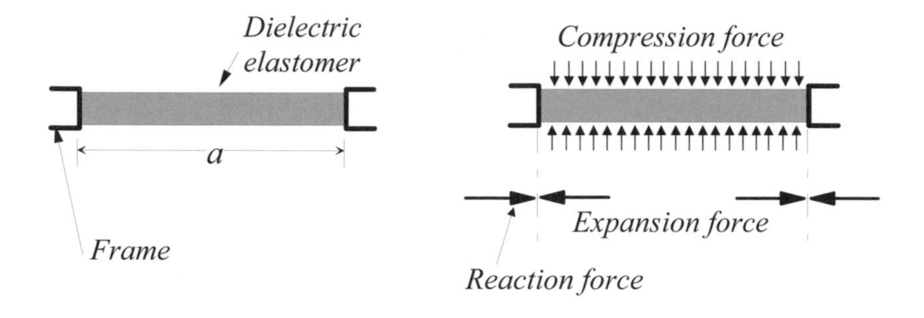


(a) initial state (b) actuated state

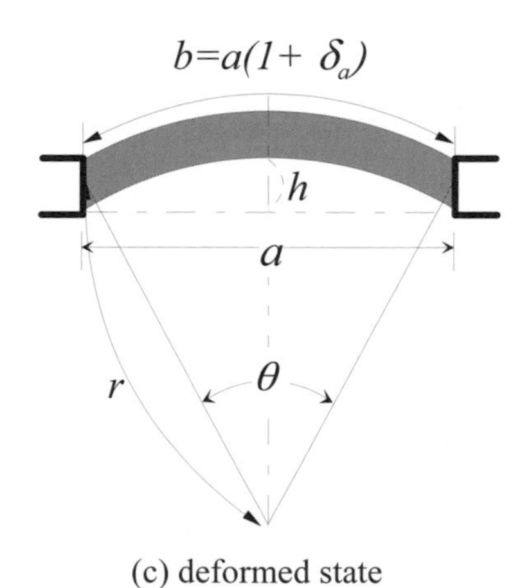


(c) deformed state

**Figure 3.27.** Basic operating concept of the nonprestrained actuator

From the Taylor series expansion,

$$\sin(\theta/2) \approx \left(\frac{\theta}{2}\right) - \frac{(\theta/2)^3}{3!} = \frac{\theta(24-\theta^2)}{48} \tag{3.30}$$

By substituting Eq. (3.30) in (3.29), angle $\theta$ can be derived as follows:

$$\theta = \sqrt{24[1 - 1/(1+\delta_a)]} \tag{3.31}$$

The strain can be derived using Eqs. (3.23), (3.24), and (3.30) so that the displacement $h$ is

$$h = r[1 - \cos(\theta/2)] \tag{3.32}$$

where

$$r = \frac{[(1+\delta_a)a]}{\theta} \tag{3.33}$$

### 3.3.1 Prototype Building and Testing of a Nonprestrained Actuator

*3.3.1.1 Actuator Prototype*

In Figure 3.28, a schematic illustration of the nonprestrained actuator construction is provided, and its actual dimensions are listed in Table 3.3. KE441(ShinEtsu) silicone that has a lower viscosity than VHB4905 is used. The spin-coated elastomer film has been coated with carbon electrodes. They are stacked to make multiple layers. The total membrane thickness is 0.75 mm and each dielectric elastomer is approximately 0.05 mm thick. To make an insulated area between electrodes, both sides of the dielectric elastomer have a nonelectrode area. The diameter of the membrane ($d$) is slightly larger than that ($d_f$) of the fixed frame and it might create either a concave or convex circular membrane that could provide more stable control of deformation in the desired direction during actuation. Figure 3.29 shows an actual fabricated prototype of a dielectric elastomer actuator.

Only the area with electrodes, $d_r$, expands when a driving voltage is applied; thus $\delta_r$ should be converted into $\delta_a$ that can be derived as

$$\delta_a = \delta_i + \delta_r \frac{d_r}{d} \tag{3.34}$$

**Table 3.3.** Dimensions of the nonprestrained actuator

| $d$ | 5.8 mm | $d_f$ | 5.7 mm |
|---|---|---|---|
| $d_r$ | 5.1 mm | $t$ | 0.75 mm |

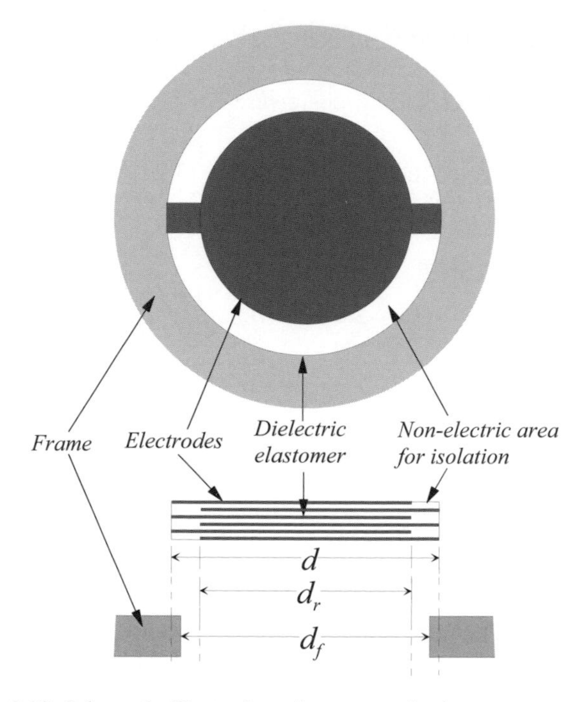


**Figure 3.28.** Schematic illustration of nonprestrained actuator construction

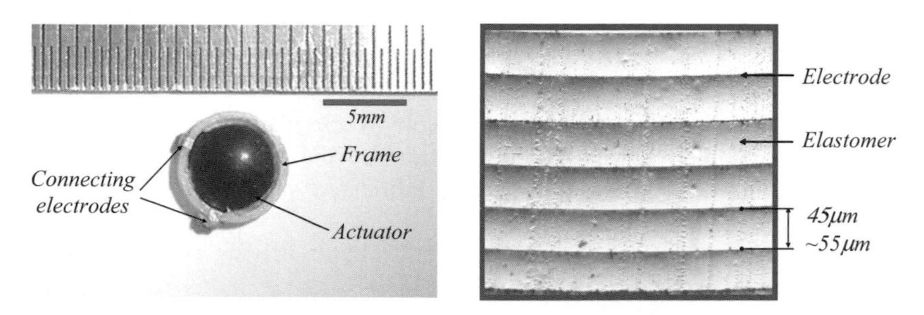


**Figure 3.29.** Prototype of a nonprestrained actuator

where $\delta_a$ denotes a converted strain, and $\delta_i$ is an initial strain given by the initial condition $\delta_i = (d/d_f - 1)$. $\delta_r$ is given by Eq. (3.25) and the vertical displacement $h$ is derived by Eq. (3.32).

*3.3.1.2 Driving Circuit*

A schematic diagram of the driving circuit for the elastomer actuator is provided in Figure 3.30. The response and output characteristic of the actuator are closely related to the charging-discharging characteristics. The duration of the charging process depends on the physical properties of the polymer and is difficult to improve electrically, whereas the discharging duration can be reduced by adding a simple switching device, as shown in the figure. By the addition of the discharging circuit, the actuator can be operated at more than 100 Hz input frequency without significant attenuation.

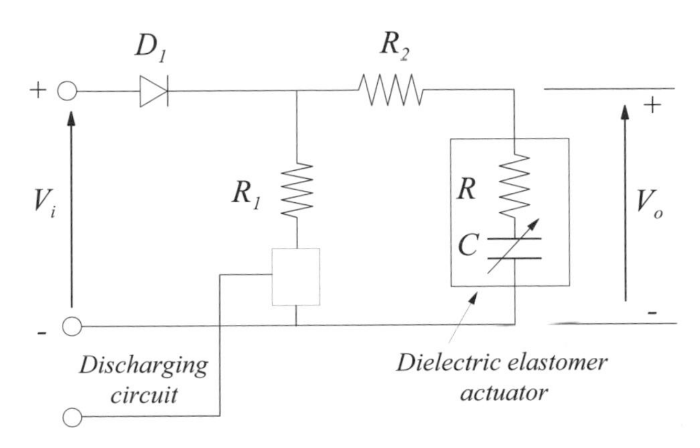


**Figure 3.30.** Driving circuit

*3.3.1.3 Simulation and Experimental Results*

A test and an analysis have been compared in Figure 3.31. The simulation and the experiments have shown good agreement. There is a small error between the calculated result and the experiment that might happen because of the disparity and difference in the thickness of each layer, the externally coated shield layer, and the fabrication process.

For complete measurement of the actuator performance, the frequency response of the actuator is also tested in both displacement and force. As shown in Figure 3.32, the soft actuator generates a fairly large displacement and force. The weight of the actuator is only 0.02 g, and its diameter is 6 mm with a 0.75 mm thickness. Besides, the actuator shows a fast response for square waveform inputs, as shown in Figure 3.33.

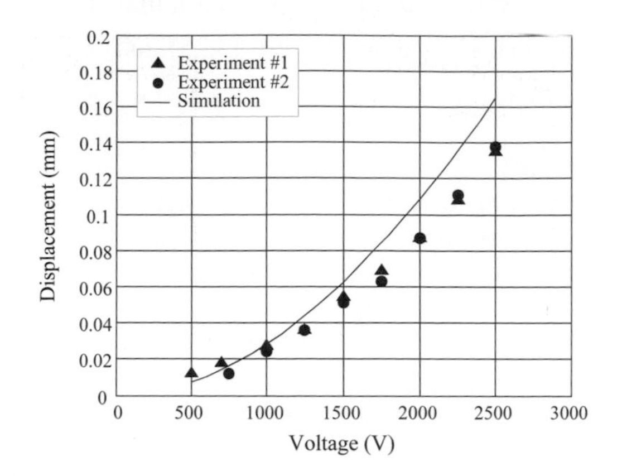


**Figure 3.31.** Comparison of displacement in analysis and test

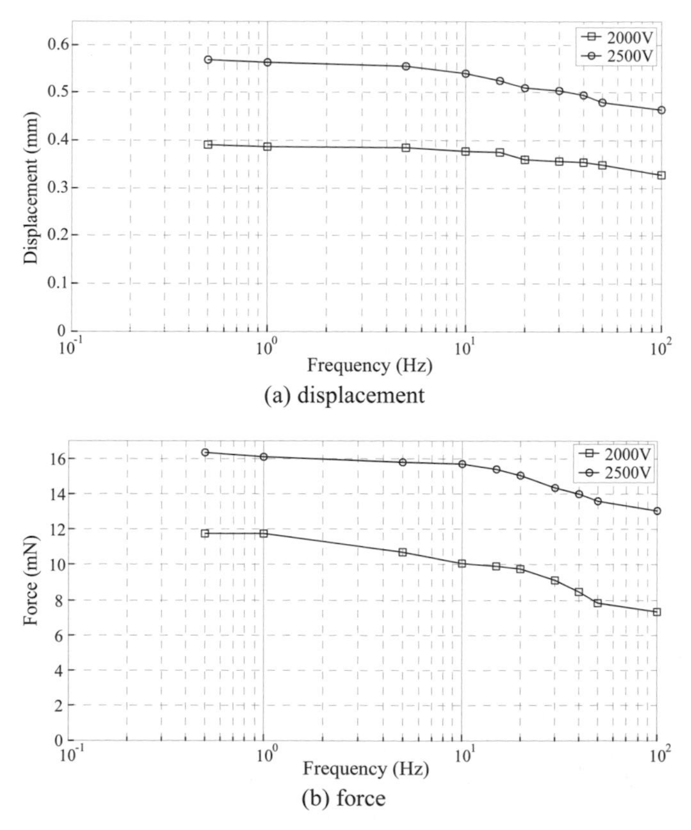


**Figure 3.32.** Frequency response of a nonprestrained actuator

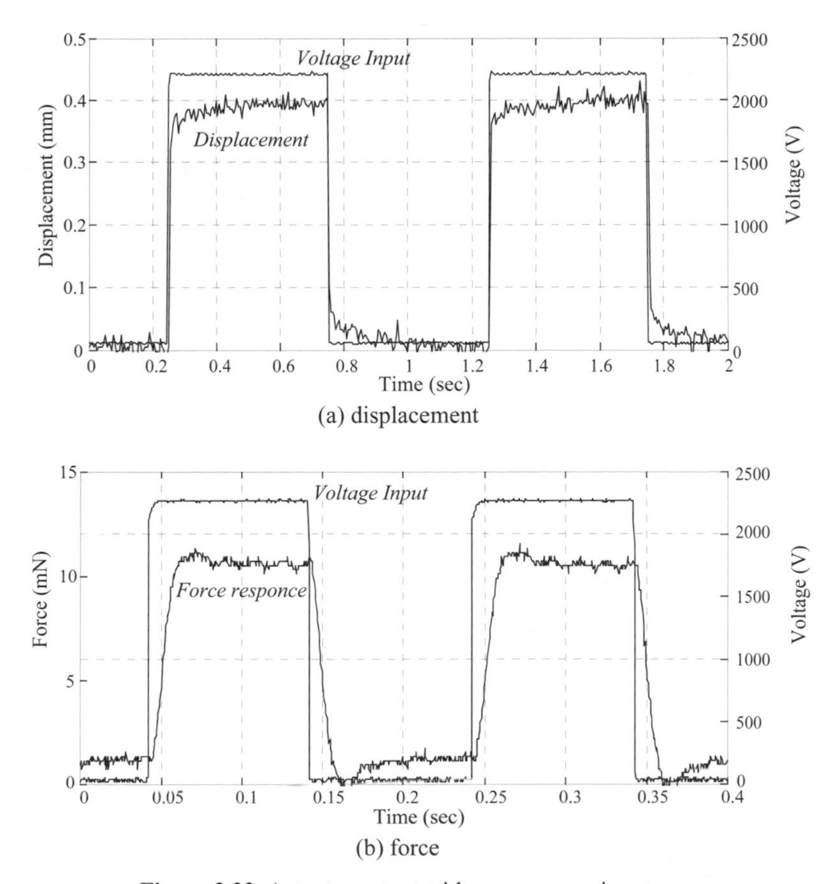


**Figure 3.33.** Actuator output with square wave inputs

### 3.3.2 Inchworm microrobot Using a Nonprestrained Actuator

An inchworm robot made with the nonprestrained actuator has been developed as an example of actuator applications. In Figure 3.34, an actuator module that has three degrees of freedom is shown. If the module is serially connected, a multi-degree-of-freedom inchworm could be constructed. The actuator module is made with 12 serially connected modules. This actuator module works as both a power plant for the movement and a body skeleton of the inchworm robot structure. In other words, the inchworm robot can be built by simply stacking the actuator modules without any additional mechanical structure.

The actuator module shown in Figs. 3.34 has a 20mm diameter, 3mm thickness and 0.4g weight. In Figure 3.35, a fully assembled inchworm robot is shown. This robot has eight actuator modules (96 actuators). Four wires for supplying electric power are connected to the each module. For connecting each module, small silicone cylinders, which have a 1mm diameter and 0.2~0.4mm height, are used to make point-to-point connections between modules and they are bonded by silicone adhesives. The inchworm robot is parted with front and rear sectors and each sector has four actuator modules. Each sector is operated sequentially to create inchworm motion.

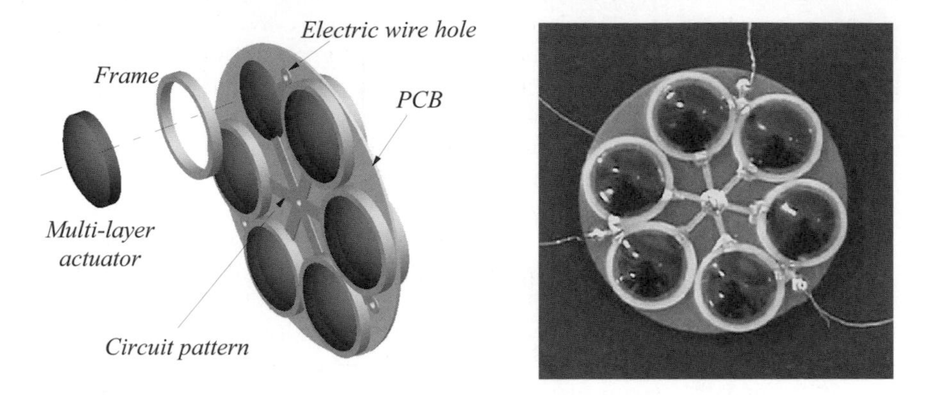


**Figure 3.34.** An actuator segment of nonpresetrained actutor

**Figure 3.35.** An inchworm robot

### 3.3.3 A Braille Display Using Nonpestrained Actuators

Although visual graphical display devices have been the dominant method for information interchange, the role of tactile sense is getting more attentions as a new the way of modern information exchange in various technical fields such as robotics, virtual reality, remote manipulation, rehabilitation, and medical engineering. For human-device interface application, a tactile display transfers information through controlled displacement or force that stimulates human skin.

Communications relying only on graphical presentations are definitely impossible for the visually impaired. For that reason, a large population in the world might be left out of Internet access that results in isolation from educational resources and cultural activities. Advances in tactile display technology for higher sensitivity and higher resolution might benefit the handicapped.

Braille is a tool for exchanging information among the visually disabled and has been extensively used to transfer textual information. It consists of six pins arranged in pattern of a 3×2 matrix (a 4×2 matrix for Chinese characters). Information is delivered by stimulating fingertips by vertical displacement of the pins. The tactile display device can be used as a refreshable dynamic braille. In particular, application of the display also can be expanded to a tablet capable of displaying textural or graphical information. With that capability, even an entire web page can be delivered in a single display step. However, it is very difficult to enable braille to deliver graphical information mainly due to the limitation of arranging massive braille dots for high spatial density. The complicated and bulky driving mechanism of a conventional tactile display hampers development of high-resolution tablet type braille. According to a physiological study for standardization of braille devices, the pin-matrix density of a tactile display is typically up to one cell per square millimeter, the actuating speed should be faster than 50Hz, and the energy density must be about 10W per square centimeter [12,13]. Although the numbers are based on experimental studies, the outcome of the display function is often deceptive because the sensitivity of responses depends on testing conditions such as speed, depth, and strength of stimulation. Meanwhile various mutated tactile display types are introduced to accommodate variable human sensitivity that normally varies from fingertips to palms. Many publications introduce several different types of tactile display devices that employ pneumatics, solenoids, voice coil, shape-memory alloy, electrostatics, or electroactive polymers.

Although previous developments deserve serious attention, most of them commonly suffer from low actuation speed due to a complex actuation mechanism. Moreover, the complicated actuator design limits expansion to the tablet type application due to high manufacturing cost and low integration density. In this section, an alternative new type of dynamic braille display is introduced. Employing a dielectric elastomer as the basis of the tactile display, it is constructed with a notably simple mechanical and electrical architecture. The proposed device is organized with a dual-layered array of tactile cells, shown in Figure 3.29, which generates vertical motion to push the braille pins up or down. These electrically driven tactile cells can generate either small-scale vibratory motion or linear displacement, and they differ from the conventional devices in softness and controllable compliance, cost effectiveness, simple manufacturability, and high actuator density. Furthermore, the small size of the design introduced enables development of a high-resolution display device. Realizing the advantages of the nonprestrained actuator cell, shown in Figures 3.28 and 3.29, a braille display device has been developed using the cells. In this section, a detailed construction procedure for the device and its electronics is presented.

A typical braille display unit is constructed with six stimulating pins that are arranged in a 3×2 array format, and an array normally represents a character as defined by the Braille alphabet. The standard braille display unit is illustrated in Figure 3.36.

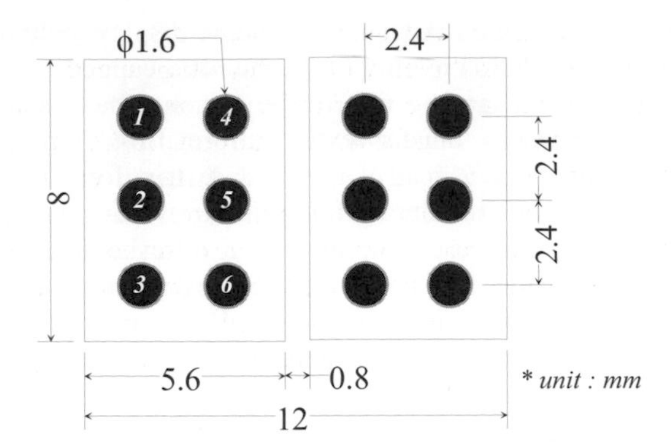


**Figure 3.36.** Standard braille cell consisting of six dots

In this section, a braille display unit is constructed with the introduced non-prestrained actuator tactile cells arranged in the format defined by the standard braille display.

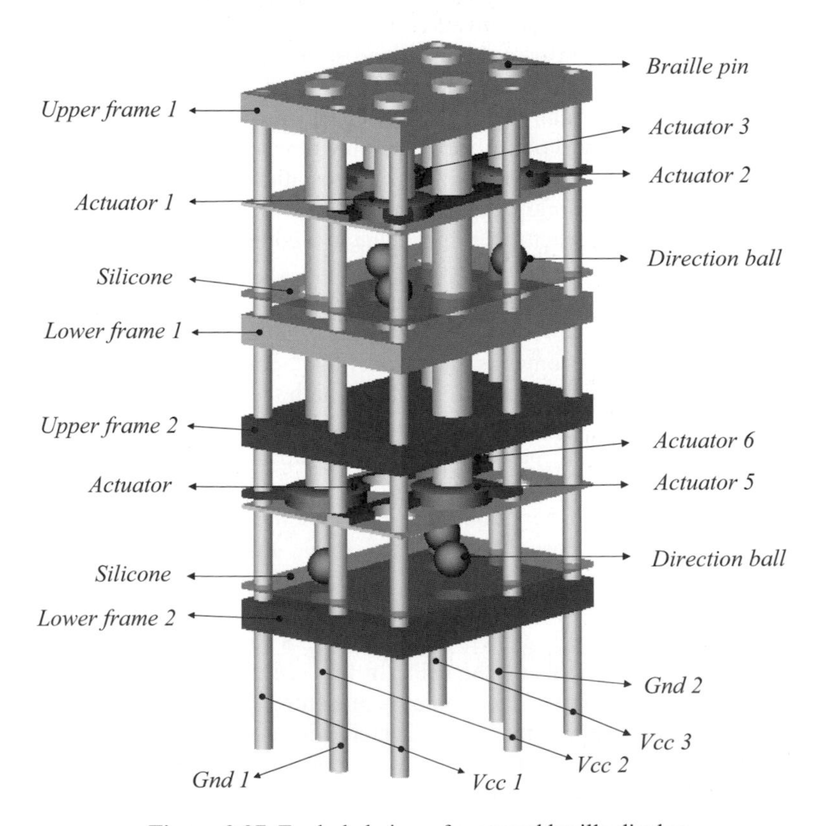


**Figure 3.37.** Exploded view of proposed braille display

The construction concept is depicted in Figure 3.37. Although the dielectric elastomer based tactile cell is driven with high-voltage electricity, users have no direct contact the actuator surface. A braille pin made of insulating material actually contacts with human fingertips. In addition to the predeformed convex feature of a cell, directional balls are placed underneath each cell to guarantee unidirectional actuation. Packaging the six pin actuators and corresponding electric wires in a constrained small space normally requires an expensive manufacturing process. Dual-layer construction is introduced to alleviate fabrication problems. By allocating three pins in each layer in a staggered pattern, interference caused by complicated wiring can be minimized. Each layer is shown in Figure 3.39.

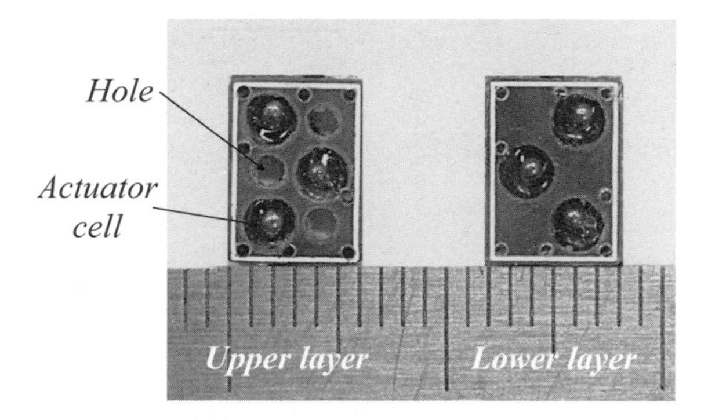


**Figure 3.38.** Upper and lower layers of a braille cell

As shown in Figure 3.39, the fully assembled device is only about 9mm high excluding the length of the terminals. Each braille cell is modularized for convenient installation, so each unit can be simply plugged onto a circuit board. With this simple drop-in feature, a number of braille cells can easily be combined so that a braille tablet may be manufactured by arranging many braille cells in a matrix format, as illustrated in Figure 3.40. A complete actuator system for a braille display unit composed of an embedded controller, high-voltage driving circuit, and a host PC is organized and its schematic description is provided in Figure 3.41. All necessary control electronic parts are embedded and packaged on a PCB and it communicates with a hosting PC through a universal serial bus. A microcontroller (AVR, Atmega 163) is used for the controller and USB 1.1 (Philips, PDIUSBD12) works for communication. A D/A converter (TI TLV5614) and OP-Amp (TI TLV4112) have been integrated in the controller for the modulation of high electric voltage. A complete circuit board is pictured in Figure 3.42.

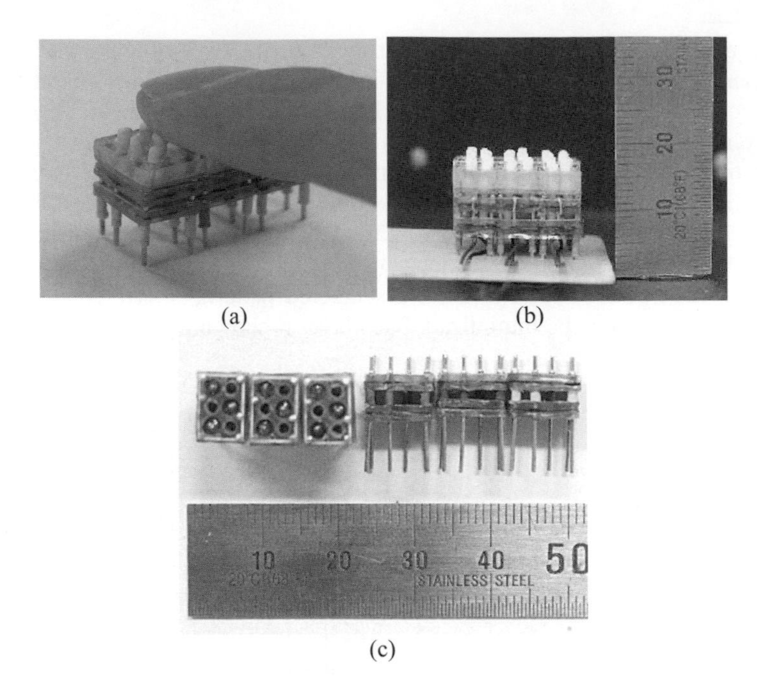


**Figure 3.39.** Assembled braille cell

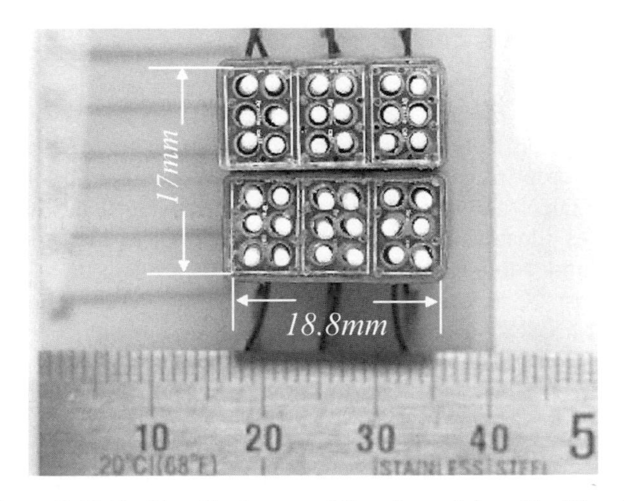


**Figure 3.40.** Braille tablet by assembling six modules of braille cells

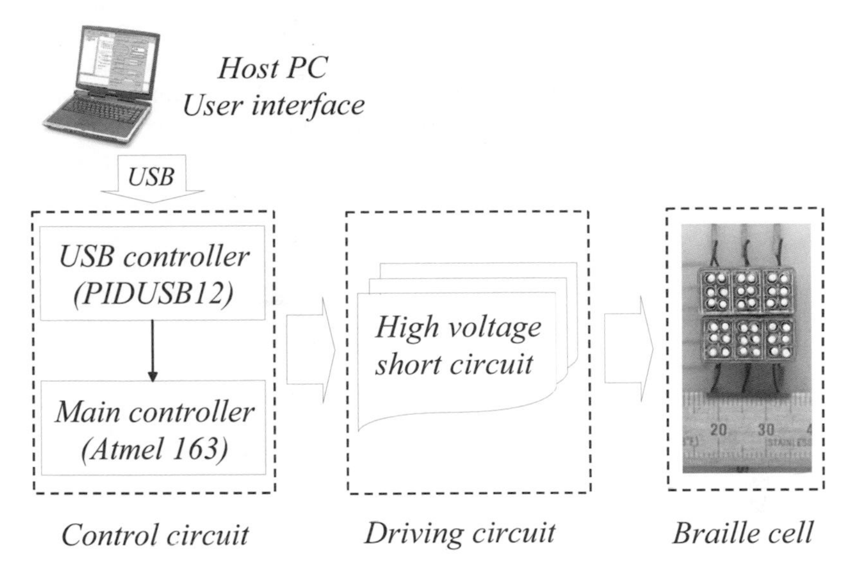


**Figure 3.41.** Schematic illustrations of a complete braille display unit

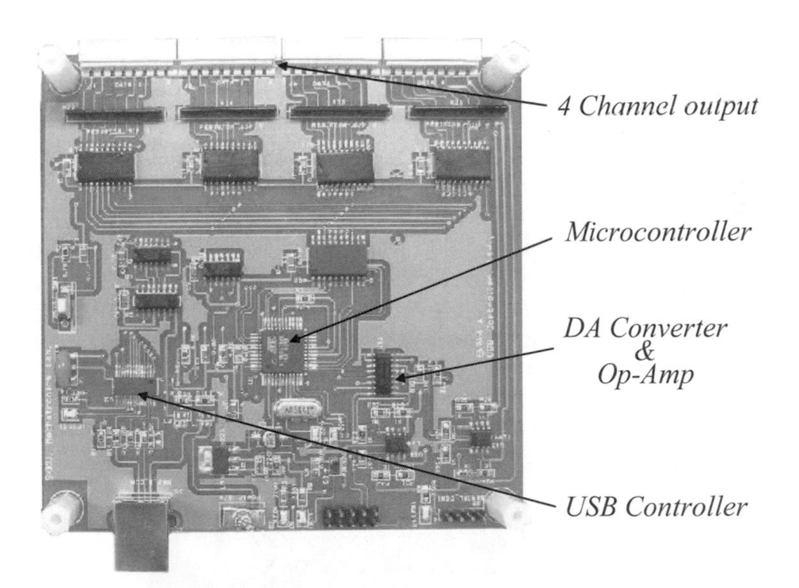


**Figure 3.42.** Control and communication circuit board

The device is organized by cramming many small-scale actuators into a tiny space, electrical wiring might be one of the key problems to be resolved. Identifying constructional and functional analogies of the braille device to a computer input keyboard, a control scheme, so-called dynamic scanning actuation(DSA), can be used here. This strategy enables easy control of n×n braille cells with only 2n electrical lines. The concept of the control is depicted in Figure 3.43. One of the lines named a data line delivers high-voltage driving electric power to each cell and the other line called a scan line functions as a ground. Therefore, each actuator cell can be driven using n lines of the data line and n lines of the scan line with alternating 'On-Off' patterns on both lines.

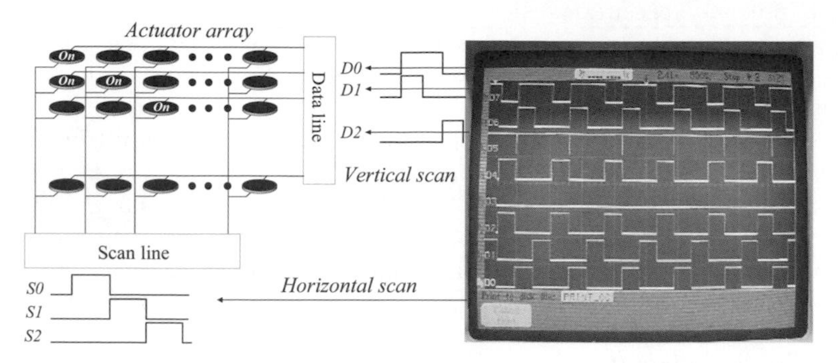


**Figure 3.43.** Concept of dynamic scanning actuation

For example, refer to Figure 3.43, although the data lines D0, D1, D2 get 'High'/'On' signal during three clocks, only one actuator located in the upper left corner operates because only one scan line S0 maintains 'High'/'On' during the time. In essence, an actuator cell moves only if those two lines are synchronized with a 'High'/'On' signal. The scan frequency is $\boldsymbol{L} \times \boldsymbol{M}$ when the number of actuators in a row is $\boldsymbol{M}$ and the desired driving frequency is $\boldsymbol{L}$.

Because a dielectric elastomer actuator normally operates with high electrical voltage of about 1~2kV, introduction of the proposed control scheme DSA provides significant benefits by reducing the number of required high-voltage sources. For example, a single braille unit composed of six tactile cells can be actuated with only a single voltage source, and the number of cells can be easily expanded. In Figure 3.44, HVSC (high-voltage switching circuit) high-voltage reed switches and photocouplers are shown. For faster actuator operation, the method shown in Figure 3.30 has been implemented in the circuit.

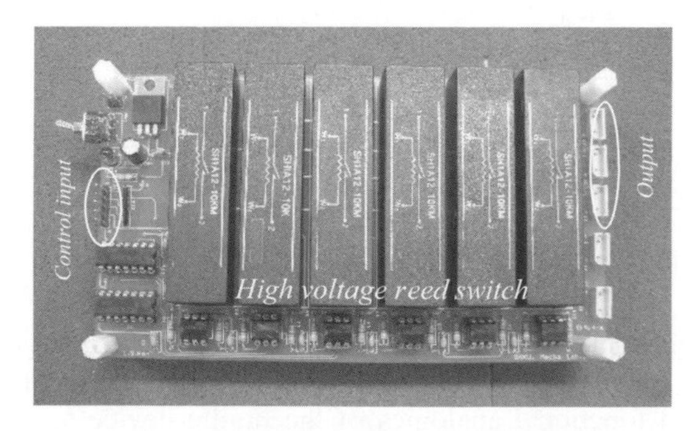


**Figure 3.44.** High-voltage switching circuit

## 3.4 Conclusions

Research and development of the dielectric elastomer actuator have produced significant progress for decades. A remarkable amount of work has focused on delivering the feasibility of the industrial application of the material. However, few commercial products equipped with polymer actuators have been introduced in the market, probably because most of the research has been dedicated either to discovering new actuation properties of the polymeric material or comparative study of the operating range of a particular traditional actuator and its polymeric substitute.

One of the implications of the term "actuator" might be a "controllable" motion generator. If a material produces just some motions, it can not be referred as an actuator unless its motions are controlled. The actuator designs introduced in this chapter have been developed under a common philosophy that actuator motions should be controllable with a reasonable amount of control actions. This clearly differs from many previous developments. The physics of the polymeric material actuation and its construction as an actuator is quite straightforward if a proper application is well sought in advance. In other words, if a good application where the polymer actuators can be used is established and the actuator functionality is well defined, application of the current dielectric elastomer actuator technology to industrial products could be accomplished before long. Future research activity may focus on development of a new energy effective material that has higher permittivity and creation of new application fields such as biomimetic robotics and tactile or braille displays.

## 3.5 References


[1] Y. Osaka and D. E. DeRossi (1999) *Polymer Sensors and Actuators,* Springer.

[2] Y. Bar-Cohen (2002) *Electroactive Polymers [EAP] Actuators as Artificial Muscles*, SPIE press.

[3] G. Kofod (2001) *Dielectric Elastomer Actuators*, Doctoral Dissertation, The Technical University of Denmark.

[4] R. Perline, R. Kornbluh, et al. (2000), High-Speed Electrically Actuated Elastomer with Strain Greater than 100 %, *Science*, Vol. 287, pp. 836-839.

[5] R. Perline, R. Kornbluh, and J. Joseph (1988), Electrostriction of Polymer Dielectrics with Compliant Electrodes as a Means of Actuation, *Sensors and Actuators*, Vol. 64, pp. 77-85.

[6] H. Choi, K. Jung, et al. (2005), Effect of Prestrain on Behavior of Dielectric Elastomer Actuator, *Proc. of 12th SPIE Conference*, San Diego, CA.

[7] H. R. Choi, S. M. Ryew, K. M. Jung, H. M. Kim, J. W. Jeon, J. D. Nam, R. Maeda and K. Tanie (2002), Soft Actuator for Robot Applications Based on Dielectric Elastomer : Quasi-static Analysis," *Proc. IEEE Int. Conf. on Robotics and Automation*, pp. 3212-3217.

[8] H. R. Choi, S. M. Ryew, K. M. Jung, H. M. Kim, J. W. Jeon, J. D. Nam, R. Maeda and K. Tanie (2002), Soft Actuator for Robot Applications Based on Dielectric Elastomer : Dynamic Analysis and Applications, *Proc. IEEE Int. Conf. on Robotics and Automation*, pp. 3218-3223.

[9] H. R. Choi, K. Jung, S. Ryew, Jae-Do Nam, J. Jeon, J. C. Koo, and K. Tanie (2005), Biomimetic Soft Actuator: Design, Modeling, Control, and Applications, IEEE/ASME Transactions on Mechatronics, Vol.10, No.5. pp.581-586.
[10] S. Guo, T. Fukuda, and K. Asaka (2003), A New Type of Fish-Like Underwater Microrobot, *IEEE/ASME Trans. on Mechatronics*, 8(1):136-141.
[11] M. Binnard and M. R. Cutkosky (2000), A Design by Composition Approach for Layered Manufacturing, *ASME Transactions, Journal of Mechanical Design*, Vol. 122, No. 1, pp. 91-101.
[12] N. Asamura, T. Shinohara, Y. Tojo, N. Koshida, and H. Shinoda (2001), Necessary Spatial Resolution for Realistic Tactile Feeling Display, *Proc. Int. Conf. on Robotics and Automation*, pp.1851-1856.
[13] D. G. Caldwell, N. Tsagarakis, and C. Giesler (1999), An Integrated Tactile/Shear Feedback Array for Stimulation of Finger Mechanoreceptor, *Proc. IEEE Int. Conf. on Robotics and Automation*, pp.287-292.

# 4

# Ferroelectric Polymers for Electromechanical Functionality

J. Su

Advanced Materials and Processing Branch
Langley Research Center
National Aeronautics and Space Administration (NASA)
Hampton, Virginia 23681, U.S.A.
ji.su-1@nasa.gov

## 4.1 Introduction

Piezoelectric and ferroelectric polymers have been recognized as a new class of electroactive materials since the significant piezoelectricity in polyvinylidene fluoride (PVDF or $PVF_2$) was discovered in 1969 by Kawai [1]. Since then, a variety of new piezoelectric polymers have been developed including copolymers of vinylidene fluoride and trifluoroethylene, p($VF_2$-TrFE), odd-numbered nylons, composite polymers, *etc.* [2–8]. These materials offer options of material selection for sensor and actuator technologies that need lightweight electroactive materials. In this chapter, the origins of piezoelectricity and the properties of several kinds of piezoelectric and ferroelectric polymeric materials are introduced for reference.

## 4.2 Dipole Moment and Polarization

### 4.2.1 Dipole and Dipole Orientation

An electric dipole can be defined as an entity in which a positive charge, $q$, is separated by a relatively short distance, $\boldsymbol{d}$, from an equal negative charge [9]. The electric dipole moment, $\boldsymbol{\mu}$, is a vector quantity and is defined as

$$\boldsymbol{\mu} \equiv q\boldsymbol{d} \tag{4.1}$$

By convention, its direction points from the negatively charged end toward the positively charged end. Traditionally, the Debye has been used as the unit of measure of the dipole moment. In international units,

$$1\ \mathbf{D} = 3.338 \times 10^{-30}\ \text{C-m} \tag{4.2}$$

A dipole moment is defined as a permanent dipole moment if it exists in the absence of an external electric field, while an induced dipole is the dipole formed by an applied electric field. Besides these, "defect dipole moments" can be introduced, which are associated with some structural heterogeneity, such as a missing lattice site, crystallite interface, crystallite-amorphous interface in solids, or an interface between two media [10]. At structural defects, the bonding energy is decreased and the radius of the defects is, correspondingly, increased. In the presence of an external electric field, a dipole may be formed at structural defects. This can be referred to as a defect dipole. Since the bonding energy is dependent on the dielectric constant of a medium, any heterogeneity in the dielectric constant should result in a corresponding change in the formation of dipoles in the presence of an external field. This change is known as the Maxwell-Wagner effect. If the sample contains regions of different dielectric constants in an external electric field, an interfacial dipole polarization builds up [11]. The electric dipoles mentioned above are schematically illustrated in Figure 4.1.

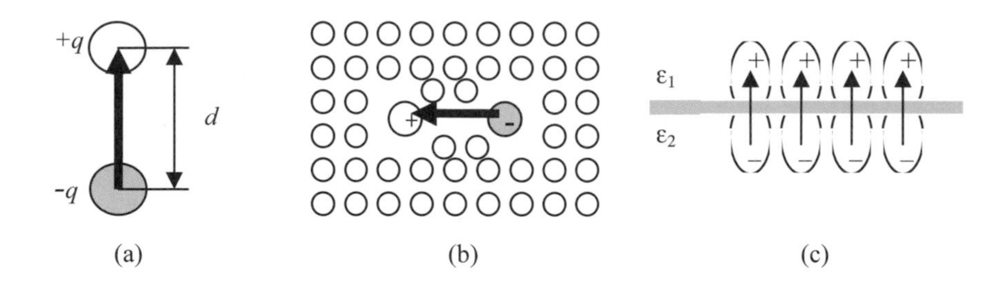


**Figure 4.1.** Schematical illustrations of the formation of a dipole. (a) a typical dipole, (b) a defect dipole, and (c) interfacial dipoles when an interface is formed by two consituents that have different dielectric constants ($\varepsilon_1 \neq \varepsilon_2$).

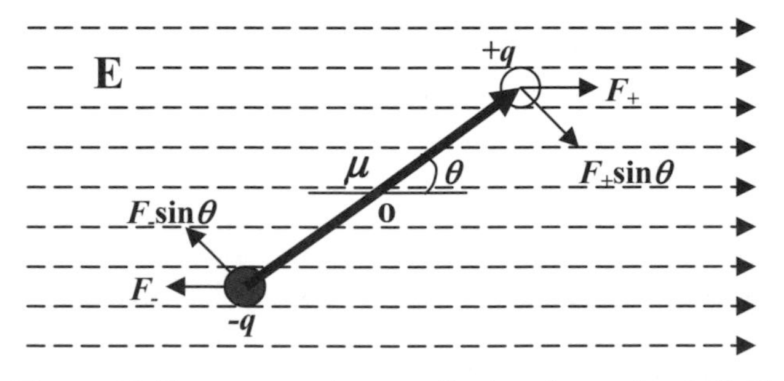


**Figure 4.2.** The torque acting on a dipole under an electric field

When an electric field is applied to a permanent dipole, the positive charge at one end of the dipole experiences a force in the direction of the applied field and the negative charge experiences an equal force in the opposite direction. Consequently, the net force on the dipole is zero; therefore, the translational movement will be zero. However, the charges are separated by a fixed displacement ***d***, so a torque (or moment) may exist. The torque consists of two components, $F_+ \sin\theta$ and $F_- \sin\theta$, acting on the dipole, as shown in Figure 4.2. The torque will tend to align, or orient, the dipole in the direction of the applied electric field [12]. As is well known, torque is a quantity which is defined as a cross product of the force ***F*** and length vector ***d***:

$$T = d \times F \tag{4.3}$$

Using scalar notation, the total torque is given by the following equation:

$$T = F_+(d/2)\sin\theta + F_-(d/2)\sin\theta = Fd\sin\theta \tag{4.4}$$

Since the force $F$ is given by $qE$, the torque is rewritten as

$$T = qEd\sin\theta \tag{4.5}$$

where $\boldsymbol{q}$ is the charge. Using the relation $\mu \equiv qd$, therefore

$$T = \mu E\sin\theta \tag{4.6}$$

The incremental work $dw$ which must be done to align the dipole by an angle $d\theta$ is

$$dw = Td\theta \tag{4.7}$$

Integrating this equation, we obtain

$$w = \int Td\theta = \mu E\int \sin\theta d\theta = -\mu E\cos\theta \tag{4.8}$$

Here, the amount of work or energy which is needed to align a dipole at an angle $\theta$ to an applied electric field ***E*** is given. Thus, a dipole, $\boldsymbol{\mu}$, inclined by an angle, $\theta$, to an electric field, ***E***, has a potential energy, $w$, given by $w = \mu E \cos\theta$, or the scalar product $w = \boldsymbol{\mu} \cdot \boldsymbol{E}$. The alignment, or orientation, of the dipole under an electric field is called polarization. However, even in the presence of an electric field, dipoles are still under the influence of thermal motion amounting to an energy, ~ $kT$, where $k$ is the Boltzman constant ($1.38 \times 10^{-23}$ J-K$^{-1}$) and $T$ is temperature. Therefore, the torque aligning the dipoles in the field direction is always counteracted by thermal energy to some extent.

### 4.2.2 Polarizability and Polarization

Polarization, **P**, is defined as a vector quantity expressing the magnitude and direction of the electric dipole moment per unit volume. In the absence of the external electric field, the polarization is equal to the charge per unit area on the surface of the material. In general, there are four fundamental components of polarization: electronic, atomic, orientational, and interfacial (or space charge) polarizations [13].

#### *4.2.2.1 Electronic Polarization*

An electric field will cause a slight displacement of the electron cloud of any atom in a molecule relative to its positive nucleus. As a result, the nucleus is no longer at the centroid of the electronic charge. This situation is termed electronic polarization. The applied electric field is usually very weak compared with the intra-atomic field at an electron, so this polarization is quite small.

#### *4.2.2.2 Atomic Polarization*

When atoms of different types form molecules, they will not normally share their electrons symmetrically, as the electron clouds will be displaced eccentrically toward the atoms having higher electronegativity. Thus atoms acquire charges of opposite polarity. When an electric field is applied, these net charges tend to change the equilibrium position of the atoms themselves. This results in atomic polarization. The magnitude of atomic polarization is usually quite small, often only one-tenth that of electronic polarization.

#### *4.2.2.3 Orientational Polarization*

If the molecules already possess a permanent dipole moment, the moment will tend to be aligned by the applied field to a net polarization in the direction of the applied electric field. The orientational polarization that develops when the field is applied is the average of the orientations of the permanent dipole moments favoring the direction of the applied field. The tendency to revert to random orientation opposes the tendency of the field to align the dipoles and thus allows for polarization to vary in proportion to the applied field.

#### *4.2.2.4 Interfacial (Space Charge) Polarization*

Besides electronic, atomic and orientational polarization, another polarization, interfacial polarization, should be introduced, especially for heterogeneous systems. Charge carriers usually exist in such systems and they can migrate through the material under an electric field. When the carriers are impeded in their motion, they become trapped at defect sites or at an interface between two media having different dielectric constants and conductivity, and they cannot be freely discharged or replaced at the electrodes. This can give rise to space charges and macroscopic field distortion [14]. These effects will create a localized accumulation of charges that will induce image charges on electrodes and give rise to a dipole moment, therefore making contributions to the total polarization of the system.

If we think of polarization on the molecular level, the polarization mechanisms can also be expressed in terms of a molecular physical quantity called polarizability $\alpha$. Polarizability is defined as the average molecular polarized dipole moment produced under the action of an electric field of unit strength [15]. Suppose that the action of the local electric field strength $\mathbf{E}_l$ on each molecule produces a polarized dipole moment, $\mu$ which is generally proportional to $\mathbf{E}_l$, or

$$\mu = \alpha \, \mathbf{E_l} \tag{4.9}$$

where the proportionality constant is the polarizability.

Assume that the total number of dipole moments per unit volume is $N_o$ and all of them are equally polarized. In that case, the polarization, **P,** will be obtained by

$$\mathbf{P} = N_o\mu = N_o \, \alpha \, \mathbf{E_l} \, . \tag{4.10}$$

Here, $\mathbf{E_l}$ is the local electric field, which is different from the applied electric field **E.** Clausius-Mossotti's calculation based on a spherical model of the Lorentz local field assumes that an imaginary sphere is drawn around a particular molecule and that the local field at that point is the sum of the fields resulting from the real charges on the electrodes which sustain the external applied field and all of the charges lying outside and inside the sphere. For a molecular arrangement that gives a zero field at the point due to the material inside the sphere, the Lorentz local field can be obtained as [9]

$$\mathbf{E_l} = \frac{\varepsilon + 2}{3} \, \mathbf{E} \tag{4.11}$$

where $\varepsilon$ is the dielectric constant, and therefore, the polarization becomes

$$\mathbf{P} = N_o \, \alpha \left(\frac{\varepsilon + 2}{3}\right)\mathbf{E} \tag{4.12}$$

In Eq. (4.12),

$$\mathbf{P} = \mathbf{P_e} + \mathbf{P_a} + \mathbf{P_o} + \mathbf{P_i} \tag{4.13}$$

where $\mathbf{P_e}$ is electronic polarization, $\mathbf{P_a}$ is atomic polarization, $\mathbf{P_o}$ is orientational polarization, and $\mathbf{P_i}$ is interfacial polarization. And

$$\alpha = \alpha_e + \alpha_a + \alpha_o + \alpha_i \tag{4.14}$$

where $\alpha_e$ is electronic polarizability, $\alpha_a$ is atomic polarizability, $\alpha_o$ is orientational polarizability, and $\alpha_i$ is interfacial polarizability [16].

As previously mentioned, the mechanisms of polarization in dielectrics are associated with the type of charge displacement, so they should be dependent on frequency. A particular mechanism of polarization gives rise to resonance phenomena in a particular frequency range. Figure 4.3 shows the relationship between polarization mechanism and frequency [9].

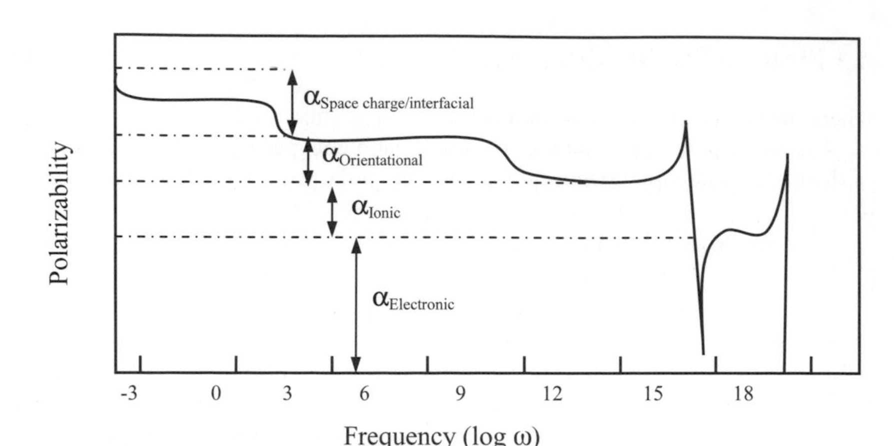


**Figure 4.3.** A schematic representation of the frequency dependence of contributions of each mechanism of the polarizability

If the polarization vector represented in terms of electric permittivity, or dielectric constant, of dielectrics under an electric field is

$$\mathbf{P} = (\varepsilon\text{-}1)\varepsilon_o\mathbf{E} \tag{4.15}$$

where $\varepsilon_0$ is the permittivity of free space; the combination of Eqs. (4.12) and (4.15) leads to

$$\mathrm{N_o}\,\alpha\left(\frac{\varepsilon+2}{3}\right)\mathbf{E} = (\varepsilon\text{-}1)\varepsilon_o\mathbf{E} \tag{4.16}$$

Rearrangement of Eq. (4.16) gives

$$\frac{\mathrm{N_o}\alpha}{3\varepsilon_o} = \frac{\varepsilon-1}{\varepsilon+2}\,. \tag{4.17}$$

By converting the relationship to a molar basis, the molar polarization is given as

$$\mathrm{P_m} = \frac{\mathrm{N}\alpha}{3\varepsilon_o} = \frac{\varepsilon-1}{\varepsilon+2}\cdot\frac{\mathrm{M}}{\rho} \tag{4.18}$$

where N is the molar number of dipole moment. Equation (4.18) is the Clausius-Mossotti equation, which expresses the relationship between the dielectric constant, $\varepsilon$, and polarizability, $\alpha$. The equation indicates that molar polarization should be proportional to the polarizability, $\alpha$, and and a function of the dielectric constant $\varepsilon$. In the equation, M is the molecular weight of the material and $\rho$ is its density.

## 4.3 Piezoelectricity and Ferroelectricity

A material is defined as piezoelectric if the short-circuit current resulting from the change in polarization when the material is stressed is large enough to be measurable; a material is defined as ferroelectric if the material is polar and the polarization is reorientable under an external electric field.

The piezoelectric response of materials was discovered by the Curies in 1880 [17]. They found that some crystals can convert mechanical energy into electrical energy. In 1881, W. G. Hankel contended that the response must obey certain laws of its own (like other intrinsic properties of materials), and Hankel proposed the term "piezoelectricity" [18], which means pressure-electricity [*piezo* means *press* in Greek], to describe the property. Later, a classical theory for single crystals was developed by W. Voigt [19] and others. According to the classical theory for single crystals, among the 32 crystal classes, 20 with noncentrosymmetry can be piezoelectrically active. Ten of these 20, which have neither a reflection plane nor a two-fold axis perpendicular to a particular direction, can possess a net intrinsic dipole moment in their crystal unit cell, and thus exhibit "pyroelectricity" which means thermal-electricity [*pyro* means *heat/thermal* in Greek]. A crystal is defined as ferroelectric when the intrinsic dipole moments in the crystal can reverse their direction, following the application of a sufficiently high external electric field [20]. Figure 4.4 shows a simple box-relationship to distinguish ferroelectricity, piezoelectricity, and pyroelectricity by the nature of crystal materials.

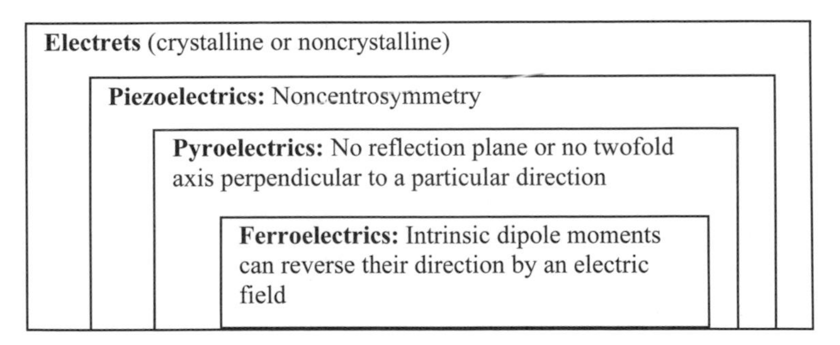


**Figure 4.4.** A simple method to distingduish ferroelectricity, piezoelectricity, and pyroelectricity by the nature of crystal materials

### 4.3.1 Definition of Piezoelectricity

Piezoelectricity can be defined thermodynamically [21]. According to the first and second laws of thermodynamics,

$$dU = dQ + dW \tag{4.19}$$

where $U$ is the internal energy density, and $Q$ is the heat given to a unit volume of a crystal from outside,

$$dQ = TdS \tag{4.20}$$

for a reversible process, where $T$ is Kelvin temperature and $S$ is the entropy. Considering $dW$ as the work done by electrical and mechanical forces on the unit volume, it can be expressed as

$$dW = Xdx + EdP \tag{4.21}$$

where $X$ is the stress, $x$ is the strain, $E$ is the electric field and, $P$ is electric polarization. If the Gibbs free energy is given by

$$G = U - TS - Xx - EP \tag{4.22}$$

Then,

$$dG = -SdT - xdX - PdE\,. \tag{4.23}$$

Since $X$ and $E$ are independent parameters of $G$, $G(X,E)$ can be expanded at $X = 0$, $E = 0$ by using the Maclaurin series and by neglecting the higher order terms:

$$G(X,E) \approx \frac{1}{2}\left(\frac{\partial^2 G}{\partial X^2}\right)_0 X^2 + \frac{1}{2}\left(\frac{\partial^2 G}{\partial E^2}\right)_0 E^2 + \left(\frac{\partial^2 G}{\partial E \partial X}\right)_0 XE \tag{4.24}$$

We define the coefficients

$$\left(\frac{\partial^2 G}{\partial X^2}\right)_0 = -\left(\frac{\partial x}{\partial X}\right)_{E,0} = -s^E \tag{4.25a}$$

$$\left(\frac{\partial^2 G}{\partial E^2}\right)_0 = -\left(\frac{\partial P}{\partial E}\right)_{X,0} = -\varepsilon_o \chi^X \tag{4.25b}$$

and

$$\left(\frac{\partial^2 G}{\partial E \partial X}\right)_0 = -\left(\frac{\partial P}{\partial X}\right)_{E,0} = -\left(\frac{\partial x}{\partial E}\right)_{X,0} = -d \tag{4.25c}$$

Thus,

$$G = -\frac{1}{2} s^E X^2 - \frac{1}{2} \varepsilon_o \chi^X E^2 - dXE\,. \tag{4.26}$$

Differentiating with respect to $X$ or $E$:

$$\left(\frac{\partial G}{\partial X}\right)_E = -x = -s^E X - dE \tag{4.27a}$$

and

$$\left(\frac{\partial G}{\partial E}\right)_X = -P = -\varepsilon_o \chi^X E - dX \tag{4.27b}$$

Thus

$$x = s^E X + d^X E \tag{4.28a}$$

and

$$P = \varepsilon_o \chi^X E + d^E X \tag{4.28b}$$

where $s^E$ is the elastic stiffness in a constant electric field, $\chi^X$ is the electric susceptibility at constant stress, and $d$ is the piezoelectric strain coefficient. Similarly, if the Gibbs free energy in terms of strain is used, the following relations can be obtained:

$$G = -\frac{1}{2}c^E x^2 - \frac{1}{2}\varepsilon_o \chi^X E^2 - e^X E \tag{4.29a}$$

$$X = c^E x + e^X E \tag{4.29b}$$

and

$$P = \varepsilon_o \chi^X E + e^E x \tag{4.29c}$$

where $c^E$is the elastic compliance in a constant electric field and $e$ is the piezoelectric stress coefficient. By comparing Eqs. (4.28b) and (4.29c), the relationship among piezoelectric strain, stress coefficients, and elastic modulus, $C$, is obtained as

$$e^E = d^E C \tag{4.30}$$

This relation can also be derived by referring to the changes in electric polarization with strain $x$ (or stress $X$).

### 4.3.2 Ferroelectricity and Polarization Reversal

Ferroelectrics are materials that possess a spontaneous electric polarization, $\boldsymbol{P_s}$, which can be reversed by applying a suitable electric field $\boldsymbol{E}$. This process is known as polarization reversal, or polarization switching, and is always accompanied by hysteresis. This characteristic property of a material was discovered by Valasek in 1921 and is known as ferroelectricity [22]. The electric field at which the polarization reversal occurs is defined as the coercive field, $\boldsymbol{E_c}$. Most ferroelectrics cease to be ferroelectric when the temperature is above a critical one, which is called the Curie transition temperature. The process is known as a ferroelectric-paraelectric transition [23].

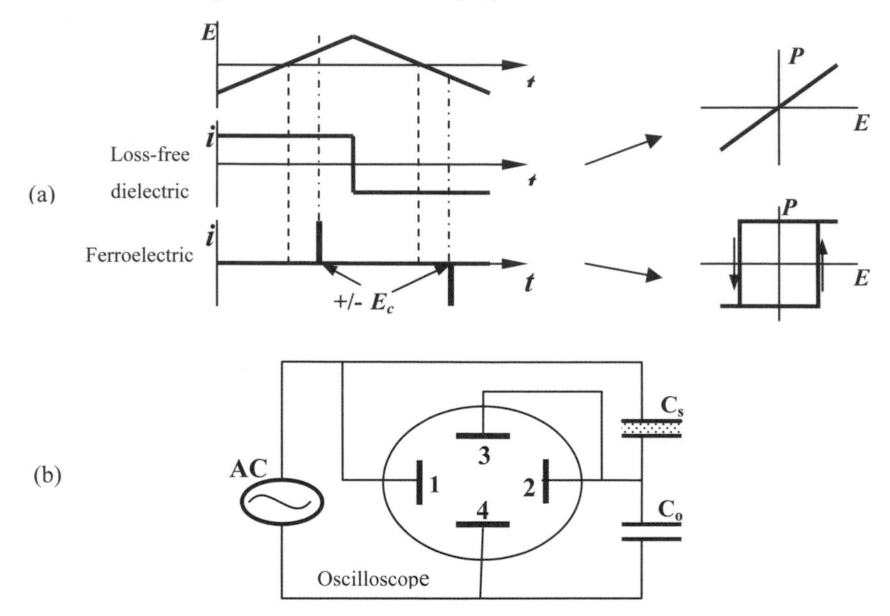


**Figure 4.5.** (a) Typical ferroelectric hysteresis characteristics and (b) the Sawyer-Tower circuit to conduct the characterization

Ferroelectrics may be distinguished by their characteristic response to a changing electric field $\boldsymbol{E}$, as indicated in Figure 4.5a, where $\boldsymbol{t}$ is the time and $\boldsymbol{i}$ is the current. $\boldsymbol{E}$ and $\boldsymbol{i}$ are a function of $\boldsymbol{t}$. Ferroelectrics show a current ($\boldsymbol{i}$) peak at the coercive field, $\boldsymbol{E_c}$ which is attributed to polarization reversal, whereas nonferroelectrics (P = 0 when E = 0) show a linear relationship of the current to the applied electric field [23]. This result can be obtained by using a (hysteresis) bridge circuit called the modified Sawyer-Tower circuit [24, 25] which is shown in Figure 4.5b. In the bridge, $C_o$ is a standard capacitor with a capacitance much larger than that of the sample $C_s$. The voltage, $V$, is applied to $C_s$ and to plates 1 and 2 of the oscilloscope, which have a small capacitance. Then, the electric charges, $Q$, stored in $C_s$ and $C_o$ should be equal. The voltage between plates 3 and 4 is then equal to $Q/C_o$. The curve of $Q/C_o$ versus the applied voltage $V$ can then be observed. For loss-free dielectrics, or nonferroelectrics, the charge stored on the sample electrodes should be proportional to $V$ and a linear relationship between them should be observed. If a material inserted as $C_s$ is ferroelectric, the charge $Q$ will be proportional to the electric displacement $\boldsymbol{D}$, ($\boldsymbol{D}_{E=0}$ = Q/A, therefore, Q = A$\boldsymbol{D}_{E=0}$, where A is a unit area), instead of to the electric field $\boldsymbol{E}$, and a hysteresis loop should be observed. The electric displacement is vectorially defined as

$$\mathbf{D} = \varepsilon_0 \mathbf{E} + \mathbf{P} \tag{4.31}$$

where $\varepsilon_0$ is the vacuum permittivity, $E$ is the applied electric field (which equals $V/d$), $V$ is the applied voltage, $d$ is the thickness of the sample, and $\boldsymbol{P}$ is the polarization magnitude. If the change in the electric displacement with the applied electric field exhibits a hysteresis loop when the applied electric field is changed alternatively, reaching a maximum in one direction, passing through a zero field and reaching a maximum in the opposite direction, the material usually possesses spontaneous polarization, $\boldsymbol{P}_s$, *i.e.* the polarization persists when the electric field is returned to zero. However, the stable polarization, or remanent polarization, $\boldsymbol{P}_r$ is usually less than $\boldsymbol{P}_s$ due to the thermal motion of molecules that form dipoles, or thermal disturbance. By the relationship shown in Eq. (4.31), the spontaneous polarization, $\boldsymbol{P}_s$, is represented by the intersection of $\boldsymbol{P}$ at $\boldsymbol{E} = 0$, whereas the field needed to switch the polarization, or coercive field, $E_c$, is represented by the intersection at $\boldsymbol{D} = 0$. Practically, the $\boldsymbol{D}$ vs. $\boldsymbol{E}$ hysteresis is considered a typical characteristic of a ferroelectric. Figure 4.6 shows a typical ferroelectric hysteresis loop with the important ferroelectric characteristics [26]. As mentioned earlier, a transition phenomenon is commonly observed in ferroelectrics. When the temperature is above the Curie transition temperature, the spontaneous polarization of a material decreases rapidly to zero, that is, the material changes from ferroelectric to become paraelectric, or nonferroelectric. Accompanying the transition, a large change in the electric permittivity, or dielectric constant, is generally observed. Curie-Weiss, in 1907, developed the relationship between the measured dielectric constant and temperature [27]. When the temperature is above the transition temperature, the dielectric constant becomes

$$\varepsilon = \varepsilon_o + \frac{C_{c-w}}{T - T_o} \tag{4.32}$$

where $\varepsilon_0$ is the permittivity of free space, $C_{c\text{-}w}$ is the Curie-Weiss constant and $T_0$ is the Curie-Weiss temperature (known as the Curie point for a ferroelectric-paraelectric transition).

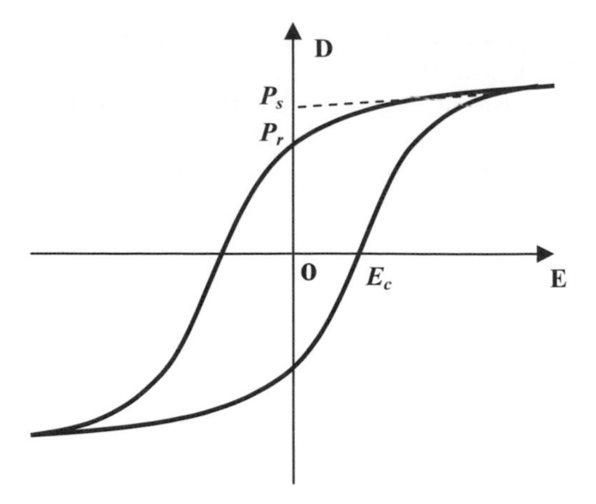


**Figure 4.6.** A typical ferroelectric hysteresis lood received by the Sawyer-Tower circuit-based characterization method. $\boldsymbol{P_r}$: remanent polarization, $\boldsymbol{P_s}$: spontaneous polarization, and $\boldsymbol{E_c}$: coesive field. (Adapted from M.E. Line and A.M. Glass)

## 4.4 Ferroelectric and Piezoelectric Polymers

Investigations of piezoelectric effects in polymers have been carried out since 1924 when Brian first studied a number of dielectrics, including some polymers [28]. However, no significant piezoelectric response had been observed until 1969 when Kawai discovered that poly(vinylidene fluoride) ($PVF_2$), prepared by a proper thermal, mechanical, and poling treatment, can show unusually large piezoelectric coefficients[1]. Thereafter, the piezoelectricity, pyroelectricity, and ferroelectricity of $PVF_2$ and related mechanisms have been extensively studied worldwide. Bergman and co-workers first reported in 1971 that $PVF_2$ was pyroelectric as well as piezoelectric and speculated on the possibility that $PVF_2$ was ferroelectric [29,30]; this was later confirmed by others [31–34]. Meanwhile, the crystal structures of $PVF_2$ had been studied by Lando *et. al.* and the polymorphism of $PVF_2$ crystal structures was reported [35–37]. Investigations of the origins of piezoelectric and pyroelectric properties and mechanisms of ferroelectricity of $PVF_2$ were carried out and reported by Wada, Broadhurst, Davis, and Kepler *et. al.* [38–44]. Based on the piezoelectricity and the ferroelectricity of the $\beta$-phase (phase I) $PVF_2$ and knowing that copolymers of vinylidene fluoride ($VF_2$) and trifluoroethylene ($VF_3$) crystallize directly from the melt into the $\beta$-phase crystal structure, $VF_2$-$VF_3$ copolymers were developed and their piezoelectricity and ferroelectricity were also extensively investigated [43–49].

Another class of ferroelectric and piezoelectric polymers is odd-numbered polyamides (nylons). Hydrogen bonds formed by –NH and –C=O groups provide essential polar elements for piezoelectricity in the polymer. The significant piezoelectric properties of nylon 11 were studied and reported in the early 1980s [50–52]. Its ferroelecticity was predicted based on study of its chemical structures and crystal phases. However, the ferroelectricity in nylons was not confirmed until 1990 when Scheinbeim *et al.* reported a typical ferroelectric hysteresis loop from the electric displacement, ***D***, vs. the applied electric field, ***E***, test of specially treated nylon 11 films [53–55].

### 4.4.1 Models and Origins of Piezoelectricity of Polymers

The origins of piezoelectricity in polymers have been being discussed for more than two decades. Several models have been introduced since 1978 for the only ferroelectric and piezoelectric semicrystalline polymer, at the time, $PVF_2$, by Broadhurst and Davis [56], Taylor and Purvis [57], Taylor and Al-Jishi [58], and Wada and Hayakawa *et. al.* [59].

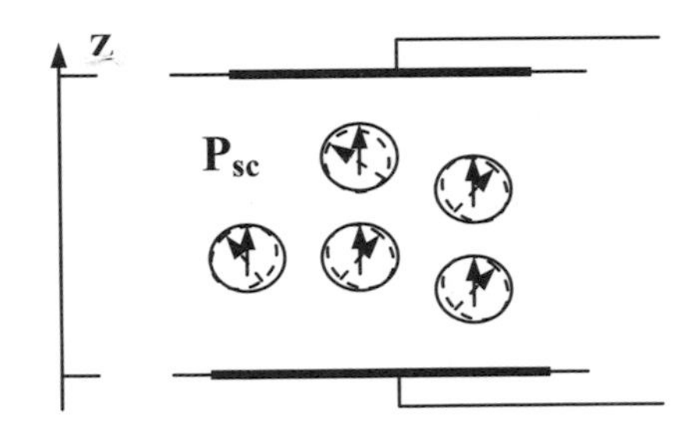


**Figure 4.7.** A schematic of the mechanism of piezoelectricity in semicrystalline polymers due to the libration of polar domains

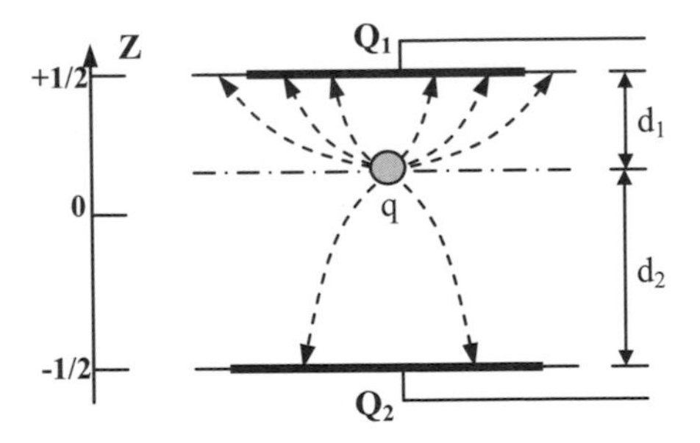


**Figure 4.8.** A schematic of the mechanism of piezoelectricity based on the analysis of response of unsymmetrical distribution of charge

In the Broadhurst and Davis model, sample dimension effects and local field effects as well as dipole libration and space charge effects are included. Figure 4.7 shows a schematic of the mechanism of dipole libration for piezoelectricity in a semicrystalline polymer. A Lorentz approximation was used to determine the local field and dipole polarizability. Through the calculation of the hydrostatic piezoelectric coefficient, it was concluded that 34% of the experimentally measured coefficient arose from the local field effect, 10% from dipole fluctuations, and 56% from dimensional changes. The Taylor-Purvis model is based on the consideration of point dipoles and an orthorhombic lattice for determination of the local field, but the dipole libration and space charge effects were not included. Later, Taylor and Al-Jishi modified the Taylor-Purvis model by introducing a better approximation for the dipoles. For the modification, the dipoles were represented by separated charges instead of by point dipoles.

Wada and Hayakawa proposed a model based on an analysis of the response of charges unsymmetrically distributed in materials. The schematic of the mechanism is shown in Figure 4.8, considering a point charge $q$ in the bulk of a polymer sample. If the electrodes are shorted, the induced charges on the electrodes, $Q_1$ and $Q_2$, are, respectively,

$$Q_1 = -q\frac{C_1}{C_1 + C_2}, \quad Q_2 = -q\frac{C_2}{C_1 + C_2} \tag{4.33}$$

where $C_1(C_2)$ is the capacitance between the upper (lower) electrode and the plane passing through $q$ parallel to the electrodes. This equation indicates that the magnitudes of $Q_1$ and $Q_2$ change when the ratio $C_1/C_2$ changes with applied strain; this results in a piezoelectric current in the external circuit. Three possibilities exist for the measured change in $Q_1$ and $Q_2$ with strain:

1)The charge $q$ is not fixed to the macroscopic plane in the material and moves out of the plane as a function of strain, that is, the displacement of the point charge is not the same as the macroscopic displacement in the material. This strain is called internal strain.

2)The charge $q$ is fixed to the material (embedded charge), but $C_1/C_2$ changes with strain owing to the heterogeneity of the materials. Since the capacitance is proportional to $\varepsilon/d$ ($d$ is the separation of the two plates constituting the capacitor), heterogeneity in the electrostriction constant and/or Poisson's ratio induces the change in $C_1/C_2$.

3)The material possesses a spontaneous polarization which changes either due to internal or external strain.

The measured piezoelectric coefficient is

$$e = \left(\frac{1}{A}\cdot\frac{\partial Q}{\partial x}\right)_{E=0} = \frac{1}{A}\cdot\frac{\partial(P_s)}{\partial x} \tag{4.34}$$

where $x$ represents mechanical strain. Based on the model and definitions, the following equation for the piezoelectric coefficient is obtained:

$$e = \langle e_i \rangle + \left\langle \left\{ \left(\frac{k}{\varepsilon} + m\right) - \left\langle \frac{k}{\varepsilon} + m \right\rangle \right\} \left\{ \int_{-1/2}^{z} \rho dz \right\} \right\rangle \tag{4.35}$$

where $k$ is electrostriction, $\varepsilon$ is the electric permittivity, $\rho$ is the charge density, $m$ is Poisson's ratio, and $z$ is the thickness of the sample. An electrostriction is a dimensional change of a material under an electric field due to the response of polar domains in the material to the field. The term $\left\langle\left\{\left(\frac{k}{\varepsilon}+m\right)-\left\langle\frac{k}{\varepsilon}+m\right\rangle\right\}\right\rangle$ is the result of sample heterogeneity, representing the piezoelectricity due to mechanism 2), whereas the intrinsic piezoelectricity appears in the average $\langle e_i \rangle$, including the contributions of mechanisms 1) and 2).

Later, Wada and Hayakawa modified the point charge model for semicrystalline polymers containing polar crystals [60]. In this model, polar crystals, as spherical particles, which possess spontaneous polarization $\boldsymbol{P}_{sc}$ and dielectric constant $\varepsilon_c$, are embedded in an isotropic amorphous matrix with dielectric constant $\varepsilon_a$. The origins of $\boldsymbol{P}_{sc}$ may be due to dipolar orientation in the crystal, dipolar orientation in the oriented amorphous phase, and dipoles arising from the trapped charges in the interfacial region. Based on this model and with the assumption of $N$ spheres of volume $v$, the following equation is derived:

$$P_s=\frac{N}{Al}\left(\frac{3\varepsilon_a}{2\varepsilon_a+\varepsilon_c}\right)\upsilon P_{sc}=\phi\frac{3\varepsilon_a}{2\varepsilon_a+\varepsilon_c}P_{sc} \tag{4.36}$$

where $A$ is the effective area, $l$ is the sample thickness, and $\phi$ is the volume fraction of the crystalline regions. According to Y. Wada's derivation [59], the piezoelectric $e$ coefficient can be expressed as

$$e=P_s\left[\left(\frac{\varepsilon_c}{2\varepsilon_a+\varepsilon}\right)\left(\frac{k_a}{\varepsilon_a}-\frac{k_c}{\varepsilon_c}g\right)-\frac{1}{l}\frac{\partial l}{\partial x}\right]+\phi\left(\frac{3\varepsilon_a}{2\varepsilon_a+\varepsilon_c}\right)e_c g \tag{4.37}$$

where $k_a$ and $k_c$ are the electrostriction constants in the amorphous and the crystal regions and $g$ is the ratio of strain in the crystal $x_c$ to that in the whole film $x$. If the piezoelectric constant of the crystal, $e_c$, is defined as:

$$e_c=\frac{1}{\upsilon}\frac{\partial(\upsilon P_{sc})}{\partial x_c}=P_{sc}\left(\frac{\partial\ln\upsilon}{\partial x}+\frac{\partial\ln P_{sc}}{\partial x_c}\right) \tag{4.38}$$

then,

$$e=P_s\left[\left(\frac{\varepsilon_c}{2\varepsilon_a+\varepsilon}\right)\left(\frac{k_a}{\varepsilon_a}-\frac{k_c}{\varepsilon_c}g\right)-\frac{1}{l}\frac{\partial l}{\partial x}\right]+\phi\left(\frac{3\varepsilon_a}{2\varepsilon_a+\varepsilon_c}\right)\left(\frac{P_{sc}}{\upsilon}\frac{\partial\upsilon}{\partial x_c}+\frac{\partial P_{sc}}{\partial x_c}\right)g \tag{4.39}$$

For a tensile strain,

$$g=\frac{\partial x_c}{\partial x}=\frac{5G_a}{3G_a+2G_c} \tag{4.40}$$

where $G_a$ and $G_c$ are the elastic moduli of the amorphous and crystal regions, respectively.

An alternative approach based on a two-phase system of crystals embedded in an amorphous matrix has been adopted by Broadhurst *et al.* In their approach, the crystals are assumed to be thin lamellae with their large surface perpendicular to the film surface rather than the spheres used by Wada *et al.* In this case, the factor $\frac{3\varepsilon_a}{2\varepsilon_a+\varepsilon_c}$ in Eq. (4.37) reduces to unity. The hydrostatic piezoelectric coefficient, $e$, is then given as

$$e_{hydro.} = -\phi P_{sc}\left(\frac{\varepsilon_c-\varepsilon_o}{3\varepsilon_o}+\frac{1}{2}\cdot\phi_o^c\gamma_g+\frac{1}{l}\cdot\frac{\partial l}{\partial S_c}\right)\frac{K_c}{K} \tag{4.41}$$

where $K_c$ and $K$ are, respectively, the compressibility of crystal and film and $\gamma_g$ is the Gruneisen constant defined by $\gamma_g=-\frac{\partial \ln\Omega}{\partial S}$ where $\Omega$ is the characteristic frequency of the dipole vibration.

From Eq. (4.39), it should be noted that the heterogeneity of the electrostriction, dimensional change, or strain dependence, Poisson's ratio, and intrinsic piezoelectricity (polarization) are considered main effects on the measured piezoelectric response. The comparison of the proposed mechanisms is summarized in Table 4.1.

The origins of the piezoelectric properties of polymers are still not completely understood. Contributions of dimensional change and polarization as well as electrostriction effects are widely accepted. It is also suggested by several researchers that space charge effects should be considered as another significant factor [29]. It is believed that the space charge plays an important role in the piezoelectric properties of polymers. Space charges can have two effects on spontaneous polarization, $P_s$: direct contribution to polarization and modification of the local field during the poling process, which gives rise to local heterogeneity. Coupled with an appreciable heterogeneity, space charges can contribute to the measured piezoelectricity of a sample (previously mentioned mechanism 2).

**Table 4.1.** Contributions from various proposed mechanisms to piezoelectricity

| | *Percentage of total response* | | | | | |
|---|---|---|---|---|---|---|
| | $d_p$ | | | $p_y$ | | |
| | **B** et al. | **W-H** | **P-T** | **B** et al. | **W-H** | **P-T** |
| Electrostriction | 34 | 22 | 45 | 27 | 0 | 76 |
| Dipole libration | 10 | - | - | 23 | - | - |
| Dimensional change of sample | 56 | 78 | 55 | 50 | 47 | 24 |
| Dimensional change of crystal | - | - | - | - | 53 | - |

*The coefficient considered was the piezoelectric stress coefficient, $e_{31}$

B et al.: Broardhurst et al., W-H: Wada and Hayakawa, and P-T: Purvis and Taylor.

Detailed discussion on the contributions of mechanisms presented here can be found in the references [56–59]. However, due to the complicated nature of the piezoelectric properties of polymeric materials, it is extremely difficult to determine exactly which theoretical model is closest to presenting the real mechanisms. However, the theoretical models proposed by researchers are useful in providing insight into the physical basis of the observed piezoelectricity.

## 4.4.2 Representative Ferroelectric and Piezoelectric Polymeric Materials

### *4.4.2.1 Polyvinylidene-Fluoride (PVDF or $PVF_2$) and its Copolymers*

The discovery of the significant piezoelectric response of poly(vinylidenefluoride) ($PVF_2$) films which were mechanically stretched followed by electrical poling initiated comprehensive investigations of structure and morphology of the polymer. The investigations led to several significant accomplishments in understanding the intrinsic mechanisms of the ferroelectric and piezoelectric properties of $PVF_2$. The understanding of the nature of ferroelectricity and piezoelectricity in $PVF_2$, including the formation of polymorphic crystal structures, the control of semicrystalline morphology, and the thermal stability of crystal phases, as well as their effects on ferroelectric and piezoelectric properties, guided the development of a series of $PVF_2$-based copolymers with trifluoroethylene. Systematic studies of this class of polymers have resulted in a more comprehensive understanding of the mechanisms of ferroelectricity and piezoelectricity of this type of polymer, especially the phase transition phenomena. Several review papers have been published separately by Wada and Hayakawa [38], Furukawa [42], Davis [40], Lovinger [61], and Kepler and Anderson [41,62].

Structures and Ferroelectricity of $PVF_2$

The repeat unit in the molecular structure of $PVF_2$ is $–CF_2–CH_2–$. The $–CF_2–$ groups are perpendicular to the molecular chain axis and are primary dipoles contributing to the remnant polarization. When the molecules crystallize, four primary crystal phases can be formed. They are α phase (II), or trans-gauche-trans-gauche' (TGTG') phase, β phase (I), or all-trans (TT) phase, δ phase (III), and γ phase (IV), or trans-trans-trans-gauche-trans-trans-trans-gauche' (TTTGTTTG'). The TGTG' conformation of α phase (II) and the molecular chain packing in the crystal result in a nonpolar unit cell. The other three phases are polar. However, β phase (I) which has an all-trans (TT) conformation is the most stable polar phase. α phase (II) and β phase (I) are the two most commonly studied and understood because α phase (II) is the primary thermodynamically stable phase when $PVF_2$ film is formed by solution processing and thermal processing techniques, whereas β phase (I) is the desired phase for electromechanical activities (ferroelectricity and piezoelectricity). Figure 4.9 shows a schematic representation of the view parallel to the chain axis in a crystal unit for a comparison of nonpolar α phase (II) and polar β phase (I). The dipole cancellation due to the opposite orientation of the polar groups happens in α phase (II), whereas the dipoles in β phase (I) are all aligned in one direction.

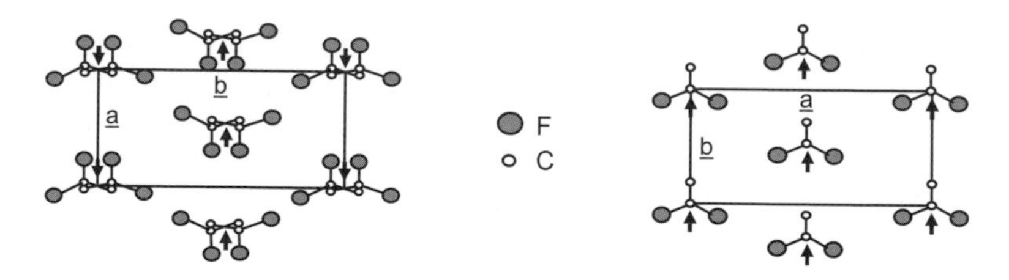


(a) α phase (II): canceling dipoles (nonpolar) (b) β phase (I): additive dipoles (polar)

**Figure 4.9.** A schematic comparison of the view parallel to the chain axis in crystal units of (a) nonpolar α phase (II) and (b) polar β phase (I)

To make β phase (I) $PVF_2$, mechanical stretching, electrical poling, and combinations of mechanical and electrical processing are commonly used to transfer thermodynamically stable and nonpolar α phase (II) $PVF_2$ to electromechanically active polar β phase (I). Figure 4.10 provides a block chart demonstrating the routes to make polar β phase (I) $PVF_2$ from nonpolar α phase (II) $PVF_2$.

In the as-mechanically-stretched $PVF_2$ film, the polar axis, *i.e.* the dipole moment axis, is randomly oriented to the stretching direction. An electrical poling treatment is needed to make the dipole axis reorient to align predominantly along the poling *z*-axis, that is, perpendicular to the film's plane. The field-induced dipole alignment in phase I crystals has been investigated by Raman spectroscopy, x-ray diffraction, and infrared dichroism [60], and all of the investigations suggest that the phase I crystal form in $PVF_2$ is ferroelectric according to the observation of polarization reversal via a cooperative switching process.

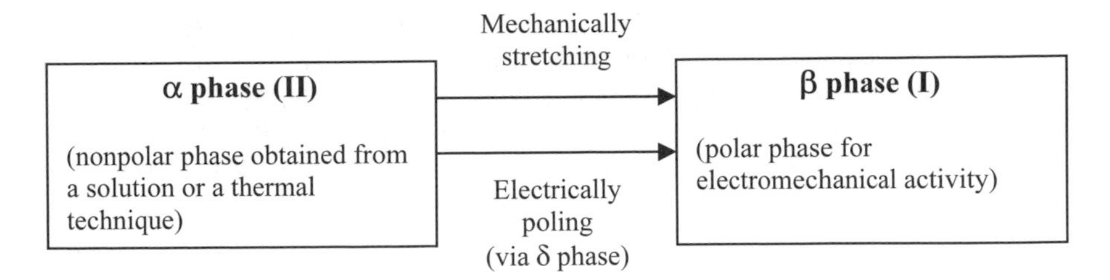


**Figure 4.10.** A block chart demonstrating the routes to make polar β phase (I) $PVF_2$ from nonpolar α phase (II) $PVF_2$

Ferroelectricity in $PVF_2$ was suggested by Bergman [29], Nakamura, and Wada [31] in 1971. Since then, various experimental findings in agreement with this speculation have been reported [63–69], particularly the hysteresis loop of spontaneous polarization $P_s$, the piezoelectric coefficient $e$, and the pyroelectric coefficient $p_y$, as well as the infrared absorbency vs. the poling field. Details of the dipole switching mechanisms in $PVF_2$ crystals occurring during poling are still not known. Suggested possible mechanisms are a) rotation of individual chains or cooperative rotation of a small number of chains around the chain axis, b) motion of domain walls, c) rotation of individual crystals, and d) intramolecular rotation (for the transition between phase II and phase IV). A theoretical model for dipole alignment in a polarizing field was first proposed by Aslasen [70]. Dvey-Aharon *et.al.* developed a theory of kink propagation as a model for poling [71,72]. Broadhurst and Davis presented a cooperative dipole orientation model. It is commonly accepted that the dipole orientation in the polymer is close to being perpendicular to the surface of the sample through a 60° switching mechanism, if the sample is sufficiently poled [32].

The ferroelectric phase I of $PVF_2$ does not show a Curie point [73] or ferroelectric-paraelectric transition in thermal characterization even though it exhibits clear typical ferroelectric D-E hysterisis when a Sawyer-Tower circuit-based ferroelectric characterization is conducted. The reason for the interesting phenomenon was not understood until it was observed that with an increase in the content of introduced defects, such as head-head (H-H) or tail-tail (T-T) joints in the chain, phase I, instead of phase II, might become the energetically stable phase [74].

Lando and Doll reported that copolymers of the $VF_2$, (–$CH_2$–$CF_2$–), and $VF_3$, (–CHF–$CF_2$–), when the $VF_3$ content is more than 17 mol %, accommodate the more extended all-trans conformation, which is similar to phase I of $PVF_2$ [75]. Yagi *et. al.* synthesized the copolymers with various relative mole compositions (50-80 mol %) in 1979 [76]. Because of the advantage of being crystallized directly into the planar zigzag (all-trans) conformation, which is similar to phase I of $PVF_2$, the ferroelectricity and piezoelectricity of these copolymers have been extensively studied [77]. Ferroelectric transition phenomena have been observed in investigations of the copolymers. Curie transition temperatures of the copolymers were found to be dependent on the relative compositions of $VF_2$ and $VF_3$. The higher the $VF_2$ content, the higher the Curie transition temperature. If the $VF_2$

content is higher than 81 mol %, the Curie transition disappears because the transition temperature of such copolymers is higher than the melting temperature, which also explains why $PVF_2$ homopolymer does not show a Curie transition [78]. Table 4.2 provides the melting temperatures and the Curie transition temperatures of various copolymers and $PVF_2$ homopolymers. If $(T_m - T_c) < 0$, the Curie transition temperature cannot be observed by a thermal test such as a differential scanning calorimeter (DSC).

**Table 4.2.** Melting and Curie transition temperatures of $PVF_2$ and its copolymers

| | *$PVF_2$* | *52 mol% ($VF_2$) copolymer* | *65 mol% ($VF_2$) copolymer* | *78 mol% ($VF_2$) copolymer* | *80 mol% ($VF_2$) copolymer* |
|---|---|---|---|---|---|
| $T_m$ (°C) | 182 | 160 | 152 | 153 | 151 |
| $T_c$ (°C) | 198* | 67 | 92 | 132 | 141 |
| $(T_m$-$T_c)$ (°C) | -16** | 93 | 60 | 19 | 10 |

*extrapolated and ** $(T_m$-$T_c) < 0$.

Piezoelectricity of $PVF_2$

The piezoelectric properties of semicrystalline polymers, such as $PVF_2$ and its copolymers, are determined by the remanent polarization in the materials and the mechanical properties. The remnant polarization depends primarily on the crystallinity, or the content, of polar crystals and the alignment of the polar crystal domains. The most popular method to increase crystallinity is thermal treatment, or annealing. The annealing temperature should be significantly above the glass transition temperature and below the melting temperature and Curie transition temperature if the Curie transition temperature is lower than the melting temperature. In the effective temperature range, the higher the temperature, the better the annealing effect. The method to generate good alignment of polar crystal domains, that is, the remnant polarization, is electrical poling. Three primary factors are important for poling the materials to achieve high remnant polarization: poling temperature, poling field, and poling time. Generally, higher poling temperature, higher poling field, and longer poling time result in a higher remnant polarization. Usually, an optimization of the three primary factors is needed for the best poling effect. A well-poled (at room temperature) uniaxially stretched (five times) $PVF_2$ film usually can offer the remanent polarization, $\boldsymbol{P_r}$, of 50 mC/m$^2$ and a piezoelectric strain coefficient, $d_{31}$, of 25 pC/N at room temperature ($d_{31}$ means that the electric signal collected in the out-of-surface direction, direction 3, is generated by the mechanical force applied in the in-surface direction, direction 1). The piezoelectric response of a poled $PVF_2$ depends on the temperature. When the temperature is below its glass transition temperature, $T_g$, which is around –50 °C, the piezoelectric strain coefficient is very small (less than 2 pC/N) above its $T_g$, the piezoelectric strain coefficient increases and reaches the maximum around room temperature. As the temperature keeps increasing, the piezoelectric coefficient starts to decrease around 50°C due to thermal depolarization [79].

#### *4.4.2.2 Odd-numbered Polyamides (Nylons)*

Another class of representative ferroelectric and piezoelectric polymers is odd-numbered polyamides (nylons). Hydrogen bonds formed by –NH and –C=O groups provide essential and stable polar elements in the polymer for piezoelectricity. The significant piezoelectric properties of nylon 11 were studied and reported in the early 1980s [50]. Its ferroelecticity was predicted based on study of its chemical structures and crystal phases. However, ferroelectricity in nylons was not confirmed until 1991 when Scheinbeim *et al.* reported a typical ferroelectric hysteresis loop from the electric displacement, *D*, vs. the applied electric field, *E,* test of a specially treated nylon 11 film. The film was produced by a melt-quench and cold-stretching method. The melt-quenched and cold-stretched nylon 11 not only exhibited the ferroelectric switching mechanism but also increased its piezoelectric property significantly compared to that previously reported, especially at an elevated temperature [53–55]. The discovery has led to a series of achievements in research on the ferroelectricity and the piezoelectricity of odd-numbered nylons, including nylon 9, nylon 7, and nylon 5, in Scheinbeim's group in the following years [82].

<u>Molecular structures and ferroelectricity of odd-numbered nylons</u>

The alpha-phase crystal structure of nylon 11 suggested by Slichter in 1959 is polar. The molecular conformation of the proposed structure is all trans, and for an odd-numbered nylon, this entails a net dipole moment per chain. Figure 4.11 shows the molecular structures of odd-numbered nylons and even-numbered nylons) [79]. As can be seen, in odd-numbered nylons, the electric dipoles formed by amide groups (H–N–C=O) with a dipole moment of 3.7 Debyes are sequenced in a way that all the dipoles are in the same direction. Therefore, a net dipole moment occurs. In even-numbered nylons, one amide group is in one direction, the next one will be in the opposite direction, alternately. This results in an intrinsic cancellation of the dipole moments, as demonstrated schematically in the figure. The ferroelectricity of nylon 11 and nylon 7 was discovered by Lee *et al.* in 1991, when typical electric displacement, ***D***, versus electric field, ***E***, hysteresis loops were obtained. The reversion of polarization for the ferroelectricity in the nylon 11 films was confirmed in molecular structure characterization by x-ray diffraction (XRD) and Fourier transform infrared (FTIR) [80]. The typical ferroelectric characteristics were also observed in the following studies on other odd-numbered nylons, including nylon 9, nylon 7, and nylon 5 films produced using the melt-quench and cold-stretching method [82].

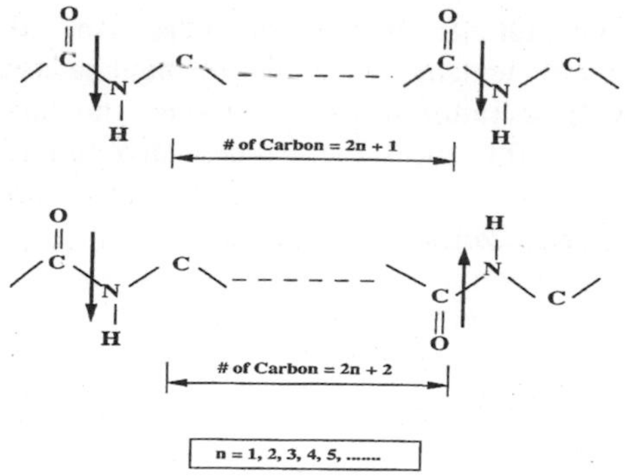

**Figure 4.11.** Schematics of molecular structures of odd-numbered nylons and even-numbered nylons (Adapted from J. Su [79])

The investigation also discovered that cold-stretching following melt-quenching is a very critical step in obtaining the polyamide chains in the parallel form needed for ferroelectric polarization switching and piezoelectricity. For instance, when a nylon 11 film is prepared using a melt-quenched, uniaxially cold draw technique, a doubly oriented structure is formed, with dipole orientation in the sheet plane parallel to the surface of the film, whereas the molecular chains are oriented in the drawing direction. There are two possible molecular arrangements: parallel (progressive) chain packing and antiparallel (staggered) chain packing, which are known as the molecular bases of the $\alpha$ form and $\beta$ form crystals, respectively. In both cases, hydrogen bonded sheets are formed. A 90° followed by an 180° polarization switching mechanism was reported for nylon 11. When an electric field is applied in the direction perpendicular to the film surface, the dipoles rotate to the field direction by 90° from a doubly oriented state to parallel to the applied electric field. In the following polarization reversals, the dipoles switch by 180° when the applied electric field reverses direction [81]. A diagram of the simplified switching mechanism is shown in Figure 4.12 and the arrows represent the orientation of the axial direction of the neighbering molecular chains (up or down alternatively as the arrows indicate).

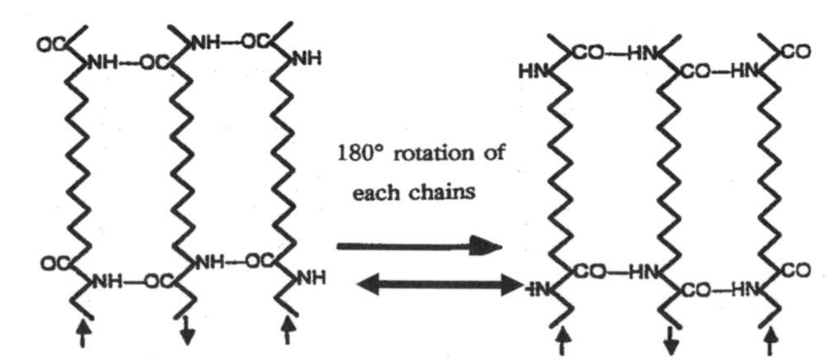


**Figure 4.12.** A schematic representation of the proposed 90°–then–180° mechanism for the ferroelectricity of the nylon 11 (Redrawn and adapted from J-W. Lee *et al.* [54])

The remnant polarization of ferroelectric odd-numbered nylons is a function of the number of carbon atoms per repeat unit, which is decisive for the electric dipole density. The smaller the number, the higher the dipole density. The interrelationship between the number and the remnant polarization is tabulated in Table 4.3 as reported by Mei *et.al.* in 1993 [82].

**Table 4.3.** Unit molecular weight, dipole density and remnant polarization of odd-numbered nylons

| | *N-3* | *N-5* | *N-7* | *N-9* | *N-11* |
|---|---|---|---|---|---|
| Molecular weight of repeat unit | 71.1 | 99.1 | 127.2 | 155.2 | 183.3 |
| Dipole density (D/100 Angstrom$^3$) | 4.30 | 2.92 | 2.12 | 1.65 | 1.40 |
| Remnant Polarization ($mC/m^2$) | 180* | 125 | 86 | 68 | 56 |

*Predicted

Piezoelectricity of some representative odd-numbered nylons

As with the piezoelectric properties of $PVF_2$, the piezoelectric properties of odd-numbered nylons, as semicrystalline polymers, are determined by the remanent polarization in the materials and the mechanical properties. The remnant polarization primarily depends on the crystallinity, or the content, of polar crystals and the alignment of the polar crystal demains. Thermal annealing is the most popular method to increase crystallinity. However, for odd-numbered nylons, the melt-quench and cold-stretching process is critical for the parallel sheet structure formed by hydrogen bonds. Parallel-sheet-structured, the odd-numbered nylons can generate typical ferroelectric property, and therefore, can result in expected piezoelectric properties after being poled. Three primary factors are important for poling the materials to achieve high remanent polarization: poling temperature, poling field, and poling time. Generally, a higher poling temperature (higher than glass temperature), higher poling field, and longer poling time generate higher remnant polarization. Usually, an optimization of the three primary factors is needed for the best poling effect. For instance, a well-poled (at room temperature) melt-quenched and uniaxially stretched (3.5 times at room temperature) nylon 11 film usually can possess a remanent polarization, $\boldsymbol{P_r}$, of ~50 $mC/m^2$ and offer a piezoelectric strain coefficient, $d_{31}$, of 2.8pC/N at room temperature. The $d_{31}$ is much less than that of $PVF_2$ at room temperature because the piezoelectric response of a poled nylon 11 also depends on the temperature, and room temperature is below its glass transition temperature (which is around 70°C). As the temperature increases above its $T_g$, the piezoelectric strain coeffiecient increases to as much as 9pC/N [79].

Lee *et.al.* reported, in 1991, that annealing of poled nylon 11 and nylon 7 results in both an enhanced piczoelectric strain coefficient and improved piezoelectric stability [53]. Annealed odd-numbered nylons (both nylon 11 and nylon 7) possess high and stable piezoelectric properties at high temperatures (up to 200°C), which is a significant advantage over $PVF_2$ which melts at ~182°C. Figure 4.13 shows a comparison of the temperature dependence of the piezoelectric strain coefficients of annealed nylon 11, nylon 7, and $PVF_2$.

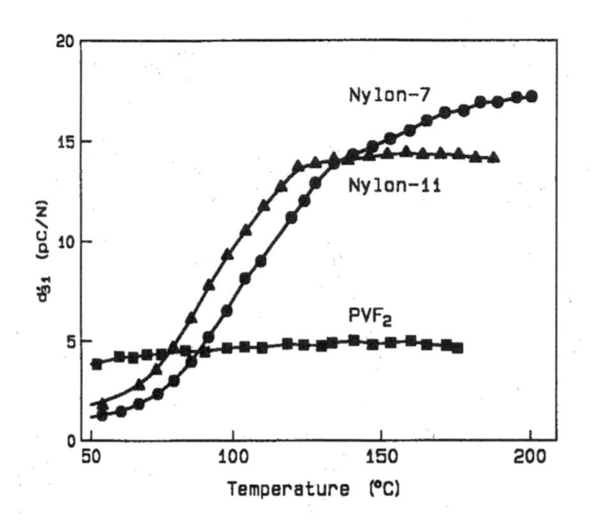


**Figure 4.13.** Temperature dependence of piezoelectric strain coefficient, $d_{31}$, of the poled and annealed nylon 11 and nylon 7 with a comparison to that of poled and annealed $PVF_2$. (Adapted from Y. Takashi *et al.*[55])

*4.4.2.3 Ferroelectric and Piezoelectric Polymer-polymer Composite Systems*

Nylon 11-$PVF_2$ bilaminates

The development of the two ferroelectric and piezoelectric polymers: $PVF_2$ and odd-numbered nylons provides the possibility of making all-polymer ferroelectric and piezoelectric composite systems. Using $PVF_2$ and nylon 11, Su, *et.al.* developed nylon 11-poly(vinylidene fluoride) bilaminates by a co-melt-pressed-stretched process in 1995 [83]. The bilaminate exhibits a typical ferroelectric ***D-E*** hysteresis loop with significantly enhanced remnant polarization, $P_r$, of 75 mC/m$^2$, which is 44% higher than those of individual nylon 11 or $PVF_2$ films made by an identical process. The results of the D-E hysteresis ferroelectric characterization of a 1:1 bilaminate are shown in Figure 4.14a with a comparison with those of individual $PVF_2$ and nylon 11. The piezoelectric strain coefficients, including the strain coefficient, $d_{31}$, the stress coefficient, $e_{31}$ and the hydrostatic coefficient, $d_h$, also show significant enhancement. The enhancement in the piezoelectricity becomes more obvious when the temperature is above the glass transition temperature of nylon 11. Figure 4.14b shows the temperature dependence of the piezoelectric strain coefficient, $d_{31}$, of a nylon 11-$PVF_2$ bilaminate having a 1:1 ratio with a comparison to that of individual nylon 11 and $PVF_2$.

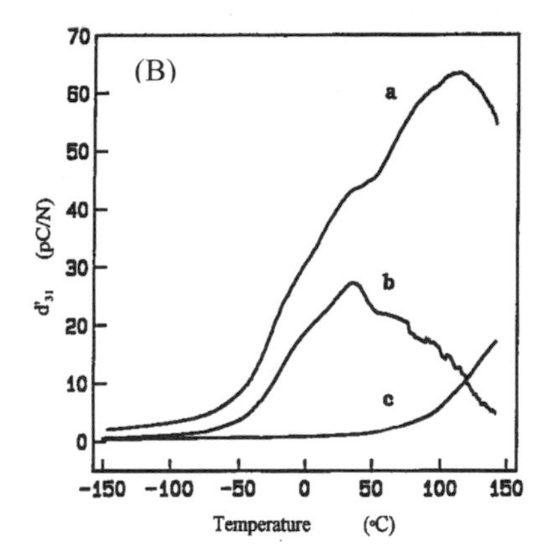


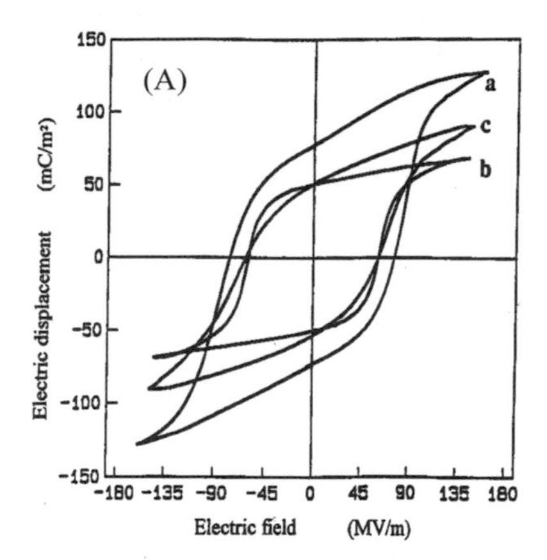


**Figure 4.14.** (A) Curves of electric field displacement, *D*, versus applied electric field, *E*, (*D-E*) and (B) temperature dependence of the piezoelectric strain coefficient, $d_{31}$, for (a) nylon 11/$PVF_2$ bilaminate, (b) $PVF_2$, and (c) nylon 11 films. (Adapted from J. Su *et al.* [83])

The enhancement is attributed to interfacial space charge accumulation and the asymmetrical distribution of the accumulated space charges in the direction across the interface between the two constituents [84]. The remnant polarization and piezoelectric coefficients of the bilaminates are a function of the fraction of the two constituents because the fraction of the two constituents decides the distribution of the effective electric field on each constituent due to the difference in their dielectric constant, therefore, the distribution of the accumulated space charges in the interfacial region.

Nylon 11-$PVF_2$ blends

In 1999, Gao *et. al.* reported the development of nylon 11-$PVF_2$ blends which also exhibit significantly enhanced remnant polarization, $\boldsymbol{P_r}$. The $\boldsymbol{P_r}$ of the blend with the 50:50 composition is 85 mC/m$^2$, which is more than 60% higher than those of individual nylon 11 and $PVF_2$ [85]. The curves of the electric field displacement, ***D***, versus the applied electric field, ***E***, are shown in Figure 4.15. The same paper also reported that the ferroelectricity of the blends depends on the fraction of the two constituents and that the enhancement might also be attributed to the space charge accumulation and distribution [86]. The piezoelectric strain coefficient, $d_{31}$, of nylon 11-$PVF_2$ blend films also shows significant enhancement compared with individual nylon 11 and $PVF_2$ films. The coefficient depends on the the fraction of the two constituents in blends.

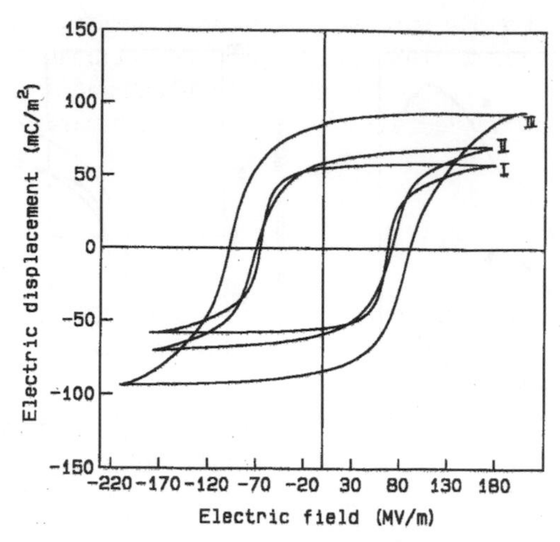


**Figure 4.15.** Curves of electric field displacement, *D*, versus applied electric field, ***E***, (***D-E***) (-I- nylon 11, -II- $PVF_2$ films and for -III- nylon 11/$PVF_2$ 50:50 blend). (Adapted from Q. Gao *et al.* [85])

### *4.4.2.4 Summary of Ferroelectric and Piezoelectric Properties of Ferroelectric and Piezoelectric Polymers*

To summarize the ferroelectric and piezoelectric properties of the ferroelectric and piezoelectric polymers discussed, some important ferroelectric and piezoelecetric parameters are tabulated in Table 4.4.

**Table 4.4.** Summary of ferroelectric and piezoelectric properties

| | *$PVF_2$* | *Nylon 11* | *Nylon 7* | *Bilminate 1:1 (nylon11: $PVF_2$) (thickness ratio)* | *Blend 20:80 (nylon 11:$PVF_2$) (weight ratio)* |
|---|---|---|---|---|---|
| $P_r$ (mC/m$^2$) | 52 | 52 | 86 | 75 | 85 |
| $d_{31}$@25°C | 25 | 3 | 2 | 41 | – |
| $d_{31}$@110°C | 12 | 9 | 7 | 62 | 34 |
| $d_{31}$@150°C | 4 | 13 | 15 | 53 | 52 |
| $d_{31}$@180°C | – | 13 | 18 | – | – |

As discussed in the previous sections, the ferroelectric and piezoelectric properties of polymeric and polymeric composite systems depend on various factors, such as crystallinity, poling conditions, glass transition temperature, and before- and after-poling treatments (electrical, mechanical, and thermal treatments). In addition to the factors mentioned above, for composite systems, laminates, or blends, the fraction of constituents and interfaces are also important. Therefore, the properties tabulated may vary due to the dependence of the ferroelectric properties and piezoelectric properties on the these factors and the dependence of the factors on material preparation methods. However, the tabulated summary should provide a reference for the selection of materials for applications or a guideline for developing new ferroelectric and piezoelectric polymeric materials.

## 4.5 Remarks

Electroactive polymers including ferroelectric and piezoelectric polymers and their applications are still relatively new research fields. Due to several advantages of electroactive polymers over electroactive ceramics (light weight, good processability, low cost, and mechanical toughness, *etc.*) these research fields have been drawing more and more attention of researchers worldwide since the significant ferroelectric and piezoelectric properties of poly(vinylidene fluoride), $PVF_2$ or PVDF, were discovred and reported in 1969. In the past three decades, various ferroelectric piezoelectric polymers have been developed. Among them, $PVF_2$ and its copolymers and odd-numbered nylons are two representative classes of polymers that have been systematically investigated, and the mechanisms of their ferroelectricity and piezoelectricity have been well understood. This chapter provides readers with primary information about these two classes of ferroelectric and piezoelectric polymers and these polymer-based ferroelectric and piezoelectric polymer-polymer composite systems. The information provided may serve readers as a guideline for understanding the nature of ferroelectric and piezoelectric polymers and for developing techniques to tailor or control the ferroelectric and piezoelectric properties as desired.

*Ackowledgement:* The author of this chapter thanks Dr. Jeffrey Hinkley, NASA Langley Research Center for his valuable discussions, comments, and suggestions.

## 4.6 References

[1] H. Kawai, *Japan. J. Appl. Phys.*, **8**, 975 (1969).
[2] Y. Wada and R. Hayakawa, *Ferroelectrics*, **32**, 115 (1981).
[3] R. G. Kepler and R. A. Anderson, *Adv. Phys.*, **41(1)**, 1 (1992).
[4] J. W. Lee, Y. Takase, B. A. Newman, and J. I. Scheinbeim, *J. Polym. Sci.: Part B: Polymer Phys.*, **29**, 279 (1991).
[5] Y. Takase, J. W. Lee, J. I. Scheinbeim, and B. A. Newman, *Macromolecules*, **24**, 6644 (1991).
[6] J. I. Scheinbeim, J. W. Lee, and B. A. Newman, *Macromolecules*, **25**, 3729 (1991).
[7] B. Z. Mei, J. I. Scheinbeim, and B. A. Newman, *Ferroelectrics*, **144**, 51 (1993).
[8] J. Su, "*Ferroelectric and Piezoelectric Polymer-Polymer Composite Systems,*" Ph.D. Dissertation, Rutgers (1995).
[9] A. von Hippel, Waves and Dielectrics, John Wiley & Sons, New York, 1954.
[10] C. Ku and R. Liepins, Electrical Properties of Polymers, Hanser, New York, 1987.

[11] P. Hedvig, Dielectric Spectroscopy of Polymers, John Wiley & Sons, New York, 1977.
[12] S. Takashima, Electrical Properties of Biopolymers and Membranes, Adam Hilger, 1989.
[13] W. D. Kingery, H. K. Bowen, and D. R. Uhlmann, Introduction to Ceramics, John Wiley & Sons, New York 1976.
[14] B. Gross, Topics in Applied Physics: Electrets, Ed. by G. M. Sessler, Springer-Verlag, 1980.
[15] A. R. Blythe, Electrical Properties of Polymers, Cambridge University Press, 1977.
[16] L. L. Hench and J. K. West, Principles of Electronic Ceramics, John Wiley & Sons, New York, 1990.
[17] J. Curie and P. Curie, *Bull. Sco. Min. de France*, **3**, 90 (1880).
[18] W. G. Hankel, *Abh. Sachs.*, **12**, 451 (1881).
[19] W. Voigt, *Abh. Gott.*, **36**, 1 (1890).
[20] Y. Takase, "Electric Properties of Polymers". Unpublished Lecture Text Notebook (1991).
[21] T. Mitsui, An Introduction to the Physics of Ferroelectrics. Gordon and Breach, New York, 1976.
[22] K. Valasek, *J. Phys. Rev.*, **17**, 475 (1921).
[23] J. C. Burfoot, Ferroelectrics: An Introduction to the Physical Principles, Van Nostrand Ltd. London, 1967.
[24] C. W. Sawyer and C. H. Tower, *Phys. Rev.*, **35**, 269 (1930).
[25] M. E. Lines and A. M. Glass, Principles and Applications of Ferroelectrics and Related Materials, Clarendon Press, Oxford, 1977.
[26] J. C. Burfoot and G. W. Taylor, Polar Dielectrics and Their Applications, University of Califonia Press, 1979.
[27] P. Weiss, *J. Physique*, **6**, 667 (1907)
[28] K. R. Brian, *Proc. Phys. Soc. London*, **36**, 81 (1924).
[29] J. B. Bergman Jr., J. H. McFee, and G. R. Crane, *Appl. Phys. Lett.*, **18**, 203 (1971).
[30] J. H. McFee, J. B. Bergman Jr., and G. R. Crane, *Ferroelectrics*, **3**, 305 (1972).
[31] K. Nakamura and Y. Wada, *J. Polym. Sci. A-2*, **9**,161 (1971).
[32] R. G. Kepler and R. A. Anderson, *J. Appl. Phys.,* **49**, 1232 (1978).
[33] J. B. Lando, H. G. Olf, and A. Peterlon, *J. Polym. Sci. A-1*, **4**, 941 (1966).
[34] A. J. Lovinger and T. T. Wang, *Polymer,* 20, 725 (1979).
[35] J. B. Lando and W. W. Doll, *J. Macromol. Sci.; Phys.*, **2**, 205 (1968).
[36] J. B. Lando and M. A. Backman, *Macromolecules*, **14**, 40, (1981).
[37] R. Hasegawa, Y. Takahashi, Y. Chatani, and H. Tadokoro, *Polym. J.*, **3**, 600 (1972).
[38] Y. Wada and R. Hayakawa, *Jpn. J. Appl. Phys.*, **15**, 2041 (1976).
[39] R. G. Kepler and R. A. Anderson, *CRC Critical Rev.: Solid State and Mat. Sci.*, **9**, 399 (1980).
[40] G. T. Davis, Polymers for Electronic and Photonic Applications, Ed. by C. P. Wong, Academic Press, 1992.
[41] R. G. Kepler and R. A. Anderson, *J. Appl. Phys.*, **49**, 1232 (1978).
[42] T. Furukawa, *Phase Transitions*, **18**, 143 (1989).
[43] M. G. Broadhurst, G. T. Davis, J. E. McKinney, and R. E. Collins, *J. Appl. Phys.*, **49**, 4992 (1978).
[44] M. G. Broadhurst and G. T. Davis, *Ferroelectrics*, **60**, 3 (1984).
[45] C. K. Purvis and P. L. Tayler, *Phys. Rev.: B*, **26**, 4547 (1982).
[46] R. Al-Jishi and P. L. Tayler, *Ferroelectrics*, **73**, 343 (1987).
[47] Y. Wada and R. Hayakawa, *Ferroelectrics*, **32**, 115 (1981).
[48] Y. Wada and R. Hayakawa, *Jpn. J. Appl. Phys.*, **15**, 2041 (1976).

[49] G. M. Sessler, D. K. Das-Gupta, A. S. DeReggi, W. Eisenmenger, T. Furukawa, J. A. Giacometti, and R. Gerhard-Multhaupt. *IEEE Trans. on Electrical Insulation, 27, 872* (1992).
[50] B. A. Newman, P. Chen, K. D. Pae, and J. I. Scheinbeim, *J. Appl. Phys.*, **51**, 5161(1980).
[51] J. I. Scheinbeim, S. C. Mathur, and B. A. Newman, *J. Polym. Sci.: Part B: Polym. Phys.*, **24**, 1791 (1986).
[52] S. C. Mathur, J. I. Scheinbeim, and B. A. Newman. *J. Polym. Sci.: Part B: Polym. Phys.*, **26**, 447 (1988).
[53] J. W. Lee, Y. Takase, B. A. Newman, and J. I. Scheinbeim. *J. Polym. Sci.: Part B: Polym. Phys.*, **29**, 273 (1991).
[54] J. W. Lee, Y. Takase, B. A. Newman, and J. I. Scheinbeim. *J. Polym. Sci.: Part B: Polym. Phys.*, **29**, 279 (1991).
[55] Y. Takase, J. W. Lee, J. I. Scheinbeim, and B. A. Newman. *Macromolecules*, **24**, 6644 (1991).
[56] M. G. Broadhurst and G. T. Davis, *Ferroelectrics*, **60**, 3 (1984).
[57] C. K. Purvis and P. L. Tayler, *Phys. Rev.: B*, **26**, 4547 (1982).
[58] R. Al-Jishi and P. L. Tayler, *Ferroelectrics*, **73**, 343 (1987).
[59] Y. Wada and R. Hayakawa, *Jpn. J. Appl. Phys.*, **15**, 2041 (1976).
[60] Y. Wada and R. Hayakawa, *Ferroelectrics*, **32**, 115 (1981); *Insulation, **28**, 243* (1993).
[61] A. J. Lovinger, Development in Crystalline Polymers I, Ed. by D.C. Bassett, London Applied Science, 1982.
[62] R. G. Kepler and R. A. Anderson, *Adv. Phys.*, **41**, 1 (1992).
[63] N. Takahashi and A. Odajima, *Ferroelectrics*, **32**, 49 (1981).
[64] A. J. Bur, J. D. Barnes, and K. J. Wahlstrand, *J. Appl. Phys.*, **59**, 2345 (1986).
[65] B. Servet, S. Ries, D. Broussoux, and F. Micheron, *J. Appl. Phys.*, **55**, 2763 (1984).
[66] M. Tamura, S. Hagiwara, S. Matsumoto, and N. Ono, *J. Appl. Phys.*, **48**, 513 (1977).
[67] D. Naegele and D. Y. Yoon, *Appl. Phys. Lett.*, **33**, 132 (1978).
[68] G. Cortili and G. Zerbi, *Spectrochim. Acta: A*, **23**, 285 (1967).
[69] M. Kobayashi, K. Tashiro, and H. Tadakoro, *Macromolecules*, **8**, 158 (1977).
[70] E. Aslasen, *J. Chem. Phys.*, **57**, 2358 (1972).
[71] H. Dvey-Aharon, T. J. Sluckin, P. L. Tayler, and A. J. Hopfinger, *Phys. Rev.: B*, **21**, 3700 (1980).
[72] H. Dvey-Aharon, P. L. Tayler, and A. J. Hopfinger, *Phys. Rev.: B*, **21**, 3700 (1980).
[73] A. J. Lovinger, G. E. Johnson, H. C. Bair, and E.W. Anderson, *J. Appl. Phys.*, **56**, 2412 (1984).
[74] T. T. Wang, J. M. Herbert, and A. M. Glass, Applications of Ferroelectric Polymers, Blakie, Glasgow, 1988.
[75] J. B. Lando and W. W. Doll, *J. Macromol. Sci. B: Phys.*, **2**, 205 (1968).
[76] T. Yagi, M. Tatemoto, and J. Sako, *Polym. J.*, **12**, 209 (1980).
[77] A. J. Lovinger, T. Furukawa, G. T. Davis, and M.G. Broadhurst, *Ferroelectrics*, **50**, 227 (1983).
[78] T. Yamada, T. Ueda, and T. Kitayama, *J. Appl. Phys.*, **52**, 948 (1981).
[79] J. Su, Ph.D. Thesis, Rutgers (1995).
[80] J. I. Scheinbeim, J. W. Lee, and B. A. Newman, *Macromolecules*, **25**, 3729 (1991).
[81] J. I. Scheinbeim and B. A. Newman, *TRIP*, **1**, 394 (1993).
[82] B. Z. Mei, J. I. Scheinbeim, and B. A. Newman, *Ferroelectrics*, **144**, 51 (1993).
[83] J. Su, Z. Y. Ma, J. I. Scheinbeim, and B. A. Newman, *J. Polym. Sci.: Part-B: Polym. Phys.*, **33**, 85 (1995).

[84] G. C. Chen, J. Su, and L. J. Fina, *J. Polym. Sci.: Part B: Polymer Physics*, **32**, 2065 (1994).
[85] G. Gao, J. I. Scheinbeim, and B. A. Newman, *J. Polym. Sci.: Part B: Polym. Phys.,* **37**, 3217 (1999).
[86] G. Gao, and J. I. Scheinbeim, *Macromolecules,* **33**, 7546 (2000).

# 5

# Polypyrrole Actuators: Properties and Initial Applications

J. D. Madden

Molecular Mechatronics Lab, Advanced Materials & Process Engineering Laboratory and Department of Electrical & Computer Engineering, University of British Columbia, Vancouver, British Columbia, V6T 1Z4 Canada
jmadden@ece.ubc.ca

## 5.1 Summary

Polypyrrole actuators are low-voltage (1–3 V), moderate to large strain (2–35%), and relatively high stress (up to 34 MPa) actuator materials. Strain rates are moderate to low, reaching 11%/s, and frequency response can reach several hertz. Faster response (> 1 kHz) is anticipated in nanostructured materials. Forces can be maintained with minimal power expenditure. This chapter reports on the current status and some of the anticipated properties of conducting polymer actuators. Applications investigated to date include braille cells, shape changing stents, and variable camber foils. Situations where low voltage operation is valuable and volume or mass are constrained favor the use of conducting polymers.

Polypyrrole and other conducting polymers are typically electrochemically driven and can be constructed in linear or bending (bilayer) geometries. Synthesis can be by chemical or electrochemical means, and raw materials are generally very low in cost. These polymers are electronically conducting organic materials. They also allow ions to diffuse or migrate within them. An Increase in the voltage applied to a polymer electrode leads to removal of electrons and an increasingly positive charge within the volume of the polymer. This charge is balanced by negative ions that enter the polymer from a neighboring electrolyte phase (or by positive ions that leave). Ion insertion is generally accompanied by expansion of the polymer. The ions, solvent, and synthesis conditions determine the extent of this expansion, which can be anisotropic. A change in modulus has also been observed as a function of the oxidation state.

Models relating charge, strain, voltage, stress, and current have been developed that allow designers to evaluate the feasibility of designs. One of these modeling approaches is presented with the aim of enabling selection of appropriate device geometry.

The field of conducting polymer actuators is developing rapidly with larger strains, stresses, cycle lifetimes, and rates reported every year. The background needed to understand these developments and to decide if polypyrrole and in general conducting polymer actuators are appropriate for use in a given application is provided.

## 5.2 Introduction and Overview

Conducting polymer actuators are relatively new, and applications are at an early stage of development. In this article, basic properties and models are presented, and a few example applications are given, in the hope that this information will guide and stimulate further applications. The introduction provides a general overview, following the format of a recent review article on electroactive polymers [1]. Many of the topics are discussed subsequently in more detail, including an overview of conducting polymer properties, a description of electrochemical synthesis of the polymers, and a brief overview of two basic device configurations. This is followed by a discussion of models that are intended to guide design.

Conducting polymers are electronically conducting organic materials featuring conjugated structures, as shown in Figure 5.1. Electrochemically changing the oxidation state leads to the addition or removal of charge from the polymer backbone, shown in Figure 5.2, and a flux of ions to balance charge. This ion flux, often accompanied by solvent, induces swelling or contraction of the material [2–8]. Insertion of ions and solvent between polymer chains likely induces the majority of the volume changes. Conformational changes in the backbone may also play a role. Changes in oxidation state and in dimension can also be chemically induced [6].

Important advantages relative to most other electroactive polymers are the high tensile strengths, which can exceed 100 MPa [9], and large peak stresses of up to 34 MPa [10]. The stiffness is also higher than in many electroactive polymers, with the modulus generally exceeding 0.1 GPa and often reaching ~ 1 GPa modulus [11–13]. Furthermore, the modulus can be a function of oxidation state, potentially enabling controllable stiffness [14]. A major advantage over piezoelectrics, electrostatic actuators, dielectric elastomers, and ferroelectric polymers is low voltage operation (~ 2 V), which is particularly useful in portable, battery-driven applications, and often enables the actuators to be driven without the need for extensive and costly voltage conversion circuitry. Initial work demonstrated only moderate strains of several percent – much greater than piezoceramics, but less than those observed in a number of other emerging actuation technologies [1]. Recent work has demonstrated that significantly larger strains can be obtained, in excess of 35% for a few cycles and routinely around 9% [15–17].

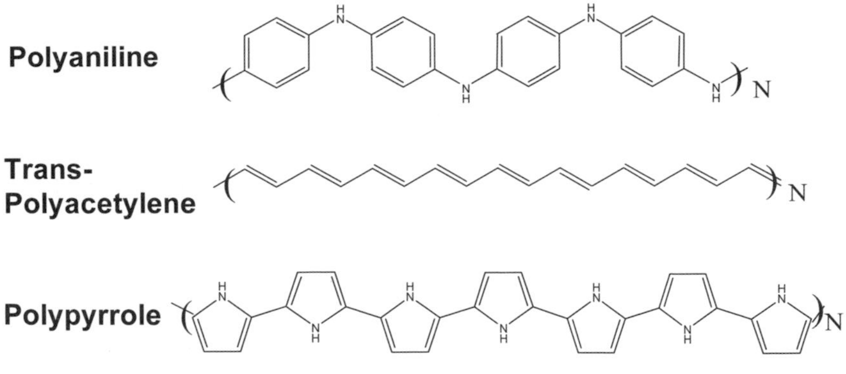

Polyaniline

Trans-Polyacetylene

Polypyrrole

**Figure 5.1.** The chemical structures of some common conducting polymers employed as actuators

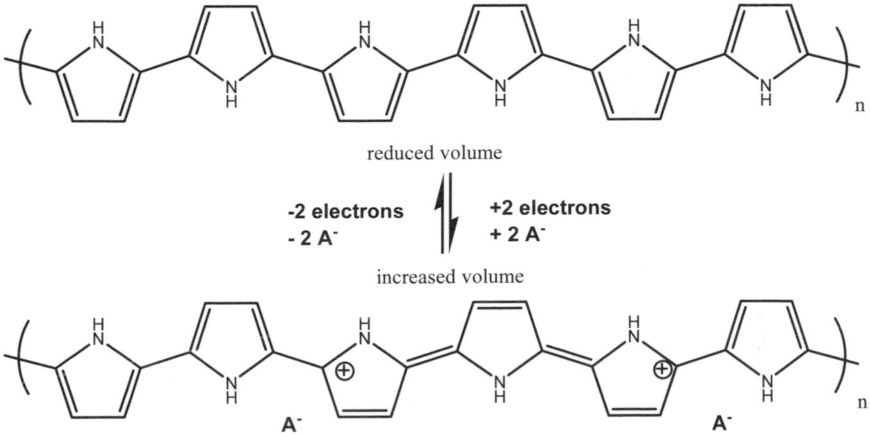

reduced volume

-2 electrons
- 2 A⁻

+2 electrons
+ 2 A⁻

increased volume

**Figure 5.2.** Electrochemical redox cycle for polypyrrole. $A^-$ represent anions, $e^-$ electrons. In general, the polymer expands when ions are inserted

The electromechanical coupling is generally less than 1%, except for small strains [18,19]. This means that only a small fraction of the input electrical energy is converted to mechanical work [1]. This does necessarily mean that the efficiency is low because much of the input electrical energy is stored and can be recovered. The principal impact is the need to shunt a lot of charge, as discussed below. This shunting can become an issue for the power supply in large devices. Encapsulation is often required to contain electrolyte, which is generally in the form of a liquid or gel. Strain rates are moderate (11%/s max to date [20]), but there is an opportunity for substantial improvement [18]. Rates are limited by internal resistance of the polymers and electrolytes and by mass transport rates of ions inside the polymer [18,21].

Experiments show that strain, $\varepsilon$, is proportional to the density of charge, $\rho$, transferred over a range of strains of about 1% or more:

$$\varepsilon = \frac{\sigma}{E} + \alpha \cdot \rho \tag{5.1}$$

where $\sigma$ is the applied stress and $E$ is the elastic modulus [4,9,11,18,19]. The strain to charge ratio, $\alpha$, is approximately $\pm 1\text{–}5 \times 10^{-10}$ $m^3 \cdot C^{-1}$ [22]. The strain to charge ratio is negative when cations dominate the ion transfer and positive when anions serve to balance charge [5]. For larger strains, this relationship continues to hold to first order. When loads exceed several megapascals, creep and stress relaxation can be significant [10].

In some systems, the volume change observed appears to be close to the volume of ions inserted [22,23], but often this model of actuation is an oversimplification. For example, the polymer can exhibit gel-like behavior, where change in oxidation state is associated with large solvent flux, large volume changes, and significant changes in modulus [24].

Rate is determined by the speed at which charge can be injected. Current is restricted by the resistance of the cell (when electrochemically activated) and by the rate at which ions are transferred within the polymer [18,19,21]. Some of the internal resistance can be compensated for, leading to substantially faster rates, as discussed below [25]. The fastest response occurs in thin, highly conducting films [12]. Changes in oxidation state have been observed at kilohertz rates [26].

**Table 5.1.** Conducting polymer actuators. Adapted from [1].

| Property | Min. | Typ. | Max. | Limit |
|---|---|---|---|---|
| Strain (%) | 1 | 3–9 | > 35 | |
| Stress (MPa) | | 1–5 | 34 | 200 |
| Work sensity ($kJ/m^3$) | | 70 | | 1000 |
| Strain rate (%/s) | | 1 | 11 | 10,000 |
| Power (W/kg) | | | 150 | 100,000 |
| Life (cycles) | | 28,000 | 800,000 | |
| Coupling | | | 0.1 | |
| Efficiency (%) | | < 1 | 18 | |
| Modulus (GPa) | 0.1 | 0.8 | 3 | |
| Tensile strength (MPa) | | 5 | 120 | 400 |
| Applied potential (V) | | 1.2 | 10 | |
| Charge transfer ($C/m^3$) | $10^7$ | | $10^8$ | |
| Conductivity (S/m) | | 10,000 | 45,000 | |
| Cost (US$/kg) | 3 | | 1000 | |

Polypyrrole and polyaniline are the most widely used conducting polymers, but actuation has also been demonstrated in polyacetylenes [3], polythiophenes, and polyethyldioxithiophene [27]. Starting materials are readily available (*e.g.*, www.aldrich.com, www.basf.com ). Electrodeposition is commonly used [28], producing thin (~0.3–40 μm thick) films with typical widths of 10 mm and lengths up to 1 m or more. An example synthesis process for polypyrrole is described below. Samples may be electrodeposited as tubes [29], often with an ingrown conductor to improve conductivity and maximize rate. Chemical synthesis may also be used in polypyrrole and other conducting polymers [30–32]. The properties of the polymers are very dependent on the solvent and salts used in deposition and also the electrolyte employed during actuation [8,11,13,33–40]. For example the cycle life (for redox cycling at least, and potentially for actuation) can be extended to approximately one million cycles from several tens of thousands by using ionic liquid electrolytes [32].

Forces can reach tens of newtons with displacements of several millimeters, and mechanical amplification increases displacements up to about 100 mm [41]. Voltages of only 1–2 V are sufficient for activation, but higher voltages help speed the response by overcoming internal resistance [42]. Currents reach hundreds of milliamperes. When not moving current is minimal, even when forces are applied. In such a case, virtually no energy is expended, and the condition is known as a catch state. This state is particularly useful where actuation is intermittent because no power is dissipated during periods where the actuator is simply holding a load.

Response under conditions other than room temperature and pressure have received little attention. Conductivity and mass transport are expected to decrease in proportion to the square root of absolute temperature, affecting rates at low temperatures. Electrolyte freezing will drastically reduce rate. Typical freezing temperatures are between –60°C and 0°C. Creep may increase substantially with temperature.

Conducting polymers are well suited to low voltage, moderate to high force, and small length scale applications. For macroscopic applications to be effective, layers of thin porous films could enable extremely fast, high power response (> 100 kW/kg) [18]. Recovery of stored electrochemical energy should enable moderate efficiencies to be achieved even at full strain. Newly designed conducting polymers promise larger strains and higher electromechanical coupling but are still at an early stage of development [17,43–46].

Data and models valuable to engineering design of conducting polymer actuators, and particularly polypyrrole based actuators, are provided by Madden *et al.* [18,19] and Mazzoldi *et al.* [13], a discussion of current status, promise and remaining challenges is given by Smela *et al.* [47,48]. Some early work on applications including braille cells, variable camber hydrofoils, and actively steerable catheters are provided by Spinks [49], Madden *et al.* [41,50] and De Rossi [51], respectively.

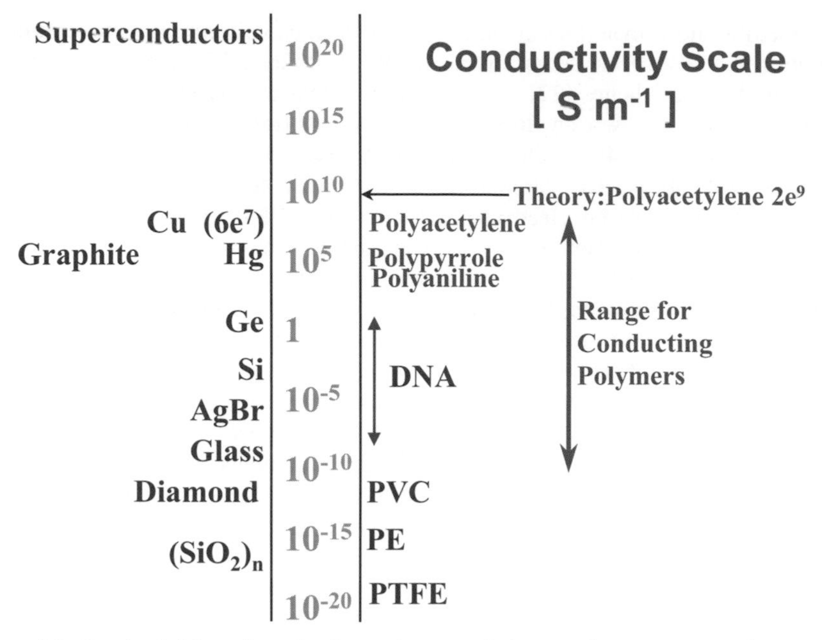


**Figure 5.3.** Conductivities of conducting polymers relative to other common polymers and inorganic materials. PE is polyethylene, PVC is polyvinylchloride, and PTFE is polytetrafluoroethylene, also known as Teflon©. The upper limits in conductivity for the conducting polymers polyaninline, polypyrrole, and polyacetylene are shown. Adapted from H.-G.Elias, Mega Molecules. Berlin: Springer-Verlag; 1987 [52]

## 5.3 Polypyrrole and Conducting Polymers – Background

What are conducting polymers, and what other properties do they have that can be useful in robotics? This section defines conducting polymers based on their molecular structure and describes their electronic properties.

Conducting polymers exhibit electronic conductivity, making them unusual among organic materials, as depicted in Figure 5.3 [52,53]. A defining feature of conducting polymers is their conjugated backbone, seen in Figure 5.1, enabling some degree of electron delocalization and hence electronic conduction, akin to the electronic behavior in benzene and graphite [54].

The polymers shown in Figure 5.1 are semiconductors. Charge carriers must overcome a band gap to become delocalized and enable conduction [55]. The band gap in polyacetylene is 1.7 eV [56]. Conductivity decreases with temperature, as the number of carriers with sufficient thermal energy to reach the conduction band is reduced. The band gap is reduced in a process analogous to doping in crystalline semiconductors [56]. Doping involves the addition or removal of charges from the polymer chain, resulting in structural changes and the creation of states in the band gap. These changes in polymer oxidation state are generally performed chemically or electrochemically. In the chemical doping process, electrons or protons are effectively donated to or withdrawn from the polymer backbone via chemical reactions with dopant molecules. In electrochemical doping, the conducting polymer is in electrical contact with an electrode in an electrochemical cell, as shown in Figure 5.4. Electrons are added to or removed from the polymer via the electrode, thereby changing the oxidation state. The electrochemical oxidation process leads to a change in the polymer charge state, which is balanced by the flux of ions to or from the electrolyte. These ions are referred to as dopants.

Although the doping process is analogous to doping in silicon and other crystalline semiconductors, there are some notable differences. Doping levels in conducting polymers are much higher, reaching one dopant for every two or three monomer units on the polymer backbone. The dopant need not be a donor or an acceptor in conducting polymers, often it is present simply to maintain charge balance. Charge carriers in the polymer are not simply electrons or holes, but are coupled with local conformational distortions in the polymer chain, among other differences [57]. In the doped state, conductivities can be appreciable, as shown in Figure 5.3.

In general, doping is chemically or electrochemically reversible. Conductivity can be switched by up to 13 orders of magnitude, an effect used in electrochemical transistors [58]. Reversible doping is used in many applications including batteries, electrochromic devices, supercapacitors, and chemical sensors, as well as being associated with volume change. The dimensional changes are used to perform mechanical work, as shown in Figure 5.4. Thus, actuation in conducting polymers is derived from dimensional changes associated with changes in the polymer oxidation state.

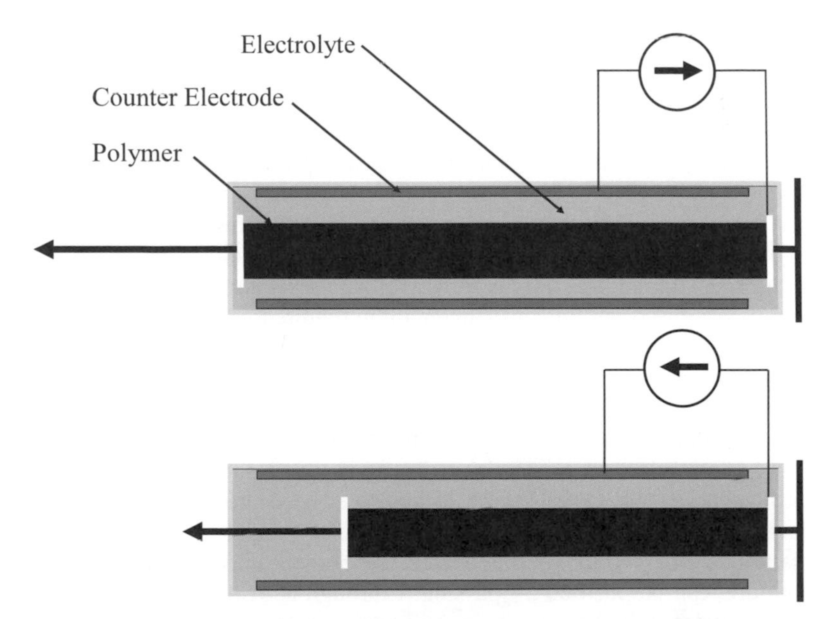


**Figure 5.4.** Electrochemical oxidation and reduction of polypyrrole. An electrochemical cell is depicted, showing the basic elements required. Current is applied leading to either oxidation or reduction of the polymer and a corresponding expansion or contraction

Conducting polymers are extremely versatile materials which have been employed to make high energy density storage devices, sensors, gas filters, generators, computational elements, display technologies, conductive wires, strong fibers, and actuators [57]. This diversity, combined with techniques for processing and patterning, make conceivable the fabrication of entire, multifunctional, and autonomous devices from a single material, as is currently being done with silicon microelectromechanical systems. Conducting polymers have a number of advantages over silicon, including much higher energy storage capability and an actuator technology that is superior in a number of respects. For example, the capacitance of > 100 F/g in polypyrrole [59, 60] is about five orders of magnitude higher than that achieved on the surface of silicon. The polymer actuators operate at low voltages, unlike the piezoelectrics and electrostatic actuators used in silicon microelectromechanical systems, offer much higher work densities than electrostatics, and ten or more times the strain offered by piezoceramics [61].

In conducting polymers, actuation is the result of molecular level interactions. The ability to tailor material properties by molecular design creates the potential to rationally design polymers that will achieve required mechanical properties. Ultimately, it may be possible to develop molecular stepping motors akin to muscle. Steps towards this goal are already being taken [1,62,63]. This article focuses on the current performance of conducting polymer actuators. This is a rapidly developing field however, in which performance limits are continually being pushed to new limits.

## 5.4 Synthesis

Properties of conducting polymer actuators are critically dependent on synthesis conditions. All polypyrrole actuators and many others reported in the literature are electrochemically synthesized, and thus an example electrodeposition proceedure is presented here. Electrochemical synthesis leads to the highest conductivity and the best mechanical properties in polypyrrole. It is also the most accessible method to nonchemists. The method presented leads to very conductive and high tensile strength films and requires only readily available solvents, salts, and monomers. A source of current is all the instrumentation required. The amount of material deposited depends only on time and current.

In this commonly used method, films are grown as summarized in Figure 5.5. A solution of 0.06 M freshly distilled pyrrole monomer (Aldrich, Milwaukee, WI, USA, www.aldrich.com) and 0.05 M tetraethylammonium hexafluorophosphate (Aldrich) in propylene carbonate is employed, following the procedure of Yamaura and colleagues [28]. The pyrrole monomer need not be distilled, but the conductivity and mechanical proerties are greatly improved by purification. If distillation is not available, the pyrrole monomer, which is a liquid at room temperature, can be purified by passing it through activated alumina or stirring in alumina powder and then removing the powder using a filter. Polypyrrole is deposited galvanostatically (constant current) onto polished glassy carbon substrates (Alfa Aesar, Ward Hill, MA, USA, www.alfaaesar.com) at current densities of between 1 and 2 $A{\cdot}m^{-2}$, resulting in a film thickness between 8 and 100 μm, depending on time. Other substrates can be used, including platinum, gold, and stainless steel. When using gold or stainless steel, it is found that thicker films can lift off the mandrel during electrodeposition, leading to films that look bubbled or warped. This liftoff may be reduced by employing a roughened surface. A copper counterelectrode is used. For best results, deposition should take place at temperatures between –30°C and –45°C in a nitrogen saturated solution. The resulting films have conductivities between 20 and 45 $kS{\cdot}m^{-1}$, densities of 1500 to 1800 $kg{\cdot}m^{-3}$ dry, and tensile strengths that can exceed 100 MPa.

The polished glassy carbon substrates take the form of either 100 mm × 100 mm × 1 mm thick plates, or a crucible. The crucible is employed to obtain films that are up to 1.5 m in length, with a width of 4 mm. Figure 5.6 shows the crucible after deposition, including the Kapton© tape that is spirally wound to mask the surface so that a continuous film is electrodeposited. Also shown is a 1.3 m long film that has been removed from the crucible.

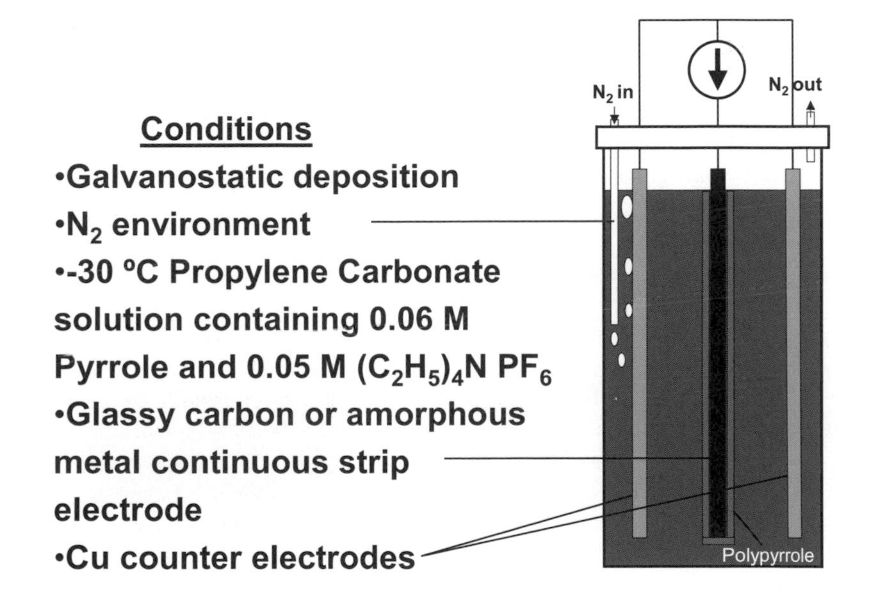


**Figure 5.5.** Polypyrrole synthesis conditions. The diagram summarizes the conditions used in the synthesis of the hexafluorophosphate doped polypyrrole films employed in this study

The reaction steps involved in electrochemical synthesis have been proposed by Baker and Reynolds [64] and are shown in Figure 5.7. It is not necessary to understand the synthesis steps, but it can help in analyzing failed depositions. When current density is too low, the solution is sometimes black, often indicating that only oligomers (short chains) have been formed. These short chains drift away from the electrode before they reach a the critical size for precipitation onto the electrode.

As synthesized, polypyrrole has a backbone configuration shown at the bottom of Figure 5.2, where roughly one-third of the monomers are charged [28,34,64-66]. The backbone charge is balanced by the presence of anions. In this state, polypyrrole is shiny black in appearance and has typical conductivities of up to ~$10^4$ S·m$^{-1}$, reaching $10^5$ S·m$^{-1}$ when stretch aligned [28]. Reduction leads to a semiconducting state, shown in Figure 5.1.

In bulk, the monomer costs ~US$1.00 per kilogram. The estimated minimum cost of electrodeposited polypyrrole is ~$1.50 per kilogram.

Variations on the method presented include growth on a platinum wire onto which a smaller wire is spirally wound [67]. The smaller wire is incorporated into the polypyrrole that is grown onto the larger wire, and the polypyrrole plus small wire are pulled off the larger wire following deposition. A key advantage of this approach is that the small wire increases electrical conductivity and hence, in many cases, the rate of actuation. It also allows the polymer to be taken deeper into its reduced (nonconducting) state, allowing larger actuation. In a variation of this approach, polymer is grown onto a spirally wound wire that does not have a core. This approach relies on the polymer growing between spiral windings but does not require removal of the large diameter wire.

**Figure 5.6.** Synthesis of long films on a crucible. At right is the glassy carbon crucible employed to electrodeposit polypyrrole films up to a meter in length. The crucible is masked using Kapton© tape, resulting in a film shown at left. Photos courtesy of Patrick Anquetil

The polypyrrole described typically produces strains in the 2–4% range. Larger strains can be obtained by varying deposition conditions. For example, Kaneto *et al.* showed a >10% strain by operating polypyrrole in an aqueous (water-based) solution of $NaPF_6$ [68], and larger strains have been claimed in other papers [16,44]. Spinks, Wallace, and others at the University of Wollongong and Santa Fe Science and Technology have shown that growth in propylene carbonate followed by redox cycling in ionic liquids can lead to increased cycle life, suggesting that it may be possible to increase the lifetime beyond 1 million cycles [32,69].

## 5.5 Actuator Configuations and Applications

Some applications require large forces, other large displacements, and sometimes a combination of both is needed. There are two common actuator configurations – linear actuators in which films or fibers contract or expand producing linear displacement, and bending actuators in which a bending motion is produced. The advantage of bending actuators is that they produce large displacements, but this is done at the expense of force. Linear actuators generate high forces – typically in the range of 1 to 10 MN per square meter of cross-sectional area–but their change in length can be relatively small – generally 1–10 % of the original length. In designing a device, the challenge is then to scale up the forces produced by bending actuators or scale up the displacements produced by linear actuators [41]. Bending actuators consist of two or more laminated films, as depicted in Figure 5.8. In the simplest case one of the films contracts leading to bending. If two sheets of polypyrrole are used, then one may expand while the other contracts. In this case, the sheets are separated by an ionically conductive medium.

The equations for the deflections and forces of bilayer and trilayer structures follows the same approach as is used in bimetallic strips [5]. For a trilayer in which both outer layers are active polymer of the same thickness and equal and opposite strain, separated by a passive layer, the equations for relative curvature, $K$, charge per unit volume, $\rho$, and force at the end of the beam, $F$, are [70]

$$F = C_{\text{spring}} \cdot K + C_{\text{charge}} \cdot \rho \tag{5.2}$$

where

$$C_{\text{spring}} = \frac{2 \cdot W \cdot E_p}{3 \cdot L} \cdot h_g^3 \left[ \left( 1 + \frac{h_p}{h_g} \right)^3 - 1 + \frac{E_g}{E_p} \right] \tag{5.3}$$

and

$$C_{\text{charge}} = \frac{E_p \cdot \alpha}{L} \cdot W \cdot h_g^2 \left[ \left( 1 + \frac{h_p}{h_g} \right)^2 - 1 \right] \tag{5.4}$$

$E_p$ and $E_g$ are the moduli of elasticity of the active polymer and separators, respectively. $H_p$ and $h_g$ are the thicknesses of each layer, $W$ and $L$ are the width and legnth of the structure, and $\alpha$ is the linear relationship between strain and charge per unit volume, introduced in Eq. (5.1) above. Maximum force is generated when bending is prevented ($K$= 0), and peak deflection occurs when the applied force is zero ($F$= 0).

**Synthesis of Polypyrrole**

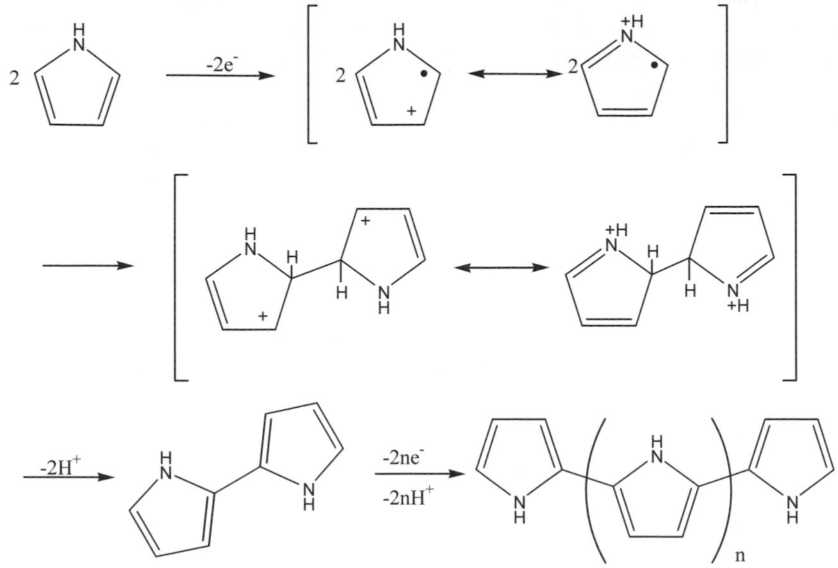

The last step is a in fact a repetion of the first steps beginning with oxidation, followed by coupling to either end of the polymer, and finally elimination of $H^+$. The electrons are either removed via an electrode (electrochemical deposition) or chemically, e.g.

$$Fe^{+++} \xrightarrow{e^-} Fe^{++}$$

Note that the polymerization does not generally result in a neutral polymer shown above, but rather the backbone is charged, as below, such that the total number of electrons transferred per monomer is $2+a$ where $a$ is generally between 0.2 and 0.5:

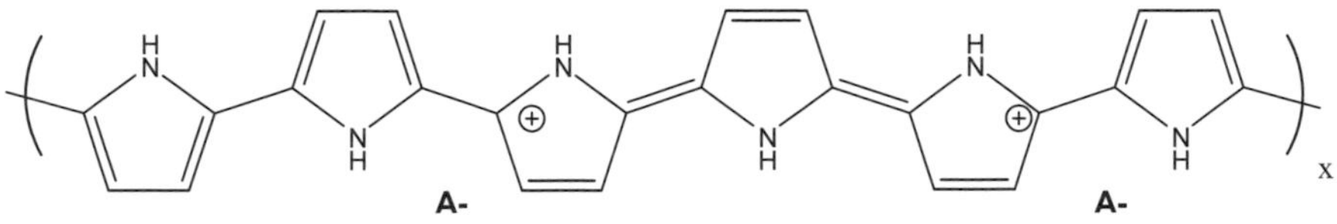

**Figure 5.7.** Mechanisms of Polypyrrole polymerization

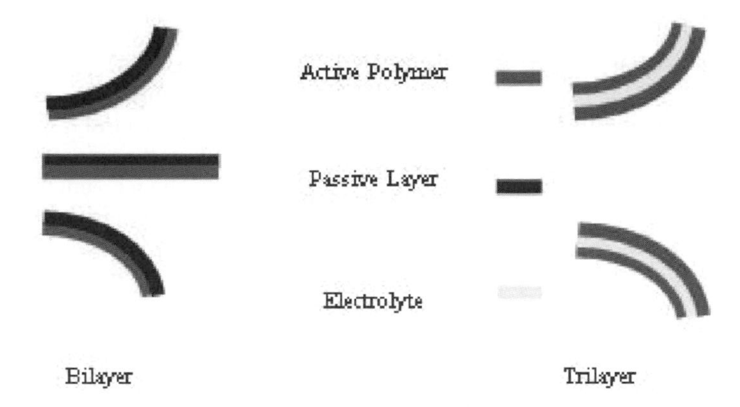


**Figure 5.8.** Bilayer and trilayer actuation configurations. In bending bilayers and trilayers, one layer expands whereas the other is passive or contracts, leading to a bending motion.

An example trilayer actuator is shown in Figure 5.9 [41]. It is employed to create a camber change in a propeller blade. The structure generates 0.15 N of force. Bilayers have been shown to be very effective for microscale actuation and have been used by Elisabeth Smela and her colleagues to create contracting fingers, cell enclosures, moveable pixels, and "micro-origami." Micromuscle.com in Sweden is working to commercialize actuated stents and steerable catheters, which appear to use the bilayer principle for operation [51].

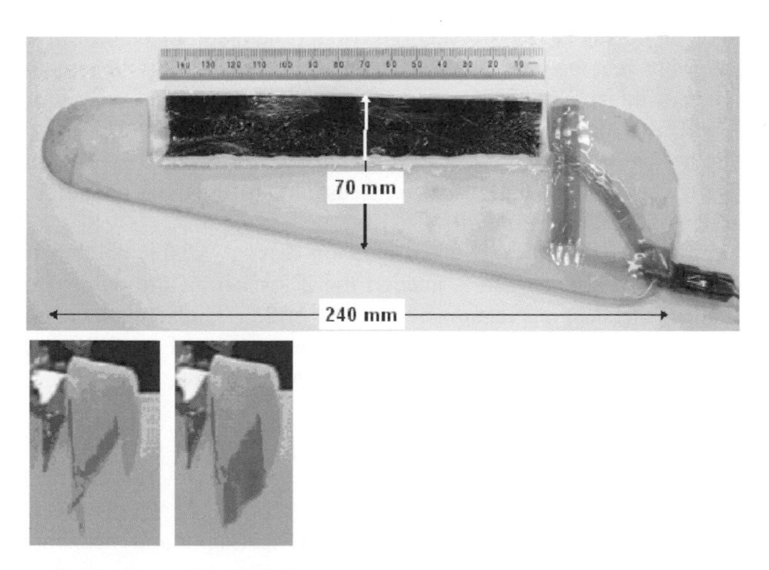


**Figure 5.9.** Trilayer acutator mounted on a propeller blade. The top image shows the geometry of the blade; the bottom two images show deflection of the structure. In the trilayers (black), two films of polypyrrole are separated by a sheet of paper soaked in gel electrolyte. A thin layer of polyethylene encapsulates the bending structure. © Journal of Oceanic Engineering, reproduced with permission [41].

Figure 5.4 shows the basic geometry of a linear actuator approach. As in all actuator configurations, a counterelectrode and an electrolyte are required. Generally, a mechanism must be available to allow transmitting force and displacement to the load. Thus, the counterelectrode, electrolyte, and any packaging must not significantly impede actuation. Also, the counterelectrode must accept a tremendous amount of charge from the polymer actuator, which stores charge within its volume with an effective capacitance of approximately 100 Farads per gram of polymer. The counterelectrode is best made of a polymer that can itself absorb a lot of charge without requiring a large voltage or degrading the electrolyte. One of the simplest solutions is to employ a conducting polymer counterelectrode.

In some cases with linear actuators, no amplification of strain is needed. One such application is the creation of braille cells for the blind. These tablets feature arrays of pins that must be actuated up or down in a pattern across a tablet so as to generate text and refresh once the reader has completed the page. A group at the University of Wollongong in collaboration with Quantum Technology of Australia is developing a braille display in which each pin is driven by a polypyrrole tube actuator [71]. The polypyrrole tube is grown on a platinum wire (~ 0.5 mm diameter) which has a smaller diameter wire wrapped around it (~ 50 μm). The polypyrrole with the small wire encapsulated in it is removed from the larger wire by sliding it free. This approach allows relatively long actuators to be produced which, when driven with large currents, have very little voltage drop along their length due to the incorporation of the platinum wire. Voltage drop needs to be prevented because a gradient in voltage leads to a different degree of strain as a function of length along the actuator. Sections of the actuator distant from the electrical contact points receive very little charge and produce negligible actuation when the resistance of the actuator is large. The spiral winding of the wire allows its mechanical stiffness to be low, minimally impeding the strain of the polypyrrole. The hollow core enables electrical and mechanical connection via a wire. Tensile strengths are not as high as in the freestanding films, but operation at several megapascals of stress is common.

Figure 5.10 shows an example of a linear-actuator-driven variable camber foil [41]. In this case, a lever mechanism is needed the 2% strain of the polypyrrole employed needs mechanical amplification by a factor of 25. The actuators shown produce 18 N of force, which is reduced to 0.7 N in the process of amplifying the displacement. The actuators are sheets of freestanding polypyrrole.

## 5.6 Modeling and Implications for Design

In this section, a model of the relationship between electrical input and mechanical output is presented and used to explore the advantages and limitations of conducting polymer actuators. The presentation is similar to that given elsewhere by the author [18,19]. In particular rate limiting factors are discussed, as well as factors that determine efficiency and power consumption. These considerations allow designers to determine the feasibility of employing conducting polymers in specific applications and then to generate designs.

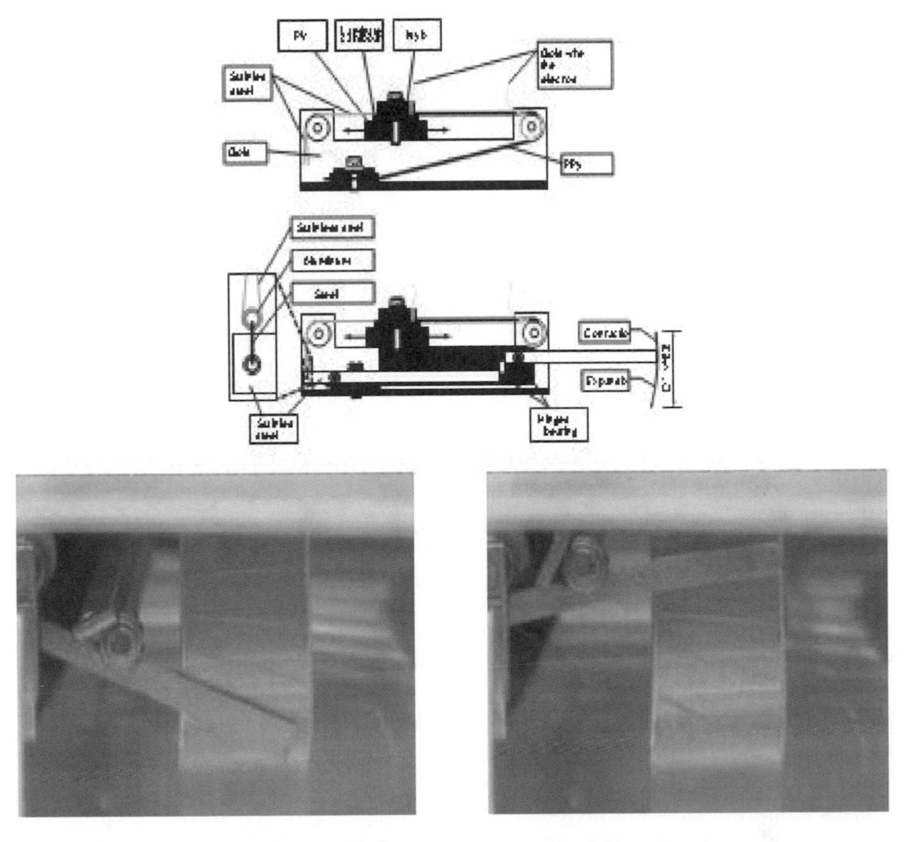


**Figure 5.10.** Linear-actuator-driven variable camber hydrofoil. The top image shows the actuator mechanism. The bottom images show the extent of deflection of the trailing edge of the foil. © Journal of Oceanic Engineering, reprinted with permission [41]

Equation (5.1), repeated again here, is a relatively simple relationship [19] between stress, $\sigma$, strain, $\varepsilon$ and charge per unit volume, $\rho$ as a function of time, $t$:

$$\varepsilon(t) = \alpha \cdot \rho(t) + \frac{\sigma(t)}{E} \tag{5.1}$$

which is found to describe the behavior of polypyrrole and polyaniline actuators to first order under a range of loads and potentials. In polypyrrole grown in $PF_6^-$ , for example, if it is operated at loads of several megapascals and below and kept within a limited potential range (~ -0.6 V to +0.2 V vs. Ag/AgCl), this equation works resonably well. The strain to charge ratio, $\alpha$, is analogous to a thermal expansion coefficient, but for charge rather than temperature. In conducting polymers, the strain to charge ratio is experimentally found to range from 0.3–$5\times10^{-10}$ $m^3 \cdot C^{-1}$ for polypyrrole and polyaniline actuators [9,13,19], and the modulus ranges between 0.1 and 3 GPa [12,13,72].

There are conditions under which Equation (5.1) does not apply. The model can be more generally expressed as

$$\varepsilon(t) = \alpha(t,V) \cdot \rho(t) + \frac{\sigma(t)}{E(t,V)} \tag{5.5}$$

*E(t, V)*, the time and voltage dependent modulus [19,22]. The modulus has been found to exhibit both time and voltage dependence, showing a viscoelastic response at higher loads, for example [18]. When taken over large potential ranges, the modulus can change significantly, leading to increases or decreases in strain as load increases, depending on whether the change in modulus adds or subtracts from the active strain [3,72]. Over long time periods (> 1000 s) at high stresses (> 10 MPa), the modulus becomes highly history dependent [73]. Also, rate of creep can be enhanced during actuation [10]. The strain to charge ratio can be load independent, but frequency and time dependence have been observed [10]. The strain to charge ratio is particularly time dependent when more than one ion is mobile, a situation that is particularly difficult to model when both positive and negative ions move, simultaneously swelling and contracting the material [5]. The strain to charge ratio can also change as a function of voltage.

### 5.6.1 Relationship Between Voltage and Charge at Equilibrium

A complete electromechanical description includes input voltage in addition to strain, stress, and charge. In conducting polymers, the relationship between voltage and charge is difficult to model because the polymer acts as a metal at one extreme and an insulator at the other. A wide range of models have been proposed. In general, the response is somewhere between that of a capacitor and a battery [13,74–80]. In a capacitor, voltage and charge are proportional, whereas in an ideal battery, the potential remains constant until discharge is nearly complete. In oxidation states where conductivity is high, it is not unusual to find that charge is proportional to applied potential over a potential range that can exceed 1 V [18], thereby behaving like a capacitor [19,22,76,79–81]. In hexafluorophosphate-doped polypyrrole, this capacitance is found to be proportional to volume [22], and has a value of $C_V = 1.3 \cdot 10^8$ F·m$^{-3}$ [19]. At equilibrium, the strain may then be expressed as

$$\varepsilon(t) = \alpha \cdot C_V \cdot V + \frac{\sigma(t)}{E}, \tag{5.6}$$

where $V$ is the potential applied to the polymer. In many conducting polymers and for extreme voltage excursions, the capacitive relationship between voltage and charge is not a particularly good approximation. These situations are difficult to model from first principles, so an empirical fit to a polynomial expansion in charge density may provide the most practical approach. The capacitative model is useful in evaluating rate-limiting factors even if the relationship between voltage and charge is complex. In such cases, the capacitance is determined by dividing the charge transferred by the voltage excursion. Before considering rate-limiting mechanisms, some considerations in choosing maximum actuator load are presented.

### 5.6.2 Position Control and Maximum Load

The designer must determine the stresses at which to operate an actuator. Conducting polymer actuators are able to actively contract at 34 Mpa [22,82]. In general, however load induces elastic deformation and creep [19,22,82–84]. To maintain position control, the actuator must be able to compensate for these effects. Over short periods, only the elastic response need be considered. The elastic strain induced by load is simply the ratio of the load induced stress, $\Delta\sigma$, and the elastic modulus, $E$. This strain must be less than or equal to the maximum active strain, $\varepsilon_{max}$, for an actuator to maintain position:

$$\varepsilon_{max} \geq \frac{\Delta\sigma}{E}. \tag{5.7}$$

The maximum strain is typically 2% in $PF_6^-$ doped polypyrrole, and the elastic modulus is 0.8 GPa, suggesting that the peak load at which elastic deformation can be compensated for is 16 MPa. At 20 MPa, the sum of the creep and the elastic deformation reaches 2% after ~ 1 hour, as shown in Figure 5.11. The designer must determine the extent of elastic deformation and creep that is acceptable and the time-scale and cycle life of the actuator. Measured creep and stress-relaxation curves and viscoelastic models will then assist in determining the appropriate upper bounds in actuator stresses.

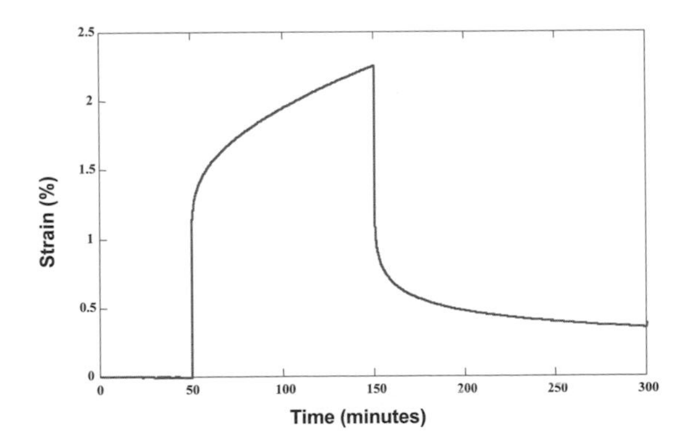


**Figure 5.11.** Creep in a polypyrrole film in response step in stress (2 MPa to 20 MPa and back to 2 MPa). The test was performed in propylene carbonate with 0.05 M tetraethylammonium hexafluorophosphate

In some designs, the maximum load will be determined by the maximum stress that can be actively generated by the actuator from an unloaded condition, or when acting against another actuator (an antagonist). In such cases, peak stress generated is simply given by the product of the modulus and the active strain:

$$\sigma = E \cdot \varepsilon \tag{5.8}$$

This peak stress will be observed at zero strain. For tetraethylammonium-doped polypyrrole this peak stress is once again 16 MPa.

### 5.6.3 Actuator Volume

How much space will a conducting polymer actuator consume? Many applications allow only limited volumes. Where a single actuator stroke is used to create motion, as in the action of the biceps muscle to displace the forearm or of a hydraulic piston on a backhoe, the amount of work performed per stroke and per unit volume, $u$, is a key figure of merit. The actuator volume required, $Vol_{\min}$, is determined based on the work , $W$, required per cycle:

$$Vol_{\min} \geq \frac{W}{u} \tag{5.9}$$

This volume is the minimum required because energy delivery, sensors, linkages, and often means of mechanical amplification generally also need to be incorporated.

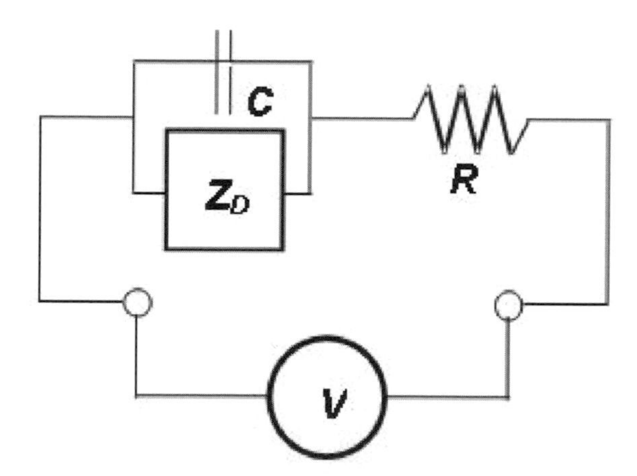


**Figure 5.12.** Equivalent circuit model of the actuator impedance. $V$ represents an external voltage source, $C$ is the double layer capacitance, $R$ is the electrolyte/contact resistance, and $Z_D$ is the diffusion impedance. © Proceedings of SPIE, reprinted with permission [18]

Work per unit volume is the integral of stress times incremental strain. The maximum stress against which mammalian skeletal muscle can maintain position is

350 kPa and typical strain at no load is 20% *in vivo* [61]. The achievable strain in muscle decreases with increasing stress, the work density is less than 70 $kJ{\cdot}m^{-3}$, the product of the peak stress and strain, and in general will be in the range [85] of 8–40 $kJ{\cdot}m^{-3}$. Work densities of 70 $kJ{\cdot}m^{-3}$ have been reported in polypyrrole [72] and exceed 100 $kJ{\cdot}m^{-3}$ in new, large strain polypyrrole [68]. Note that unlike muscle, conducting polymers can perform work both under compression and tension, and therefore can generate a further doubling in work per volume where this property is used.

### 5.6.4 Rate and Power

Generally, in any given application, a certain rate of response or output power is required. This section is dedicated to presenting a number of factors that determine the rate of response and estimating how rate will be a function of geometry in such cases. Other factors affecting rate are polymer and electrolyte conductivities, diffusion coefficients, and capacitance. The equations presented enable the designer to determine the physical and geometrical constraints necessary to achieve the desired performance.

As discussed, conducting polymers respond electrically as batteries or supercapacitors with enormous quantities of charge stored per unit volume. The capacitance can exceed 100 F/g [22]. Given that strain is proportional to charge, high strain rates and powers require high currents. Although other factors could also limit rate, including inertial effects and drag, the generally moderate to low rates of actuation in conducting polymers to date suggest that such situations are unusual.

The factors that limit current in conducting polymer actuators are the same as those that limit the discharging rate in batteries and supercapacitors. Internal resistance is one factor. To charge and discharge a battery, the time limit due to internal resistance is the product of the total amount of charge multiplied by the resistance and divided by the voltage. In a capacitor, the time constant is determined by the product of the internal resistance and the capacitance. The two are essentially equivalent because the capacitance is the ratio of the charge and the voltage. The second major limiting factor that will be discussed is the rate of transport of ions into the polymer. Another factor that could also limit response time is the rate at which electrons are exchanged between the conducting polymer and the contacting metal electrode (kinetics). This will not be discussed because it is considered not to be a significant factor, providing that the conducting polymer is in a relatively highly conducting state. More complex effects than those discussed here can also arise due to changes in ionic and electronic conductivity as a function of voltage, which often lead to the advancing of sharp fronts of oxidation state through the material rather than a concentration gradient observed in diffusionlike behavior. Models for these effects are being evaluated and are likely important when the oxidation state of the polymer is brought substantially down toward the completely reduced state [21, 86]. Mechanical relaxation and solvent swelling may also be important [21]. These cases will not be covered here.

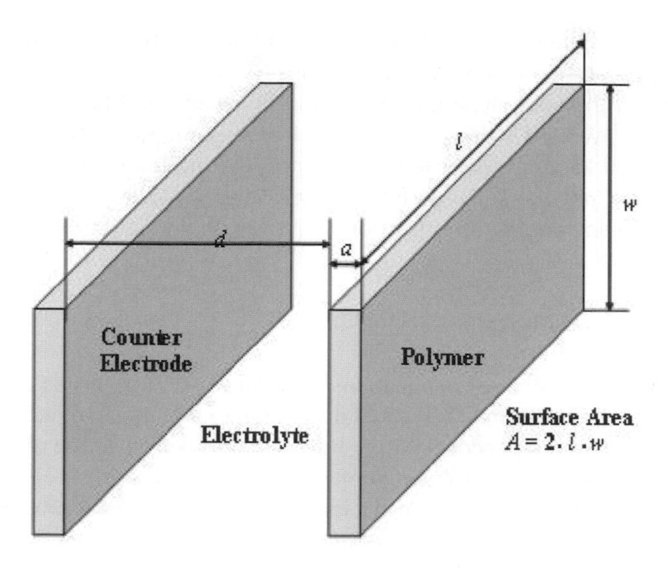


**Figure 5.13.** Dimensions of the polymer actuator. Electrical contact to the polymer is made at intervals of length *l*. © Proceedings of SPIE, reprinted with permission [18]

Current and potential are related via impedance or its inverse, admittance. Figure 5.12 is an impedance model from which rate-limiting time constants are derived. $R$ represents the electrolyte resistance, $C$ is the double layer capacitance at the interface between the polymer and the electrolyte [87], and $Z_D$ is a diffusion element, modeling mass transport into the polymer. For planar geometry, as in Figure 5.13, $Z_D$ is expressed in the Laplace or frequency domain as:

$$Z_D(s) = \frac{\delta}{\sqrt{D} \cdot C \cdot \sqrt{s}} \cdot \coth\left(\frac{a}{2} \cdot \sqrt{\frac{s}{D}}\right) \tag{5.10}$$

$D$ is the diffusion coefficient, $\delta$ is the double layer thickness, $C$ is the double layer capacitance, and $a$ is the polymer film thickness. At low frequency (lim $s \rightarrow 0$), the diffusion impedance reduces to

$$Z_D = \frac{\delta}{a \cdot C \cdot s} = \frac{1}{C_V \cdot Vol \cdot s} \tag{5.11}$$

behaving as a capacitance. The right-hand expression restates the impedance in terms of the polymer capacitance per unit volume, $C_V$, and the polymer volume, $Vol$. Details of the derivation, assumptions, and physical significance of the variables are provided elsewhere [12,19]. In essence it assumes that the rate is determined by either RC charging or the time it takes for ions to go through the polymer. It also assumes that ionic mobiility or the diffusion coefficient does not change significantly over the range of potentials employed and that the electrical conducitivity within the polymer is sufficiently high to eliminate any potential drop. A time constant accounting for potential drop along the polymer due to finite electrical conductivity is discussed below.

The model provides a reasonable description of hexafluorophosphate-doped polypyrrole impedance over a 2 V range at frequencies between 100 μHz and 100 kHz [19,22]. It also suggests the rate-limiting factors for charging and actuation. One is the rate at which the double layer capacitance charges, which is limited by the internal resistance, $R$. A second is the rate of charging of the volumetric capacitor, which is determined by the slower of the rate of diffusion of ions through the thickness and the $R \cdot C_V \cdot Vol$ charging time. These time constants and their implications are discussed further below.

The impedance model represented in Figure 5.12 is very general in the sense that the time constants derived from it are present in all conducting polymer systems. As a result, it provides a good basis for describing all systems. A more general model will require the addition of finite and changing electronic and ionic conductivities, kinetics effects, and material anisotropies.

Ion transport within the polymer can be the result of diffusion or convection through pores, molecular diffusion, or field-induced migration along pores [13,22,76,77,79–81,88,89]. The mass transport model described by Eq. (5.10) appears to represent only a diffusion response. Eq. (5.10) mathematically describes all of these effects, not just diffusion. The equivalent circuit for the diffusion element is shown in Figure 5.14. It is identical to the equivalent circuit used to describe migration and convection and thus these effects are indistinguishable based on the form of the frequency response alone. It is quite likely that the diffusionlike response is due to a combination of internal resistance (ionic or electronic) and internal capacitance.

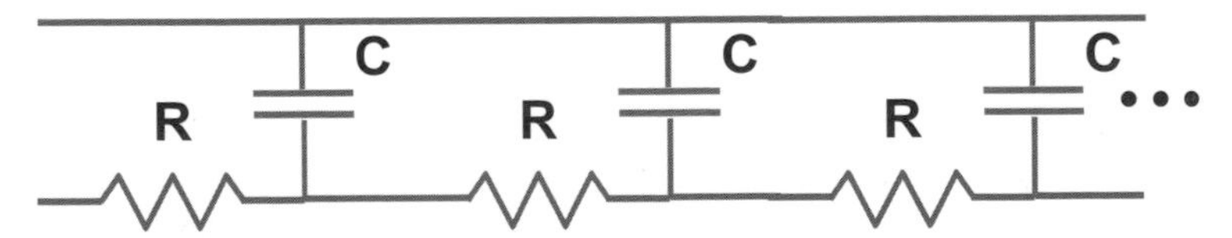


**Figure 5.14.** Diffusionlike response represented by a transmission line model. The resistors may represent solution resistance, or fluid drag, and the capacitors double layer charging or electrolyte storage. This model also represents the charging of a polymer film whose resistance is significant compared to that of the adjacent electrolyte. In this case, the resistance is that of the polymer, and the capacitance is the double layer capacitance or the volumetric capacitance. © Proceedings of SPIE, reprinted with permission [18]

*5.6.4.1 Polymer Charging Time*

In conducting polymers, charging occurs throughout the volume. Independent of the nature of the charge-voltage relationship (*e.g.,* battery or capacitor), the charge density is delivered through an internal resistance. There are two primary sources of resistance – the electrolyte and the polymer. To minimize electrolyte resistance, the electrolyte ideally covers the polymer surface area on both sides, $A = 2 \cdot l \cdot w$, with as small an electrode separation, $d$, as possible (refer to Figure 5.13 for dimensions). In a liquid electrolyte, the conductivity, $\sigma_e$, can be a high as 1 to

10 S·m$^{-1}$. For capacitorlike behavior, as in hexafluorophosphate-doped polypyrrole, the time constant for volumetric charging, $\tau_{RCV}$, is

$$\tau_{RCV} = R \cdot C_V \cdot Vol = \frac{d \cdot a}{2 \cdot \sigma_e} \cdot C_V, \tag{5.12}$$

where electrolyte resistance is large compared to the polymer resistance. To charge a 10 μm thick film in 1 s and given an electrolyte conductivity of 10 S·m$^{-1}$, the electrolyte dimension, *d*, must be less than 20 mm.

The polymer resistance can dominate RC charging time in long films, poorly conductive polymers, or over a range of potentials where the conducting polymer is no longer in its quasi-metallic state. In such a case, the combination of the film resistance and the volumetric capacitance forms a transmission line, as depicted in Figure 5.14, with the polymer resistance along the length of the film. The charging time constant can be reexpressed in terms of the polymer conductivity, $\sigma_p$, and the film length, *l*:

$$\tau_{RCVP} = R_p \cdot C_V = \frac{l^2}{4 \cdot \sigma_p} \cdot C_V. \tag{5.13}$$

The factor of 4 is appropriate only when both ends of the film are electrically connected. If the electrical connection is only from one end, the four is replaced by one. A 20 mm long film having a conductivity of $10^4$ S·m$^{-1}$ (typical of hexafluorophosphate-doped polypyrrole) has a time constant $\tau_{RCVP} = 1$ s.

The keys to improving R·C response times are to reduce the distance, *l*, between contacts, the distance between electrodes, *d*, and the polymer thickness, *a*. Maximizing electrolyte and polymer conductivities is also important. Finally, if the volumetric capacitance can be reduced without diminishing strain, the charge transfer is reduced. New polymers are being designed and tested whose strain to charge ratio is much larger and capacitance is lower [90]. These polymers promise to charge faster while also developing greater strain, compared to polypyrrole.

#### *5.6.4.2 Resistance Compensation*

Higher current in a circuit can generally be achieved by applying higher voltage. However, extreme voltages applied to the polymer lead to degradation. There is a simple case in which the application of a high voltage for a short amount of time will lead to faster actuation without degradation. If the solution resistance (and any contact resistance) is large compared to the polymer resistance, then immediately after the application of a step in potential, nearly all the potential will be across this series resistance. This is essentially the same as a series RC circuit, in which initially all the potential drop is across the resistor. At long times, as in the series RC circuit, all potential drop is across the capacitor. If we can prevent the resistance across the polymer from exceeding a threshold value, then we can avoid degradation while having increased the initial rate of charging and hence the actuation rate [19].

The polymer diffusion impedance, $Z_D$, and double layer capacitance, $C$, in Figure 5.12, act as a low-pass filter. The electrolyte resistance is then easily identified (*e.g.*, using a step, impulse, or high-frequency sinusoidal input). The product of the current and the resistance, $I{\cdot}R$, is the drop across the electrolyte. Subtracting this voltage from the total applied potential, $V$, provides an estimate of the double layer potential, $V_{\mathrm{dl}}$. If the double layer potential should reach but not exceed a voltage, $V_{\mathrm{dl}}^{\mathrm{max}}$, then the controller must simply maintain the input voltage such that the applied potential $V$:

$$V \leq V_{dl}^{\max} + I \cdot R \,. \tag{5.14}$$

This method effectively eliminates the rate-limiting effects of the resistance, $R$, in Figure 5.12 [42,91]. The remaining rate-limiting factors are then due to polymer resistance and mass transport, which are now discussed. The method does not work if polymer resistance is similar in magnitude or larger than the cell resistance.

*5.6.4.3 Mass Transport*

Ions move in the polymer to balance charge during oxidation and reduction. This can occur by molecular diffusion [19,22], conduction, or diffusion through electrolyte filled pores [76,77,79–81,88,89] and convection through pores [13]. The mathethematical form of the solutions is represented by Equation (5.10). The mass transport related charging time constant is

$$\tau_D = \frac{a^2}{4 \cdot D}, \tag{5.15}$$

where $D$ is the effective diffusion coefficient and $a$ is the film thickness. The factor of 4 is removed if ions have access only from one side of the film. Diffusion coefficients in hexafluorophosphate-doped polypyrrole appear to range[19,22] between 0.7 and $7 \times 10^{-12}$ $\mathrm{m^2{\cdot}s^{-1}}$. To obtain a 1 s response time, the polymer film thickness must be less than approximately 5 μm.

### 5.6.5 Summary of Rate-Limiting Effects

Volumetric charging times, ($\tau_{RCV}$ and $\tau_{RCVP}$) and diffusion time ($\tau_D$) are key rate-limiting factors that can be minimized by reducing the distance between electrodes, $d$, the spacing between contacts, $l$, and the film thickness, $a$. Reduction in film thickness and length are particularly effective because diffusion time and volumetric charging time are proportional to the squares of these lengths.

How fast could a conducting polymer film actuate? A 10 nm thick polypyrrole actuator that is several micrometers long is predicted to exhibit a diffusion time constant of 3 μs. Charge and discharge curves from similarly thin electrochemically driven transistors and electrochromic devices approach such a rate [26, 58]. The need for small dimensions to achieve fast response suggests that conducting polymer actuators are well suited for micro- and nanoscale applications. The next section takes a general look at what applications and scales are currently attractive for conducting polymer actuators.

## 5.7 Opportunities for Polypyrrole Actuators

Established actuator technologies used in robotics [92] for which artificial muscle technologies might offer an alternative include internal combustion engines, high-revving electric motors, direct drive electric motors, and piezoelectric actuators. Conducting polymers are not ready to compete with the internal combustion engine and high-revving electric motors in high power propulsion systems. They are appropriate for intermittent or aperiodic applications with moderate cycle life requirements and could replace solenoids, direct drive electric motors, and some applications of piezoceramics. This section follows the format of the introduction in a paper on the application of polypyrrole in variable camber foils [93] which seeks to answer the general question where conducting polymer actuators are useful. The challenges in creating large, high power actuators based on conducting polymers are then discussed in Section 5.8 using the example of a simulated biceps muscle.

### 5.7.1 Propulsion

The internal combustion engine with its 1000 W/kg power and fuel energy of ~43 MJ/kg is hard to beat for high-speed propulsion of automobiles and ships. Electric motors achieve specific power similar to the internal combustion engine and efficiencies of >90% [92]. Fuel cells and hybrid engines provide energy sources that make such high revving electric motors feasible. Of the available actuator technologies, only shape-memory alloys and piezoceramic actuators clearly surpass the power to mass of internal combustion engines and electric motors [61], but there are efficiency and cycle life issues with shape-memory alloys and mechanical amplification challenges with piezoelectrics. Can conducting polymers offer any advantages in propulsion?

The power densities of conducting polymers and carbon nanotubes are within a factor of four of the combustion engine [94]. Their musclelike nature may make polymers and nanotubes more suitable for biomimetic propulsion – such as robot walking, swimming, or flying. However, the low electromechanical coupling of conducting polymers (ratio of mechanical work output to electrical energy input) means that either the efficiency will be low, or additional energy recovery circuitry will be required. No polymer-based actuator technology has yet demonstrated cycle lifetimes [90] of more than $10^7$ and thus lifetime is generally not long enough for sustained propulsion. For example, continuous operation at 10 Hz may lead to failure after 10 days or less. Conducting polymer and other novel polymer actuators do not yet offer compelling alternatives to electric motors and combustion engines for high power, continuous propulsion.

### 5.7.2 Intermittent Actuation

Motions such as the grasping of parts by a robot arm, the opening of a valve, and the adjustment of a hydrodynamic control surface are difficult to perform using high revving electric motors and combustion engines. Direct drive electric motors are often used instead. Direct drive motors suffer from relatively low torque and force. Honda's elegant servomotor driven Asimo, for example, is limited in walking speed to ~ 2 km/hr due to the low torque of its muscles (http://world.honda.com/ASIMO/P3/spec/). Also electromagnetic actuators expend energy to hold a force even without displacement, wasting energy. Conducting polymers expend minimal energy while holding a force, feature high work density, and produce high stresses and strains, making them well suited for discontinuous, aperiodic tasks such as the motion of a robotic arm or the movement of a fin [85]. Some challenges are encountered in using the current properties of conducting polymers for moderate to large scale applications, as explained in the next section.

## 5.8 Challenges in Fabrication and Energy Delivery: Example – Biceps Muscle

To contract quickly, conducting polymer actuators need to be thin. They also have relatively low electromechanical coupling. Low electromechanical coupling does not necessarily mean low efficiency, as energy can be recovered from conducting polymers due to their capacitive nature. However, it does imply that a much larger amount of electrical energy needs to be transferred to the polymer than the amount of work performed, this energy then is either dissipated or recovered. Conducting polymer actuators operate at low voltages and the coupling is low, the implication is that currents will be very high in large, high power devices. Thin films and relatively large currents are less of a concern in micro and nano devices where dimensions are small, heat transfer is good, and batteries are readily available. However in scaling up even to moderate size, the challenge can be considerable, as made clear in the example of creating a biceps muscle.

The biceps muscle does about 45 J of work in one stroke [18]. Using Eq. (5.9) relating work and work density to the minimum actuator volume and assuming a work density of about 100 $kJ/m^3$, the volume of conducting polymer required to produce this work is approximately 500 ml. For the contraction to occur in one second, then based on Eq. (5.15), the polymer thickness must be about 5 μm. Assuming a biceps length of 150 mm and a square cross section that is 55 mm across, 11,000 layers of conducting polymer are needed to achieve the desired speed. Equation (5.12) predicts that the counterelectrode must be placed within 40 mm of the working electrode, and thus will have to be integrated with it. Equation (5.13) suggests that the electrical contacts to the polymer must be spaced by less than 20 mm. The need for the multiple thin layers and separators suggests a considerable fabrication challenge. This may be eased by using porous polymers in which mass transport is enhanced by allowing ions to travel into the polymer via liquid filled capillaries [59]. Nevertheless the challenge is significant.

The current needed to charge a conducting polymer bicep is substantial. Assuming a capacitance of 130 farads per ml (based on 100 F/g), it will take about 65 000 C, which needs to be delivered within 1 second. Normally such huge currents would be unthinkable. Fortunately, beacuse conducting polymers act as energy storage devices, enough energy for one cycle can be stored in the counter-electrode (which likely will also be polymer, and act as the antagonistic muscle – *e.g.*, triceps). 90% or more of the input energy is recoverable [19,22], a combination of batteries and capacitors can be used to store the additional 10 % of energy needed for each cycle. The fabrication and control are not trivial, however.

As the size scales down, the currents also drop dramatically. For example, a 1 $mm^3$ actuator needs only about 130 mA of current. At present, moderate to small size actuator applications are the most promising for conducting polymer actuators. Larger devices require substantial attention to the details of design and fabrication. New materials are being developed [63,95] which promise to dramatically increase electromechanical coupling and reduce the amount of charge transferred, easing scaling issues [90].

## 5.9 Summary of Properties

The chapter has described the current state of synthesis, modeling, application, and analysis of polypyrrole actuators. Polypyrrole and other conducting polymer actuators are evolving rapidly, with strains increasing dramatically over the past two years alone, for example [96,97]. It is likely that further improvments will be made in cycle life [98] and possibly also in load bearing capacity. Currently, the advantages and limitations of polypyrrole actuators are

*Limitations:*

Mass transport of ions and the high capacitance limits rates of actuation; high rates are several hertz, though these can potentially reach kilohertz frequencies using microstructured electrodes.
Low electromechanical coupling and low voltage operation mean that very high currents are needed to drive large actuators at moderate speeds, making scaling a challenge. In principle, high currents can be delivered by polymer supercapacitors, solving the problem.
Cycle life is still only moderate, limiting the range of applications unless regeneration or replacement strategies can be used.
If a liquid electrolyte is to be used, some encapsulation is often required.

*Advantages:*

Low voltage operation (several volts),
Low cost, flexible materials,
High work density (~100 $kJ/m^3$),
Moderate stress (1–5 MPa typical),
Moderate to high strain (2 – >20%),
Catch state (no work expended to hold a load),

Miniature devices are expected to be fast,
Versatile materials from which electronics, energy storage, structural elements, sensors as well as actuators can all be constructed.

## 5.10 Acknowledgment

The formulation of the contents of this chapter was made possible by years of interaction with and inspiration from Professor Ian W. Hunter and members of his laboratory.

## 5.11 References

[1] Madden, J. D. W., Vandesteeg, N. A., Anquetil, P. A., Madden, P. G. A., Takshi, A., Pytel, R. Z., Lafontaine, S. R., Wieringa, P. A., and Hunter, I. W. Artificial muscle technology: Physical principles and naval prospects. Oceanic Engineering, IEEE Journal of 29(3), 706–728 (2004).
[2] Baughman, R.H. Conducting polymer artificial muscles. Synthetic Metals **78**, 339–353 (1996).
[3] Baughman, R.H., Shacklette, R.L., and Elsenbaumer, R.L. Micro electromechanical actuators based on conducting polymers. In Lazarev, P.I. (ed.) *Topics in Molecular Organization and Engineering, Vol.7: Molecular Electronics*. Kluwer, Dordrecht (1991).
[4] Otero, T.F. Artificial muscles, electrodissolution and redox processes in conducting polymers. In Nalwa, H.S. (ed.) Handbook of organic and Conductive Molecules and Polymers. John Wiley & Sons, Chichester (1997).
[5] Pei, Q. and Inganas, O. Electrochemical applications of the beam bending method; a novel way to study ion transport in electroactive polymers. Solid State Ionics **60**, 161–166 (1993).
[6] Herod, T.E. and Schlenoff, J.B. Doping induced strain in polyaniline: Stretchoelectrochemistry. Chemistry of Materials **5**, 951–955 (1993).
[7] Kaneto, K., Kaneko, M., and Takashima, W. Response of chemomechanical deformation in polyaniline film on variety of anions. Japanese Journal of Applied Physics **34, Part 2**, L837–L840 (1995).
[8] Pei, Q. and Inganas, O. Electrochemical application of the bending beam method. 1. Mass transport and volume changes in polypyrrole during redox. Journal of Physical Chemistry **96**, 10507–10514 (1992).
[9] Baughman, R.H. Conducting polymer artificial muscles. Synthetic Metals **78**, 339–353 (1996).
[10] Madden, J.D., Madden, P.G., Anquetil, P.A., and Hunter, I.W. Load and time dependence of displacement in a conducting polymer actuator. Materials Research Society Proceedings **698**, 137–144 (2002).
[11] Mazzoldi, A., Della Santa, A., and De Rossi, D. Conducting polymer actuators: Properties and modeling. In Osada, Y. and De Rossi, D.E. (eds.) Polymer Sensors and Actuators. Springer Verlag, Heidelberg (1999).
[12] Madden, J.D., Madden, P.G., and Hunter, I.W. Conducting polymer actuators as engineering materials. In Yoseph Bar–Cohen (ed.) Proceeding of SPIE Smart Structures and Materials 2002: Electroactive Polymer Actuators and Devices. SPIE Press, Bellingham, WA (2002).

[13] Mazzoldi, A., Della Santa, A., and De Rossi, D. Conducting polymer actuators: Properties and modeling. In Osada, Y. and De Rossi, D.E. (eds.) Polymer Sensors and Actuators. Springer Verlag, Heidelberg (1999).
[14] Spinks, G., Liu, L., Wallace, G., and Zhou, D. Strain response from polypyrrole actuators under load. Advanced Functional Materials **12**, 437–440 (2002).
[15] Smela, E. and Gadegaard, N. Volume change in polypyrrole studied by atomic force microscopy. Journal of Physical Chemistry B **105**, 9395–9405 (2001).
[16] Hara, S., Zama, T., Takashima, W., and Kaneto, K. Gel–Like polypyrrole based artificial muscles with extremely large strain. Polymer Journal **36**, 933–936 (2004).
[17] Anquetil, P.A., Rinderknecht, D., Vandesteeg, N.A., Madden, J.D., and Hunter, I.W. Large strain actuation in polypyrrole actuators. Smart Structures and Materials 2004: Electroactive Polymer Actuators and Devices (EAPAD). 5385, 380–387. 2004–. San Diego, CA, SPIE.
[18] Madden, J.D., Madden, P.G., and Hunter, I.W. Conducting polymer actuators as engineering materials. In Yoseph Bar–Cohen (ed.) Proceeding of SPIE Smart Structures and Materials 2002: Electroactive Polymer Actuators and Devices. SPIE Press, Bellingham, WA (2002).
[19] Madden, J.D., Madden, P.G., and Hunter, I.W. Characterization of polypyrrole actuators: Modeling and performance. In Yoseph Bar–Cohen (ed.) Proceedings of SPIE $8^{th}$ Annual Symposium on Smart Structures and Materials: Electroactive Polymer Actuators and Devices. SPIE, Bellingham WA (2001).
[20] Hara, S., Zama, T., Takashima, W., and Kaneto, K. Free–Standing polypyrrole actuators with response rate of 10.8% $s^{-1}$. Synthetic Metals **149**, 199–201 (2005).
[21] Wang, X.Z., Shapiro, B., and Smela, E. Visualizing ion currents in conjugated polymers. Advanced Materials **16**, 1605–+ (2004).
[22] Madden, J.D. Conducting Polymer Actuators.Ph.D. Thesis. Massachusetts Institute of Technology, Cambridge, MA (2000).
[23] Kaneko, M., Fukui, M., Takashima, W., and Kaneto, K. Electrolyte and strain dependences of chemomechanical deformation of polyaniline film. Synthetic Metals **84**, 795–796 (1997).
[24] Spinks, G.M., Zhou, D.Z., Liu, L., and Wallace, G.G. The amounts per cycle of polypyrrole electromechanical actuators. Smart Materials & Structures **12**, 468–472 (2003).
[25] Madden, J.D., Cush, R.A., Kanigan, T.S., and Hunter, I.W. Fast contracting polypyrrole actuators. Synthetic Metals **113**, 185–193 (2000).
[26] Lacroix, J.C., Kanazawa, K.K., and Diaz, A. Polyaniline: A very fast electrochromic material. Journal of the Electrochemical Society **136**, 1308–1313 (1989).
[27] Vandesteeg, N., Madden, P.G.A., Madden, J.D., Anquetil, P.A., and Hunter, I.W. Synthesis and characterization of EDOT–based conducting polymer actuators. Smart Structures and Materials 2003: Electroactive Polymer Actuators and Devices (EAPAD). 5051, 349–356. 2003–. San Diego, CA, SPIE.
[28] Yamaura, M., Hagiwara, T., and Iwata, K. Enhancement of electrical conductivity of polypyrrole film by stretching: Counter ion effect. Synthetic Metals **26**, 209–224 (1988).
[29] Ding, J. *et al.* High performance conducting polymer actuators utilizing a tubular geometry and helical wire interconnects. Synthetic Metals 8 (in press ).
[30] Kaneto, K., Min, Y., MacDiarmidm, and Alan, G. Conductive polyaniline laminates . 96. 94.
[31] Gregory, R.V., Kimbrell, W.C., and Kuhn, H.H. Conductive textiles. Synthetic Metals **28**, C823–C835 (1989).
[32] Lu, W. *et al.* Use of ionic lquids for pi–conjugagted polymer electrochemical devices. Science **297**, 983–987 (2002).

[33] Yamaura, M., Sato, K., and Iwata, K. Memory effect of electrical conductivity upon the counter–anion exchange of polypyrrole films. Synthetic Metals **48**, 337–354 (1992).
[34] Sato, K., Yamaura, M., and Hagiwara, T. Study on the electrical conduction mechanism of polypyrrole films. Synthetic Metals **40**, 35–48 (1991).
[35] Yamaura, M., Sato, K., and Hagiwara, T. Effect of counter–anion exchange on electrical conductivity of polypyrrole films. Synthetic Metals **39**, 43–60 (1990).
[36] Yamaura, M., Hagiwara, T., and Iwata, K. Enhancement of electrical conductivity of polypyrrole film by stretching: counter ion effect. Synthetic Metals **26**, 209–224 (1988).
[37] Maw, S., Smela, E., Yoshida, K., Sommer–Larsen, P., and Stein, R.B. The effects of varying deposition current on bending behvior in PPy(DBS)–actuated bending beams. Sensors and Actuators A 89, 175–184. 2001.
[38] Shimoda, S. and Smela, E. The effect of pH on polymerization and volume change in PPy(DBS). Electrochimica Acta **44**, 219–238 (1998).
[39] Kaneko, M., Fukui, M., Takashima, W., and Kaneto, K. Electrolyte and strain dependences of chemomechanical deformation of polyaniline film. Synthetic Metals **84**, 795–796 (1997).
[40] Pei, Q. and Inganas, O. Electrochemical application of the bending beam method. 1. Mass transport and volume changes in polypyrrole during redox. Journal of Physical Chemistry **96**, 10507–10514 (1992).
[41] Madden, J.D.W., Schmid, B., Hechinger, M., Lafontaine, S.R., Madden, P.G.A., Hover, F.S., Kimball, R., and Hunter, I.W. Application of polypyrrole actuators: Feasibility of variable camber foils. Oceanic Engineering, IEEE Journal of *29*(3), 738–749. 2004.
[42] Madden, J.D., Cush, R.A., Kanigan, T.S., and Hunter, I.W. Fast contracting polypyrrole actuators. Synthetic Metals **113**, 185–193 (2000).
[43] Anquetil, P.A., Rinderknecht, D., Vandesteeg, N.A., Madden, J.D., and Hunter, I.W. Large strain actuation in polypyrrole actuators. Smart Structures and Materials 2004: Electroactive Polymer Actuators and Devices (EAPAD). 5385, 380–387. 2004–. San Diego, CA, USA, SPIE.
[44] Hara, S., Zama, T., Tanaka, N., Takashima, W., and Kaneto, K. Artificial fibular muscles with 20% strain based on polypyrrole–metal coil composites. Chemistry Letters **34**, 784–785 (2005).
[45] Smela, E. and Gadegaard, N. Surprising volume change in PPy(DBS): An atomic force microscopy study. Advanced Materials **11**, 953–957 (1999).
[46] Smela, E. and Gadegaard, N. Surprising volume change in PPy(DBS): An atomic force microscopy study. Advanced Materials **11**, 953–957 (1999).
[47] Smela, E., Kallenbach, M., and Holdenried, J. Electrochemically driven polypyrrole bilayers for moving and positioning bulk micromachined silicon plates. Journal of Microelectromechanical Systems. **8**, 373 (1999).
[48] Smela, E., Inganas, O., and Lundstrom, I. Conducting polymers as artificial muscles: hallenges and possibilities. Journal of Micromechanics & Microengineering **3**, 203–205 (1993).
[49] Wallace, G. *et al.* Ionic liquids and helical interconnects: bringing the electronic braille screen closer to reality. Proceedings of SPIE Smart Structures and Materials (in press), (2003).
[50] Madden, J.D. Actuator selection for variable camber foils. Smart Structures and Materials 2004: Electroactive Polymer Actuators and Devices (EAPAD). 5385, 442–448. 2004–. San Diego, CA, USA, SPIE.
[51] DellaSanta, A., Mazzoldi, A., and DeRossi, D. Steerable microcatheters actuated by embedded conducting polymer structures. Journal of Intelligent Material Systems and Structures **7**, 292–300 (1996).

[52] Elias, H.–G. Mega Molecules. Springer–Verlag, Berlin (1987).
[53] Kohlman, R.S. and Epstein, A.J. Insulator–metal transistion and inhomogeneous metallic state in conducting polymers. In Skothcim, T.A., Elsenbaumer, R.L., and Reynolds, J.R. (eds.) Handbook of Conducting Polymers. Marcel Dekker, New York (1998).
[54] Atkins, P.W. Physical Chemistry. W.H. Freeman, New York (1990).
[55] Kittel, Charles. Introduction to Solid State Physcis. 66., John Wiley & Sons, New York
[56] Roth, S. One–Dimensional Metals. Springer–Verlag, New York (1995).
[57] Noda, A. and Watanabe, M. *Electrochimica Acta* **45**, 1265–1270 (2000).
[58] Jones, E.T., Chao, E., and Wrighton, M.J. Preparation and characterization of molecule–based transistors with a 50 nm separation. Journal of the American Chemical Society **109**, 5526–5529 (1987).
[59] Izadi–Najafabadi, A., Tan, D.T.H., and Madden, J.D.W. Towards high power polypyrrole–carbon capacitors. Synthetic Metals **152**, 129–132 (2005).
[60] Arbizzani, C., Mastroagostino, M., and Sacrosati, B. Conducting polymers for batteries, supercapacitors and optical devices. In Nalwa, H.S. (ed.) Handbook of Organic and Conductive Molecules and Polymers. John Wiley & Sons, Chichester (1997).
[61] Hunter, I.W. and Lafontaine, S.A comparison of muscle with artificial actuators. *Technical Digest IEEE Solid State Sensors and Actuators Workshop.* 178–185. 92. IEEE.
[62] Anquetil, P.A., Yu, H., Madden, J.D., Swager, T.M., and Hunter, I.W. Recent advances in thiophene–based molecular actuators. Smart Structures and Materials 2003: Electroactive Polymer Actuators and Devices (EAPAD). 5051, 42–53. 2003–. San Diego, CA, SPIE.
[63] Marsella MJ, Reid RJ, Estassi S, and Wang LS. Tetra[2,3–thienylene]: A building block for single–molecule electromechanical actuators. Journal of the American Chemical Society **124 (42)**, 12507–12510 (2002).
[64] Baker, C.K. and Reynolds, J.R. A quartz microbalance study of the electrosynthesis of polypyrrole. Journal of Electroanalytical Chemistry **251**, 307–322 (1988).
[65] Yamaura, M., Sato, K., and Iwata, K. Memory effect of electrical conductivity upon the counter–anion exchange of polypyrrole films. Synthetic Metals **48**, 337–354 (1992).
[66] Yamaura, M., Sato, K., and Hagiwara, T. Effect of counter–anion exchange on electrical conductivity of polypyrrole films. Synthetic Metals **39**, 43–60 (1990).
[67] Ding, J. *et al.* High performance conducting polymer actuators utilizing a tubular geometry and helical wire interconnects. Synthetic Metals **138**, 391–398 (2003).
[68] Zama, T., Hara, S., Takashima, W., and Kaneto, K. Comparison of cxonducting polymer actuators based on polypyrrole doped with Bf4(–), Pf6(–), Cf3so3–, and Clo4–. Bulletin of the Chemical Society of Japan **78**, 506–511 (2005).
[69] Spinks, G.M., Xi, B.B., Zhou, D.Z., Truong, V.T., and Wallace, G.G. Enhanced control and stability of polypyrrole electromechanical actuators. Synthetic Metals **140**, 273–280 (2004).
[70] Madden, P.G.A. Ph. D. Thesis: Development and modeling of conducting polymer actuators and demonstration of a conducting polymer–based feedback loop. MIT, Cambridge, MA (2003).
[71] Spinks, G.M. *et al.* Ionic liquids and polypyrrole helix tubes: Bringing the electronic Braille screen closer to reality. Proceedings of SPIE Smart Structures and Materials **5051**, 372–380 (2003).
[72] Spinks, G.M. and Truong, V.T. Work–per–cycle analysis for electromechanical actuators. Sensors and Actuators A–Physical **119**, 455–461 (2005).

[73] Madden, J.D. , Rinderknecht, D., Anquetil, P.A., and Hunter, I.W. Cycle life and load in polypyrrole actuators. Sensors and Actuators A (2005).
[74] Penner, R.M. and Martin, C.R. Electrochemical investigations of electronically conductive polymers. 2. Evaluation of charge–transport rates in polypyrrole using an alternating current impedance method. Journal of Physical Chemistry **93**, 984–989 (1989).
[75] Penner, Reginald M., Van Dyke, Leon S., and Martin, Charles R. Electrochemical evaluation of charge–transport rates in polypyrrole. Journal of Physical Chemistry 92, 5274–5282. 88.
[76] Mao, H., Ochmanska, J., Paulse, C.D., and Pickup, P.G. Ion transport in pyrrole–based polymer films. Faraday Discussions of the Chemical Society **88**, 165–176 (1989).
[77] Bull, R.A., Fan, F.–R.F., and Bard, A.J. Polymer films on electrodes. Journal of the Electrochemical Society **129**, 1009–1015 (1982).
[78] Tanguy, J. and Hocklet, M. Capacitive charge and noncapacitive charge in conducting polymer electrodes. Journal of the Electrochemical Society: Electrochemical Science and Technology 795–801 (1987).
[79] Posey, F. A. and Morozumi, T. Theory of potentiostatic and galvanostatic charging of the double layer in poirous electrodes. Journal of the Electrochemical Society 113(2), 176–184. 66.
[80] Yeu, T., Nguyen, T.V., and White, R.E. A mathematical model for predicting cyclic voltammograms of electrically conductive polypyrrole. Journal of the Electrochemical Society: Electrochemical Science and Technology 1971–1976 (1988).
[81] Tanguy, J., Mermilliod, N., and Hocklet, M. Capacitive charge and noncapacitive charge in conducting polymer electrodes. Journal of the Electrochemical Society: Electrochemical Science and Technology 795–801 (1987).
[82] YuH., Anquetil, P.A., Pullen, A.E., Madden, J.D., Madden, P.G., Swager, T.M., and Hunter, I.W. Molecular Actuators. 2002.
[83] Della Santa, A., Mazzoldi, A., Tonci, C., and De Rossi, D. Passive mechanical properties of polypyrrole films: A continuum poroelastic model. Materials Science and Engineering C **5**, 101–109 (1997).
[84] Della Santa, A., Mazzoldi, A., and De Rossi, D. Journal of Smart Material Systems and Structures **7**, 292–300 (1999).
[85] Madden, J.D.W., Vandesteeg, N.A., Anquetil, P.A., Madden, P.G.A., Takshi, A., Pytel, R.Z., Lafontaine, S.R., Wieringa, P.A., and Hunter, I.W. Artificial muscle technology: physical principles and naval prospects. Oceanic Engineering, IEEE Journal of 29(3), 706–728. 2004.
[86] Otero, T.F. Artificial muscles, electrodissolution and redox processes in conducting polymers. In Nalwa, H.S. (ed.) Handbook of Organic and Conductive Molecules and Polymers. John Wiley & Sons, Chichester (1997).
[87] Bard, A.J. and Faulkner, L.R. Electrochemical Methods, Fundamentals and Applications. John Wiley & Sons, New York (1980).
[88] Ren, X. and Pickup, P.G. The origin of the discrepancy between the low frequency AC capacitances and voltammetric capacitances of conducting polymers. Journal of Electroanalytical Chemistry 372, 289–291 (1994).
[89] Kim, J.J., Amemiya, T., Tryk, D.A., Hashimoto, K., and Fujishima, A. Charge transport processes in electrochemically deposited poly(pyrrole) and poly(N–methylpyrrole) thin films. Journal of Electroanalytical Chemistry **416**, 113–119 (1996).
[90] Madden, J.D. *et al.* Artificial muscle technology: Physical principles and naval prospects. IEEE Journal of Oceanic Engineering 24 pages (2004).
[91] Barisci, J.N., Spinks, G.M., Wallace, G.G., Madden, J.D., and Baughman, R.H. Increased actuation rate of electromechanical carbon nanotube actuators using potential

pulses with resistance compensation. Smart Materials & Structures **12**, 549–555 (2003).

[92] Hollerbach, J., Hunter, I.W., and Ballantyne, J. A comparative analysis of actuator technologies for robotics. In Khatib, O., Craig, J., and Lozano–Perez (eds.) The Robotics Review 2. MIT Press, Cambridge, MA (1992).

[93] Madden, J. D. W., Schmid, B., Hechinger, M., Lafontaine, S. R., Madden, P. G. A., Hover, F. S., Kimball, R., and Hunter, I. W. Application of polypyrrole actuators: feasibility of variable camber foils. Oceanic Engineering, IEEE Journal of 29(3), 738–749. 2004.

[94] Madden, J.D. *et al.* Artificial muscle technology: Physical principles and naval prospects. IEEE Journal of Oceanic Engineering 24 pages (2004).

[95] Anquetil, P.A., Yu, H., Madden, J.D., Swager, T.M., and Hunter, I.W. Recent advances in thiophene–based molecular actuators. Smart Structures and Materials 2003: Electroactive Polymer Actuators and Devices (EAPAD). 5051, 42–53. 2003–. San Diego, CA, USA, SPIE.

[96] Anquetil, P.A., Rinderknecht, D., Vandesteeg, N.A., Madden, J.D., and Hunter, I.W. Large strain actuation in polypyrrole actuators. Smart Structures and Materials 2004: Electroactive Polymer Actuators and Devices (EAPAD). 5385, 380–387. 2004. San Diego, CA, SPIE.

[97] Nakashima, T. *et al.* Enhanced electrochemical strain in polypyrrole films. Current Applied Physics **5**, 202–208 (2005).

[98] Lu, W. *et al.* Use of ionic liquids for pi–conjugated polymer electrochemical devices. Science 297, 983–987 (2002).

# 6

# Ionic Polymer-Metal Composite as a New Actuator and Transducer Material

K. J. Kim

Active Materials and Processing Laboratory, Mechanical Engineering Department (MS 312), University of Nevada, Reno, Nevada 89557, U.S.A.

## 6.1 Introduction

Ionic polymer-metal composites (IPMCs) are a unique polymer transducer that when subjected to an imposed bending stress exhibits a measurable charge across the chemically and/or physically placed effective electrodes of the electroactive polymer. IPMCs are also known as bending actuators capable of large bending motion when subjected to a low applied electric field (~10 kV/m) across the metalized or conductive surface (Figure 6.1). The voltage found across the IPMC under an imposed bending stress is one to two orders of magnitude smaller than the voltage required to replicate the bending motion input into the system. This leads to the observation that the material is quite attractive by showing inclination for possible transduction as well as actuation [1–25]. In 1993, an IPMC was first reported as an active polymeric material by Oguro and his co-workers [21]. Since then, much attention has been given to IPMCs with the hope that they can be used as a soft actuator and sensor/transducer material for new opportunities in future engineering. IPMCs have been considered promising actuator materials, in particular for biorobotic applications.

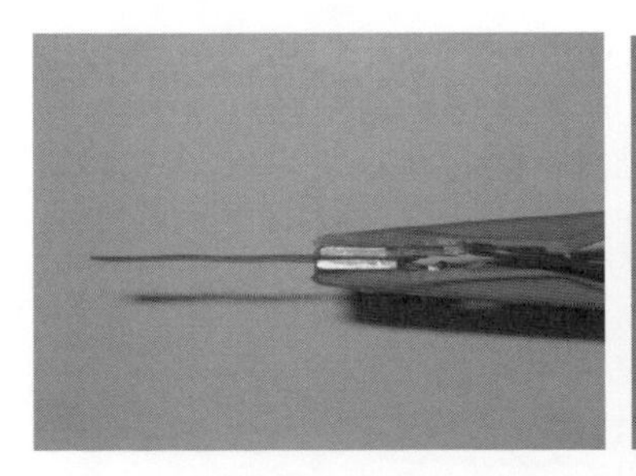
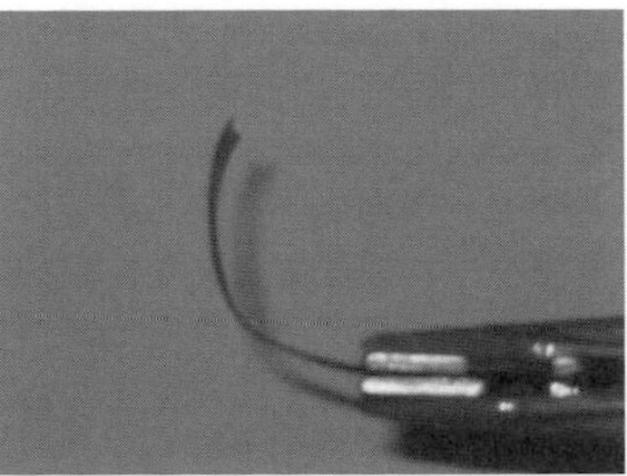

**Figure 6.1.** Actuation of a typical IPMC (from [8])

Part of what makes an IPMC so unique are its inherent transductive/sensing properties in addition to its actuation capabilities. Similar to piezo materials, IPMCs can show displacements under an applied electric field and can also engender a current from an imposed bending moment that is applied to the material. The voltage can be as high as in the 10's of millivolts range for larger imposed bending displacements. This makes IPMCs possibly effective for large motion sensor or damper applications if their behavior can be properly controlled.

## 6.2 How IPMC Works

In 2000, De Gennes *et al.* [18] presented a set of coupled equations based on linear irreversible thermodynamics to describe the behavior of a typical IPMC. The model is a compact description of the transduction and actuation principles inherent in the IPMC by defining it in the linear regime and in static conditions.

They introduce the linear irreversible thermodynamic relationship for charge transport (with a current density $J$ normal to the membrane) and solvent transport (with a flux $Q$) to describe this electromechanical coupling of ionic gels (*i.e.* IPMCs, see Figure 6.2 [18]). The standard Onsager relations for the system have the form,

$$J = \sigma E - L_{12} \nabla p \tag{6.1}$$

$$Q = L_{21} E - K \nabla p \tag{6.2}$$

Equations (6.1) and (6.2) couple the electric field, $E$, as well as the mechanical pressure gradient, $\nabla p$, the driving forces for the phenomenon involved. These equations can be elaborated upon to explain the direct effect (actuation) as well as the inverse effect (sensing or transduction) of IPMCs. Note that these Onsager relations developed by De Gennes *et al.* are for a static model. Note also that $\sigma$, $L_{12}$ ($=L_{21}$), and K are electric conductance, cross-coefficient, and permeability, respectively.

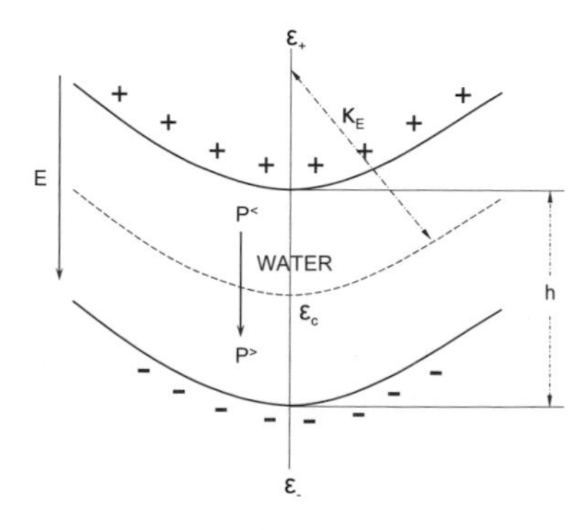


**Figure 6.2.** Principle of the bending motion. $K_E$, $h$, $E$, and $P$ are the curvature, specimen thickness, electric field, and ionic pressure of the system.

IPMCs as transducers have been modeled and developed for actuation, sensing. and control applications by several investigators. Their work can be found elsewhere [19–25].

## 6.3 IPMC Manufacturing Techniques

### 6.3.1 Metal Reduction Technique

The current state-of-the-art IPMC manufacturing technique [6] incorporates two distinct preparation processes: *initial compositing process* and *surface electroding process*. Due to different preparation processes, the morphologies of precipitated platinum are significantly different. The initial compositing process requires an appropriate platinum salt such as $Pt(NH_3)_4HCl$ for chemical reduction processes. The principle of the compositing process is to metalize the inner surface of the material (usually, Pt nano–particles, in a membrane shape) by a chemical-reduction means such as $LiBH_4$ or $NaBH_4$. The ion-exchange polymer is soaked in a salt solution to allow platinum-containing cations to diffuse through *via* the ion-exchange process. Later, a proper reducing agent such as $LiBH_4$ or $NaBH_4$ is introduced to platinize the materials by molecular plating. It has been experimentally observed that the platinum particulate layer is buried a few microns deep (typically 1–10 μm) within the IPMC surface and is highly dispersed. A TEM image near the boundary region of an IPMC strip on the penetrating edge of the IPMC shows a functional particle density gradient where the higher particle density is toward the surface electrode. Figure 6.3 describes Ni-doped IPMC manufacturing developed at the Active Materials and Processing Laboratory of the University of Nevada, Reno.

### 6.3.2 Physical Loading Technique

Although the traditional metal reduction processes described above are known to be effective in manufacturing IPMCs, one may realize that one drawback of using these processes is their relatively high cost due to the use of noble metals (platinum, gold, palladium, *etc.*) and associated complex chemical processes. For IPMCs to be successfully adopted as industrial actuators or sensors, one should be able to reduce their manufacturing cost significantly. One way to do so is to simplify the compositing and electroding processes.

The principal idea of processing this new IPMCs is first to physically load a conductive primary powder into the polymer network forming a dispersed layer which can function as a major conductive medium near boundaries and, subsequently, to further secure such a primary particulate medium within the polymer network with smaller particles (Pd or Pt in this case) via a chemical plating process so that both primary and smaller secondary particles can be secured within the polymer network. Furthermore, an electroplating process can be applied to integrate the entire conductive phase intact, serving as an effective electrode. For more details, readers are referred to recent work done by Shahinpoor and Kim [15].

This physical loading tcchnique has been elaborated by Leo and his co-workers [25]. They used a high surface area-to-volume ratio of metal particulates to achieve high capacitance at low frequencies.

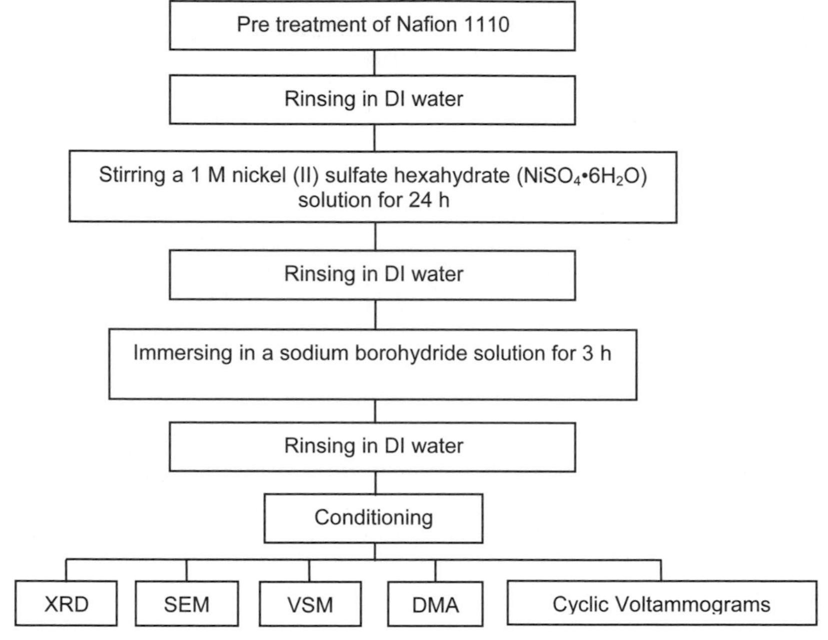


**Figure 6.3.** Experimental procedure for Ni-doped IPMCs using the ion-exchange and precipitation method

## 6.4 Engineering Properties of Interest

In this section, important engineering properterties of typical IPMCs are presented.

### 6.4.1 Mechanical Properties

Figure 6.4 shows the results of dynamic material analysis (DMA) tests for a pristine Nafion$^{TM}$ film and a Ni-doped IPMC in tensile. The experiment was performed in air. In the tensile mode, Ni-doped IPMCs have higher storage modulus ($E$') with regard to stiffness than pristine Nafion$^{TM}$ film.

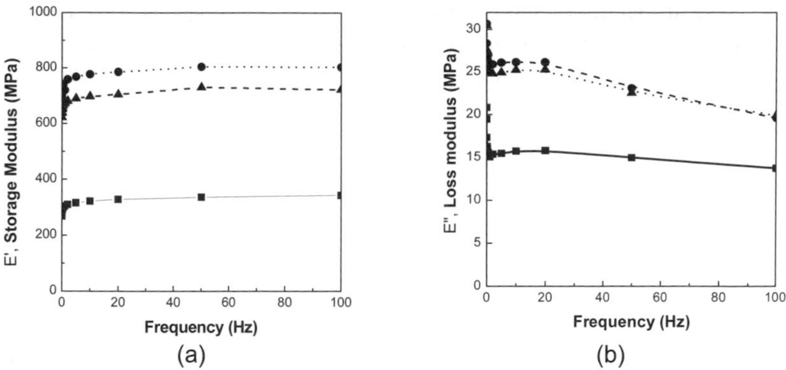


**Figure 6.4.** DMA results of the pristine Nafion film (■) and metal-doped IPMCs (● and ▲). The results are storage modulus (a) and loss modulus (b) with a frequency range from 0.01 to 100 Hz in the tensile mode

## 6.4.2. Electrical and Electrochemical Properties

*6.4.2.1 General Electrical Properties*

To assess the electrical properties of an IPMC, the standard AC impedance method that can reveal the equivalent electric circuit has been adopted. A typical measured impedance plot, provided in Figure 6.5, shows the frequency dependency of the impedance of the IPMC. It is interesting to note that the IPMC is nearly resistive in the high–frequency range and fairly capacitive in the low–frequency range.

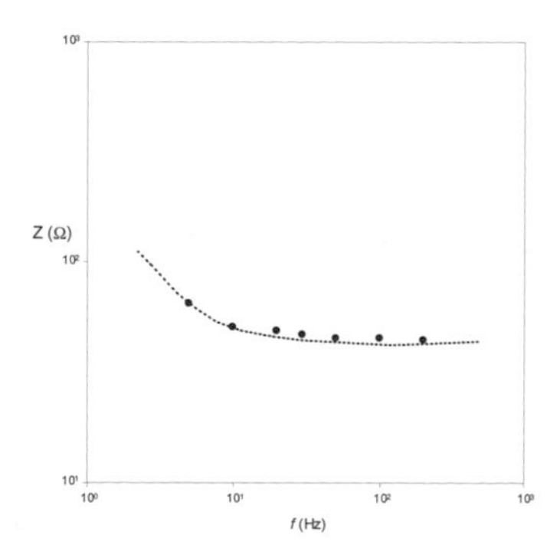


**Figure 6.5.** The measured AC impedance spectra (magnitude) of an IPMC sample [6]

Based upon the above findings, a simplified equivalent electric circuit of the typical IPMC can be considered, such as the one shown in Figure 6.6. In this approach, each single unit-circuit (i) is assumed to be connected in a series of arbitrary surface-resistance ($R_{ss}$) in the surface. This approach is based upon the experimental observation of the considerable surface-electrode resistance. We assume that there are four components to each single unit-circuit: the surface-electrode resistance ($R_s$), the polymer resistance ($R_p$), the capacitance related to the ionic polymer and the double layer at the surface-electrode/electrolyte interface ($C_d$), and an impedance ($Z_w$) due to charge transfer resistance near the surface electrode. For the typical IPMC, the importance of $R_{SS}$ relative to $R_s$ may be interpreted from $\Sigma R_{SS} / R_S \approx L / t >> 1$, where notations $L$ and $t$ are the length and thickness of the electrode, respectively.

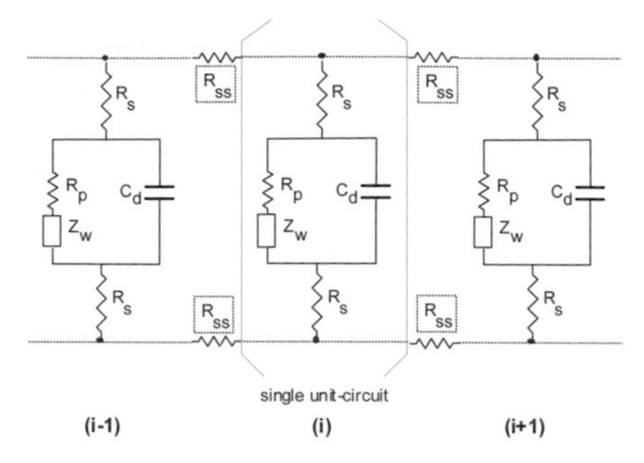


**Figure 6.6.** A possible equivalent electric circuit of a typical IPMC membrane [6]

### *6.4.2.2 Electrochemical Properties*

Figure 6.7 shows cyclic voltammograms of an IPMC with platinum electrodes. Potentiostat/galvanostat (PGZ40, Voltalab) was used for the cyclic voltammetry as well as AC impedance. By examining the voltammogram of the IPMC, it is clear to see the polycrystalline characteristics of the platinum that has been significantly altered by the presence of the base polymeric material within the testing specimen showing a unique behavior. This exhibits the importance of the surface properties of the electrodes of an IPMC. The electrochemical behavior at the surface electrodes is yet to be determined and is currently under investigation.

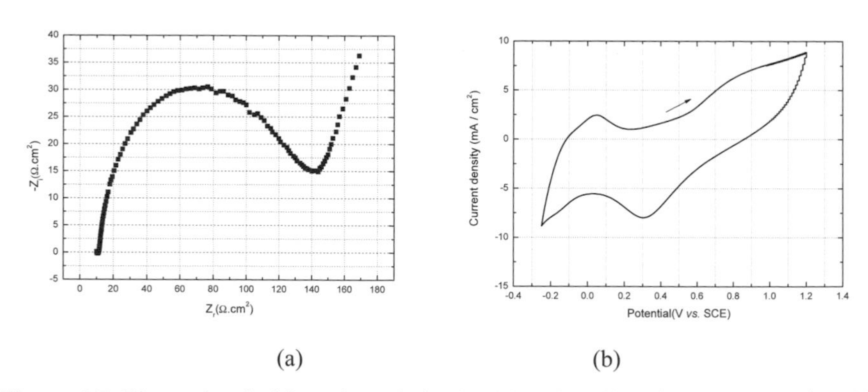


**Figure 6.7.** Electrochemical impedance behavior (a) and cyclic voltammograms (b) with a scan rate of 50 mV/s in 0.5M sulfuric acid of an IPMC [5]

### *6.4.2.3 Measurement of the Force-displacement Relationship*

The electromechanical properties considered are determined by the force–displacement relationship of an IPMC actuator. The method used to measure these properties is graphically depicted in Figure 6.8. An IPMC actuator is cantilevered at one end, and the other end is constrained, as shown in Figure 6.8a. The reaction force (or actuation force) at the right end of the actuator is generated by an electrical field. We measure the reaction force with a small force transducer. After the right-end constraint is moved up with amount of the displacement $s$, the same test is conducted. In this way, the actuation force corresponding to the end displacement $s$ can be measured, as illustrated in Figure 6.8b. Finally, without the constraint, the free end displacement can be determined. Following this procedure, the force-displacement relationship was obtained as shown in Figure 6.9. Figure 6.9 shows the measured force-displacement relationship for an IPMC actuator for two- and three-volt inputs across the IPMC. Regions **A** and **B** in Figure 6.9 include the maximum actuation forces and the maximum displacements, respectively. The specimen tested was a Nafion™-based IPMC in $Li^+$ form and plated with platinum. The length of the IPMC actuator was 20 mm with a width of 5 mm and a thickness of 0.3 mm.

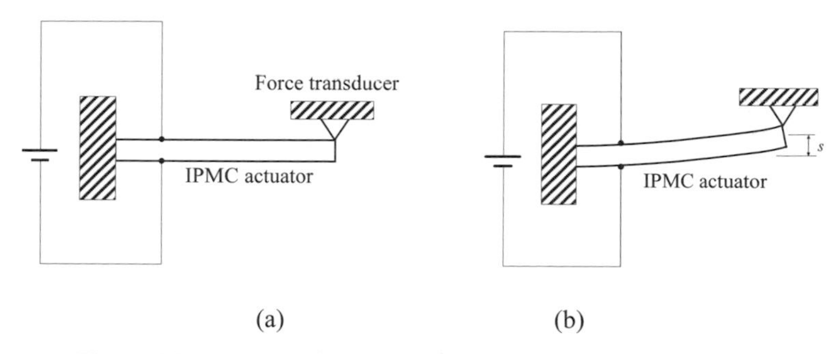


**Figure 6.8.** Test setup for the force-displacement relationship ([4])

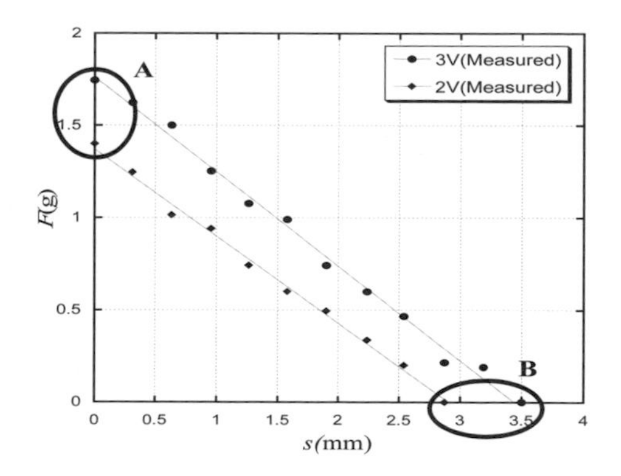


**Figure 6.9.** Force-displacement relationship of an IPMC actuator ([4])

Thus, the force $F$ for $s = 0$ (region **A** in Figure 6.9) is the reaction force for the case shown in Figure 6.8a, and the measured displacement $s$ when $F = 0$ (region **B** in Figure 6.9) stands for the tip displacement without the right-end constraint.

## 6.5 Robotic Flapping Wings

Pneumatic and motor-driven actuators have been widely adopted and used in aerospace applications as well as in other industrial robot systems. However, these actuators are not feasible for use in small or microscale flying and locomotive vehicles due to their large payload and system complexity. Furthermore, they are not suitable for mimicking the flapping motion of bird or insect wings. Using microrobots to create flapping flight is attractive due to their maneuverability that could not be obtained by conventional, fixed or rotary wing aircraft. An electroactive polymer, IPMC is a good candidate for the flapping motor because it is lightweight and can create a large deformation under low electric voltage input. Bird/insect wings can generate lift and thrust at the same time during flapping motion because the wing can flap and twist during the flapping motion [26]. To mimic the motion, the artificial flapping mechanism should also be able to create flap and twist simultaneously. Also, the width of a bird wing tip is pointed compared with the remaining parts of the wing. This reduces drag during the up-/down-strokes of the wing and also strengthens the tip of the vortex. Thus, the actuation mechanism and shape are both important for successfully mimicking a bird wing. The IPMC can generate this particular motion if it has a specially designed plan form.

Since the flapping wing must create a twisting motion as well as a bending up and down motion for thrust generation, the IPMC actuators have nonsymmetric-shapes, as shown in the two wings in Figure 6.10. The wing shapes and dimensions of the wings are also shown in Figure 6.10. Note that the areas of the IPMC actuators in the two wings are kept the same for fair comparison in the actuation displacement analysis. The wing itself is made of a thin plastic film.

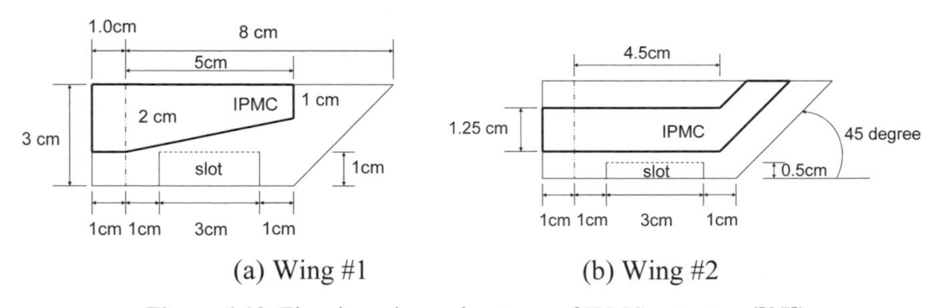


(a) Wing #1 (b) Wing #2

**Figure 6.10.** Flapping wing and patterns of IPMC actuators ([27])

The numerical deformation analysis has been conducted to determine the shape of the IPMC actuator such that the designed wing can produce maximum bending and twisting motion at the same time. Deformation of the wing has been estimated by using the equivalent bimorph beam model and MSC/ NASTRAN with the thermal analogy. For finite-element modeling, QUAD4 elements were used for both Wing #1 and #2 as shown in Figure 6.11a and Figure 6.12a. Material properties and thicknesses for the calculations are shown in Table 6.1.

**Table 6.1.** Material properties and thicknesses

| | Young's modulus (GPa) | Poisson's ratio | $d_{31} = d_{32}$ (m/V) | $t$ (mm) |
|---|---|---|---|---|
| IPMC | 1.158* | 0.487 | $1.750\times10^{-7}$ | 0.3 |
| Plastic film | 0.1 | 0.3 | N/A | 0.1 |

*Pt (~6%) heavy IPMC

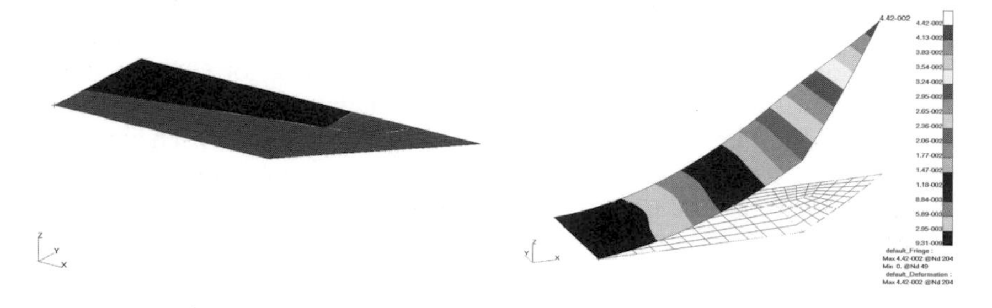

(a) Finite-element model (b) Deformed shape

**Figure 6.11.** Flapping simulation for flapping wing #1 ([27])

(a) Finite-element model (b) Deformed shape

**Figure 6.12.** Flapping simulation for flapping wing #2 ([27])

The flap-up displacement and twisting angle at the tip under 3 V (*i.e.*, $E_3 = 10$ V/mm) are calculated as 4.42 cm and 3.4°, for Wing #1, and 4.68 cm and 9.1°, for Wing #2. The deformed shapes are shown in Figure 6.11b and Figure 6.12b for Wing #1 and #2, respectively. Wing #2 is the better design for the flapping wing in terms of a twisting angle. Noted that our analysis is based on linear elasticity and thus may not accurately predict the actuation displacement. However, the present approach provides a simple but effective design tool to determine the shape of an IPMC actuator for a specific purpose.

## 6.6 Summary and Acknowledgments

In this chapter, the fundamental properties and a brief summary of recent progress in ionic polymeric-metal composites (IPMCs) as smart biomimetic sensors, actuators, and artificial muscles were presented. In the following Chapters 7, 8, 9, and 10, detailed robotics-related work on IPMCs is presented by many investigators from RIKEN, AIST, the Tokyo Institute of Technology, Tohoku University, the University of Nevada, Las Vegas, and the University of Nevada, Reno.

Also, I would like to extend my special thanks to many earlier investigators including Drs. K. Oguro (Osaka National Research Institute), M. Shahinpoor (University of New Mexico), K. Asaka (Research Institute for Cell Engineering, AIST), Y. Bar-Cohen (NASA Jet Propulsion Laboratory), S. Nemat-Nasser (University of California, San Diego), and D. Leo (Virginia Tech). Their dedicated work toward IPMCs is invaluable. The research work regarding IPMCs performed by Drs. H.C. Park, S.K. Lee, I.S. Park and, Mr. D. Kim is also appreciated.

## 6.7 References

[1] K.J. Kim, W. Yim, J.W. Paquette, and D. Kim, Ionic Polymer-Metal Composites for Underwater Operation, Journal of Intelligent Materials Systems and Structures (JIMSS), (2006, in print).

[2] D. Kim and K.J. Kim, Experimental Investigation on Electrochemical Properties of Ionic Polymer-Metal Composite, Journal of Intelligent Materials Systems and Structures (JIMSS), (2006, in print).

[3] D.Y. Lee, M.-H. Lee, K.J. Kim, S. Heo, B.-Y. Kim, and S.-J. Lee, Effect of Multiwalled Carbon Nanotube (M-CNT) Loading on M-CNT Distribution Behavior and the Related Electromechancial Properties of the M-CNT Dispersed Ionomeric Nanocomposites, Surface Coatings and Technology, Vol. 200(5-6), pp. 1916-192 (2005).

[4] S-K. Lee, H.C. Park, and K.J. Kim, Equivalent Modeling for Ionic Polymer-Metal Composite Actuators Based on Beam Theories, Smart Materials and Structures, Vol. 14, pp. 1363-1368 (2005).

[5] J.W. Paquette, K.J. Kim, D. Kim, and W. Yim, The Behavior of Ionic Polymer-Metal Composites in a Multi-Layer Configuration, Smart Materials and Structures, Vol. 14, 881-888 (2005).

[6] M. Shahinpoor and K.J. Kim, Ionic Polymer-Metal Composite-IV: Industrial and Mechanical Applications, Smart Materials and Structures, Vol. 14, 197-214 (2005); M. Shahinpoor and K.J. Kim, Ionic Polymer-Metal Composites: III. Modeling and Simulation as Biomimetic Sensors, Actuators, Transducers, and Artificial Muscles, Smart Materials and Structures, Vol. 13, pp. 1362-1388 (2004); K.J. Kim and M. Shahinpoor, Ionic Polymer-Metal Composites - II. Manufacturing Techniques, Smart Materials and Structures, Vol. 12, No. 1, pp. 65-79 (2003); M. Shahinpoor and K.J. Kim, Ionic Polymer-Metal Composites – I. Fundamentals, Smart Materials and Structures, Vol. 10, pp. 819-833 (2001).

[7] J.-D. Nam, J.H. Lee, J.H. Lee, H. Choe, K.J. Kim, and Y.S. Tak, Water Uptake and Migration Effects of Electroactive IPMC(Ion-Exchange Polymer-Metal Composite) Actuator, Sensors and Actuators A: Physical, Vol. 118, pp. 98-106 (2005).

[8] J.W. Paquette, K.J. Kim, and D. Kim, Low Temperature Characteristics of Ionic Polymer-Metal Composite Actuators, Sensors and Actuators A: Physical, Vol. 118, pp. 135-143 (2005).
[9] J. Paquette and K.J. Kim, Ionomeric Electro-Active Polymer Artificial Muscle for Naval Applications, IEEE Journal of Oceanic Engineering (JOE), Vol. 29, No. 3, pp. 729-737 (2004).
[10] J. Paquette, K.J. Kim, J.-D. Nam, and Y. S. Tak, An Equivalent Circuit Model for Ionic Polymer-Metal Composites and Their Performance Improvement by a Clay-Based Polymer Nano-Composite Technique, Journal of Intelligent Materials Systems and Structures (JIMSS), Vol. 14, pp. 633-642 (2003).
[11] J.-D. Nam, H.R. Choi, Y.S. Tak, and K.J. Kim, Novel Electroactive, Silicate Nanocomposites Prepared to Be Used as Actuators and Artificial Muscles, Sensors and Actuators: A. Physical, Vol. 105, pp. 83-90 (2003).
[12] M. Shahinpoor, K.J. Kim, and D. Leo, Ionic Polymer-Metal Composites as Multifunctional Materials, Polymer Composites, Vol. 24, No. 1, pp. 24-33 (2003).
[13] M. Shahinpoor and K.J. Kim, Experimental Study of Ionic Polymer-Metal Composites in Various Cation Forms: Actuation Behavior, Science and Engineering of Composite Materials, Vol. 10, No. 6, pp. 423-436 (2002).
[14] M. Shahinpoor and K.J. Kim, Mass Transfer Induced Hydraulic Actuation in Ionic Polymer-Metal Composites, Journal of Intelligent Materials Systems and Structures (JIMSS), Vol. 13, No. 6, pp. 369-376 (2002).
[15] M. Shahinpoor and K.J. Kim, A Novel Physically-Loaded and Interlocked Electrode Developed for Ionic Polymer-Metal Composites (IPMCs), Sensors and Actuator: A. Physical, Vol. 96, No. 2/3, pp. 125-132 (2002).
[16] K.J. Kim and M. Shahinpoor, Development of Three Dimensional Ionic Polymer-Metal Composites as Artificial Muscles, Polymer, Vol. 43(3), pp. 797-802 (2002).
[17] M. Shahinpoor and K.J. Kim, The Effect of Surface-Electrode Resistance on the Performance of Ionic Polymer-Metal Composite (IPMC) Artificial Muscles, Smart Materials and Structures, Vol. 9, No. 4, pp. 543-551 (2000).
[18] P.G. de Gennes, K. Okumura, M. Shahinpoor, and K.J. Kim, Mechanoelectric Effects in Ionic Gels, Europhysics Letters, Vol. 50, No. 4, pp. 513-518 (2000).
[19] S. Nemat-Nasser, Micromechanics of Actuation of Ionic Polymer-Metal Composites, Journal of Applied Physics, Vol. 92, No. 5, pp. 2899-2910 (2002).
[20] S. Nemat-Nasser and C.W. Thomas, Ionomeric Polymer-Metal Composites, in Electroactive Polymer (EAP) Actuators as Artificial Muscles, Reality, Potential, and Challenges, ed. Bar-Cohen, Y., SPIE Press, Washington, (2001).
[21] K. Oguro, K. Asaka, and H. Takenaka, Actuator Element, U.S. Patent #5,268,082, (1993).
[22] S. Tadokoro, T. Takamori, T., and K. Oguro, Modeling IPMC for Design of Actuation Mechanisms, in Electroactive Polymer (EAP) Actuators as Artificial Muscles, Reality, Potential, and Challenges, ed. Bar-Cohen, Y., SPIE Press, Washington, U.S.A., (2001).
[23] K. Asaka, K. Oguro, Y. Nishimura, M. Mizuhata, and H. Takenaka, Bending of Polyelectrolyte Membrane-Platinum Composites by Electric Stimuli, I. Response Characteristics to Various Waveforms, Polymer Journal, Vol. 27, No. 4, pp. 436-440 (1995).
[24] B.J. Akle, M.D. Bennet, and D. Leo, High Strain Ionomeric-Ionic Liquid Electroactive Actuators Sensors and Actuators A (in press, 2006).
[25] M.D. Bennett and D.J. Leo, Ionic Liquids as Solvents for Ionic Polymer Transducers, Sensors and Actuators A: Physical, Vol. 115. pp. 79–90 (2004).
[26] D.E. Alexander, Nature's Flyers, Chapter 4. London, The Johns Hopkins University Press (2003).

[27] H.C. Park, S. Lee, and K.J. Kim, Equivalent Modeling for Shape Design of IPMC (Ionic Polymer-Metal Composite) as Flapping Actuator, Key Engineering Materials, Vol. 297-300, pp. 616-621 (2005).

# 7

# Biomimetic Soft Robots Using IPMC

Y. Nakabo[1], T. Mukai[2], K. Asaka[3]

[1] Bio-Mimetic Control Research Center, RIKEN,
2271-130 Anagahora, Shimoshidami, Moriyama, Nagoya 463-0003, Japan and
Intelligent Systems Institute, National Institute of AIST,
1-1-1 Umezono, Tsukuba, Ibaraki 305-8568, Japan
nakabo-yoshihiro@aist.go.jp
[2] Bio-Mimetic Control Research Center, RIKEN
mukai@bmc.riken.jp
[3] Research Institute for Cell Engineering, National Institute of AIST,
1-8-31 Midorigaoka, Ikeda, Osaka 563-8577, Japan and
Bio-Mimetic Control Research Center, RIKEN
asaka-kinji@aist.go.jp

## 7.1 Introduction

### 7.1.1 Ionic Polymer-Metal Composite (IPMC)

A bimorph-type soft polymer gel actuator, which we call an artificial muscle, is an ionic polymer-metal composite (IPMC) consisting of a perfluorosulfonic acid membrane with chemically plated gold or platinum as electrodes on both sides [18, 1]. It has some excellent characteristics for robotic applications compared with other soft polymer actuators [2] as follows.

Driven with low voltage (<3 V).
Low power consumption.
Fast response (>10 Hz in water).
Mechanically and chemically durable and stable.
Soft and compliant.
Works in water (or in wet conditions).

Moreover, it has an advantage for miniaturization with its simple actuator structure. With these characteristics, we expect to apply it to micromanipulations in the bioengineering or medical fields. Our goal is to realize bioinspired soft robots, for example, a snake-like swimming robot [15, 16, 17] or a multi-degree-of-freedom (DOF) microrobot manipulator [13, 14].

The snake-like motion of a swimming robot sweeps a smaller area than a simple bending motion. It is easy to miniaturize the actuator because of its simple structure. In the future, we may be able to make robots that can swim in thin tubes or in blood vessels, or various kinds of micromanipulators in the biomedical field.

### 7.1.2 Multi-DOF Motion of IPMC

To realize a snake-like or a multi-DOF bending motion, we laser cut special electrode patterns on the surface of the actuator to control each segment individually.We have developed a variety of motions from this patterned actuator, including a snake-like motion. A kinematic modeling of the manipulator simply describes various multi-DOF motions of the artificial muscle. This model is applied to visual feedback control of the manipulator system using a Jacobian control method. For the feedback control, we have developed a visual sensing system using a 1 ms high-speed vision system which has a fast enough response to capture the fast actuator motion.

We have also measured the propulsion speed generated by the snake-like motion. By changing voltages and phases to each segment, we can control the direction of the propulsion. Finally, we have made the robot swim freely forward and backward by finding the optimal voltage, phase, and frequency. In this report, we show some results from simulations of the proposed manipulator control method and experimental results from visual sensing of the bending motion and snake-like swimming of the actuator.

## 7.2 Multi-DOF Microrobot Manipulator

### 7.2.1 Concept of Multi-DOF Microrobot Manipulator

Micromanipulations in bioengineering or in the medical field can be one of the applications of the artificial muscle. Our goal is to realize a multi-DOF microrobot manipulator that is automatically controlled by visual feedback, as shown in Figure 7.1.

In previous research studies on manipulators using an IPMC, Tadokoro *et al.* have fabricated a multi-DOF micromanipulator using a "3D-EFD" element [19] that consists of two arch-shaped IPMC components placed crosswise and works with a mechanism similar to that of parallel-link manipulators. Guo *et al.* have developed a multi-DOF microcatheter with active guide wires [6]. Using this catheter at a diverging point of a vessel, the direction of the catheter can be selected by the bending motion of an IPMC. However, up to this point, the kinematic modeling of multi-DOF bending motions or automatic visual feedback control have not yet been attempted. Automatic control will be required for applications of micromanipulators.

On the other hand, Mallavarapu *et al.* have implemented feedback control for dynamic bending motions in IPMC membranes [10]. They measured the bending motions of IPMC using a laser vibrometer to identify the dynamic models of bending responses to electrical stimuli and controlled them by sensor feedback. However, a laser vibrometer can only measure a distance from the sensor to a certain point of an IPMC; it cannot measure various shape changes in multi-DOF motion. For this, we need a better vision sensor.

In this research, we propose to use visual information for automatic feedback control of the IPMC. Our 1 ms vision system [12] enables high-speed and contact-free measurement of fast shape changes in an IPMC. Also, we propose a new type of multi-DOF manipulator that is compactly designed using a patterned artificial muscle. The patterning of electrodes on an IPMC enables the multi-DOF bending motions of a manipulator with a simple structure.

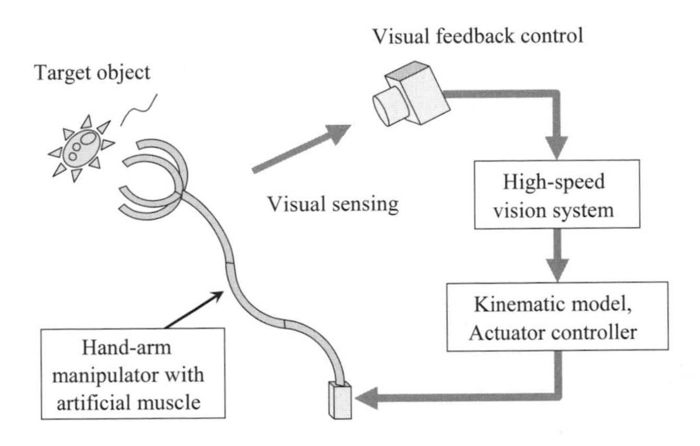


**Figure 7.1.** Application of multi-DOF micromanipulator with artificial muscle and visual feedback control

In the next section, we will describe details of our manipulator and how the patterning is carried out on the electrodes. The kinematic modeling for visual control is described in Section 7.4. The visual sensing system and proposal of feedback control are shown in Section 7.5. In Section 7.6, the experimental results of visual sensing are presented.

## 7.3 Patterned Artificial Muscle

### 7.3.1 Patterning with a Laser Beam

One way to realize this multi-DOF motion is to connect some of them using joints. The process of connecting them together is, however, complex, and it is difficult to make the joints flexible. So we propose another method to realize multiple degrees of freedom. In this method, we make special electrode patterns separated electrically by laser-cutting the gold-plated surface of the IPMC to control each segment individually. This method makes the IPMC bend as required without losing flexibility, and it is suitable for miniaturization because the process is simple.

The IPMC used in this study is a Nafion 117 membrane (by DuPont) five times chemically plated with gold. The thickness of the IPMC is 200 μm. After the plating, we cut a pattern on both sides of the membrane using a laser beam. By laser patterning, the thin gold layer is removed, and insulation between segments of the electrodes is achieved. By optimizing the conditions of the bursts of the laser beam, sufficient insulation was achieved at the minimum groove size of 50 μm wide and approximately 20 μm deep.

Photomicrographs of the laser-cut pattern are shown in Figure 7.2. The left photograph shows a cross section of the membrane with a groove and a gold layer. The right one shows a closeup of the segmented electrodes seen from above the membrane. In both photographs, plated gold remained in the bright areas and was removed in dark areas. The photographs have different magnifications, but the minimum width of insulation grooves is the same (50 μm). From the photographs, the sharp lines cut by the laser can be seen.

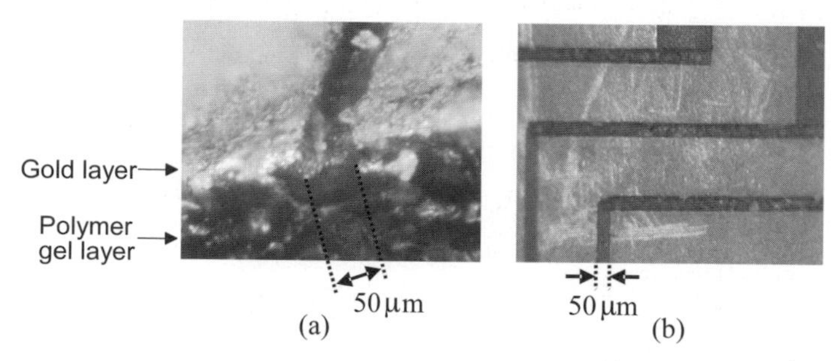


**Figure 7.2.** Results of laser beam patterning. (a) cross section of membrane with groove and plated gold layer, (b) patterned lines viewed from above

### 7.3.2 Patterning for a Multi-DOF Manipulator

Using the above laser cutting method, we have formed the electrode pattern shown in Figure 7.3. The pattern on the opposite side of the membrane has small differences only at the interconnections of electrodes.

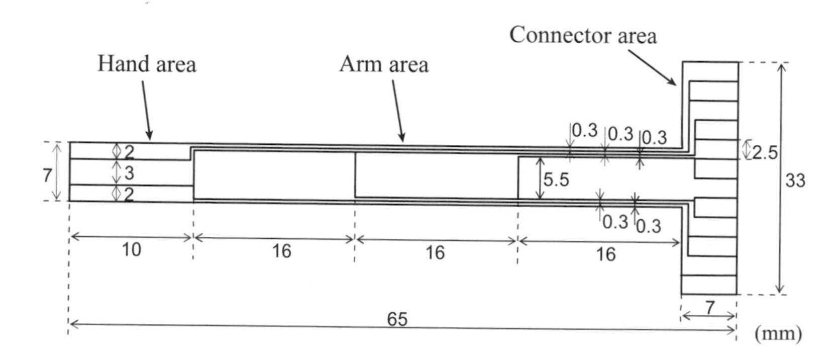


**Figure 7.3.** Pattern of electrodes on IPMC

The pattern on the membrane consists of three areas, a connector area, an arm area, and a hand area. The connector area is for providing electricity and mounting the manipulator to a stable base through an electric connector. The arm area is composed of three wide insulated segments for multi-DOF arm bending and many thin lines for electrical interconnections. Finally, the hand area is composed of three segments that are mechanically split so that it can grab a target object, as shown in Figure 7.4. A photograph of the patterned IPMC is shown in Figure 7.5. Note that three beads are attached to the joints of the arm segments, which are used for visual sensing, as explained later.

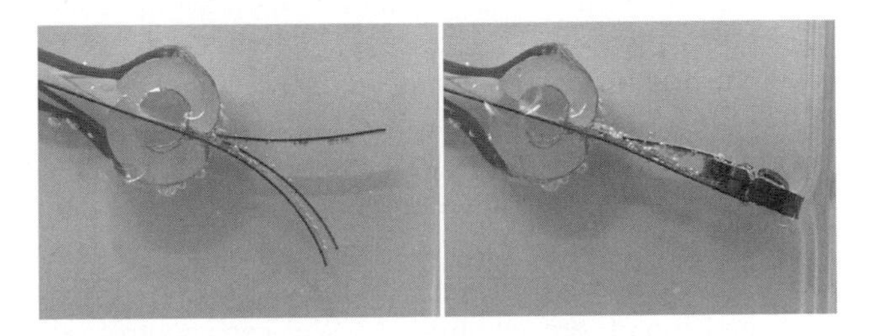

**Figure 7.4.** Hand with its open and grabbing position

**Figure 7.5.** Patterned IPMC manipulator

## 7.4 Kinematic Modeling

### 7.4.1 Modeling of One-Link Kinematics

Next we propose the kinematic model of an IPMC. Our approach is similar to the modeling of conventional serial-link multi-DOF robot manipulators, but has a marked difference in that instead of joint rotation, the link itself bends on an IPMC. We treat each segment of an electrode as a link of a manipulator that can be bent independently. Figure 7.6 shows the kinematic model of a one-link bending motion. We assume that a link bends with a constant curvature. Although this assumption is not true for very large bending angles in real IPMC membranes, it is the first step in the approximation in kinematic modeling, and for small values of curvature, this is a good approximation.

Either from the kinematic model or from real IPMC bending motion, it is important to note that the position and direction of an end point of the link is restricted to one constant trajectory, that is, one link can make only one DOF motion.

### 7.4.2 Modeling of Multi-Link Kinematics

To realize the multi-DOF motion of a manipulator, we can connect several links in series in the same way as in conventional serial-link manipulators. In this study, as a first step in our research, we restrict the motion of the manipulator to a two-dimensional space, in which two DOF for position and one DOF for the tangential direction of the end point are controlled. We can control all three DOF independently by controlling the three links of an IPMC.

We now define coordinate systems $\Sigma_i, (i = \{0, 1, 2, 3\})$ one at the origin, one at each joint of the multi-link system, and one at the end point of an arm, as shown in Figures 7.6 and 7.7. A homogeneous transfer matrix ${}^{i}A_{i+1}$ from $\Sigma_{i+1}$ to $\Sigma_i$ is written as

$$
{}^{i}A_{i+1} = \begin{pmatrix} \cos(\theta_i + \alpha_i) & -\sin(\theta_i + \alpha_i) & -\frac{l_i}{\theta_i}(1 - \cos\theta_i) \\ \sin(\theta_i + \alpha_i) & \cos(\theta_i + \alpha_i) & \frac{l_i}{\theta_i}\sin\theta_i \\ 0 & 0 & 1 \end{pmatrix} \tag{7.1}
$$

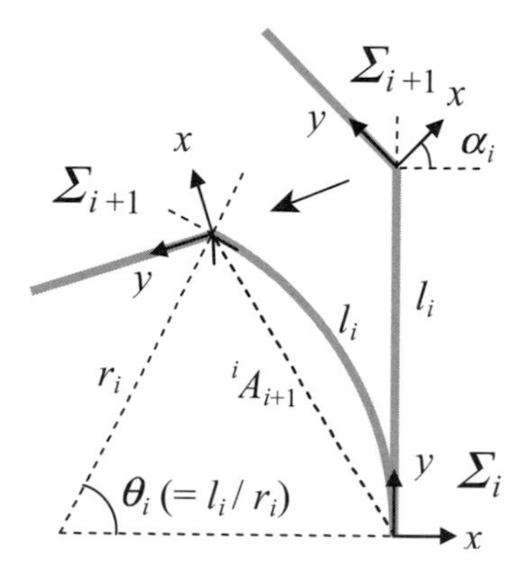


**Figure 7.6.** One-link kinematic modeling (note that usually $\alpha_i = 0$, as in this study)

where the certain length of a link is $l_i$, the curvature is $\theta_i (= r_i / l_i)$, and if the link has a certain bending angle at the joint, this angle is referred as $\alpha_i$. We consider that $l_i$ and $\alpha_i$ are known (also $\alpha_i = 0$ in this study).

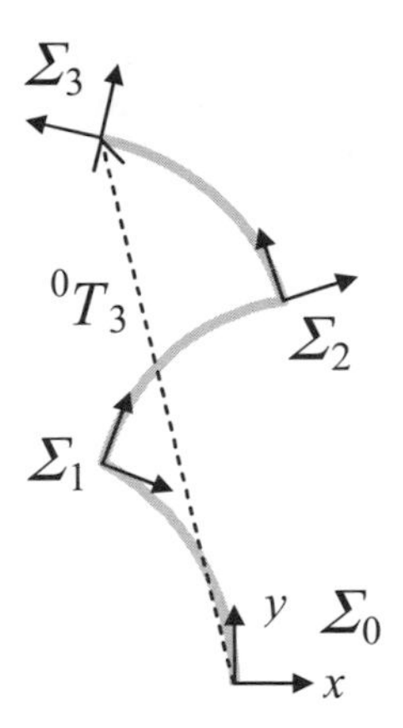


**Figure 7.7.** Multi-link coordinate system (note that joints are usually not bent, as in this study)

A transfer matrix from the base to the end point is calculated by multiplying each matrix of the link from the first to the last using

$$ {}^{0}T_3 = {}^{0}A_1 {}^{1}A_2 {}^{2}A_3 \tag{7.2} $$

which is shown in Figure 7.7. From this equation, we can obtain the position vector ${}^{0}P_e$ and rotation angle ${}^{0}\Theta_e$ of the end point in the base coordinate system $\Sigma_0$ as follows (where $\boldsymbol{P_e} = \left({}^{0}P_e, {}^{0}\Theta_e\right)^t = \left(P_{ex}, P_{ey}, P_{e\theta}\right)^t$ ):

$$
\begin{aligned}
P_{ex} &= \frac{l_2}{\theta_2}\cos(\theta_0+\theta_1+\theta_2+\alpha_0+\alpha_1) - \frac{l_2}{\theta_2}\cos(\theta_0+\theta_1+\alpha_0+\alpha_1) \\
&\quad + \frac{l_1}{\theta_1}\cos(\theta_0+\theta_1+\alpha_0) - \frac{l_1}{\theta_1}\cos(\theta_0+\alpha_0) + \frac{l_0}{\theta_0}(\cos\theta_0 - 1) \\
P_{ey} &= \frac{l_2}{\theta_2}\sin(\theta_0+\theta_1+\theta_2+\alpha_0+\alpha_1) - \frac{l_2}{\theta_2}\sin(\theta_0+\theta_1+\alpha_0+\alpha_1) \\
&\quad + \frac{l_1}{\theta_1}\sin(\theta_0+\theta_1+\alpha_0) - \frac{l_1}{\theta_1}\sin(\theta_0+\alpha_0) + \frac{l_0}{\theta_0}\sin\theta_0 \\
P_{e\theta} &= \theta_0+\theta_1+\theta_2+\alpha_0+\alpha_1+\alpha_2
\end{aligned}
\tag{7.3}
$$

### 7.4.3 Position and Orientation Control Based on a Kinematic Model

To control the position and orientation of the hand of the manipulator, we have to solve an inverse kinematic problem. More specifically, we have to solve $\hat{\boldsymbol{P}}_e = \boldsymbol{P_e}(\boldsymbol{\theta})$ by $\boldsymbol{\theta} = (\theta_0, \theta_1, \theta_2)^t$ from Eq. (7.3), where $\hat{\boldsymbol{P}}_e$ denotes both the objective position and orientation. However, it is quite difficult to solve Eq. (7.3); thus we use Jacobian methods for controlling and define control variables to converge to objective values. The Jacobian of $\boldsymbol{P}_e$ with $\boldsymbol{\theta}$ is defined as

$$\boldsymbol{J} = \left[\boldsymbol{J}_0\, \boldsymbol{J}_1\, \boldsymbol{J}_2\right] \tag{7.4}$$

$$\boldsymbol{J}_i = \frac{\partial}{\partial \theta_i} \boldsymbol{P}_e, \quad (i = \{0, 1, 2\})$$

Using the Jacobian, the following control, which is an inverse Jacobian method, will induce the control variable to converge to an objective value $\hat{\boldsymbol{P}}_e$ exponentially.

$$\dot{\boldsymbol{\theta}} = \lambda \boldsymbol{J}^{-1}\left(\hat{\boldsymbol{P}}_e - \boldsymbol{P}_e\right) \tag{7.5}$$

Here, $\lambda$ is a given constant.

The following control, which is a transposed Jacobian method, will also induce convergence to an objective:

$$\dot{\boldsymbol{\theta}} = \lambda' \boldsymbol{J}^t\left(\hat{\boldsymbol{P}}_e - \boldsymbol{P}_e\right) \tag{7.6}$$

This is possible if an appropriate value of a constant $\lambda'$ is chosen. $\boldsymbol{J}$ and $\boldsymbol{J}^{-1}$ can be calculated explicitly, but we omit such calculations because of space restrictions.

### 7.4.4 Simulations

We performed some simulations to verify our control methods. A block diagram of a control system simulation is shown in Figure 7.8. Results obtained by the inverse Jacobian method are shown in Figure 7.9, and those obtained by the transposed Jacobian method in Figure 7.10.

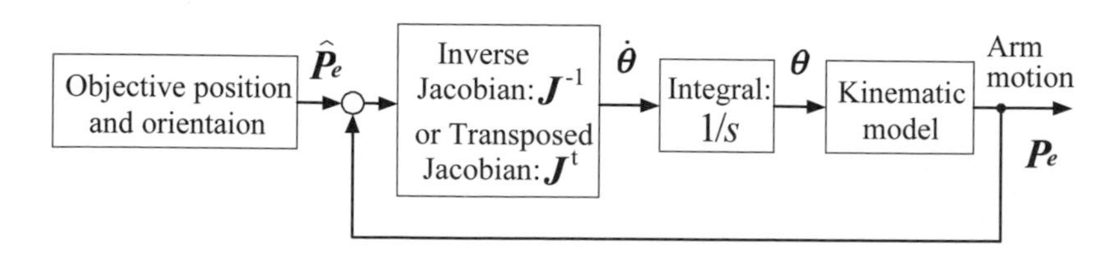


**Figure 7.8.** Block diagram of simulation of control system

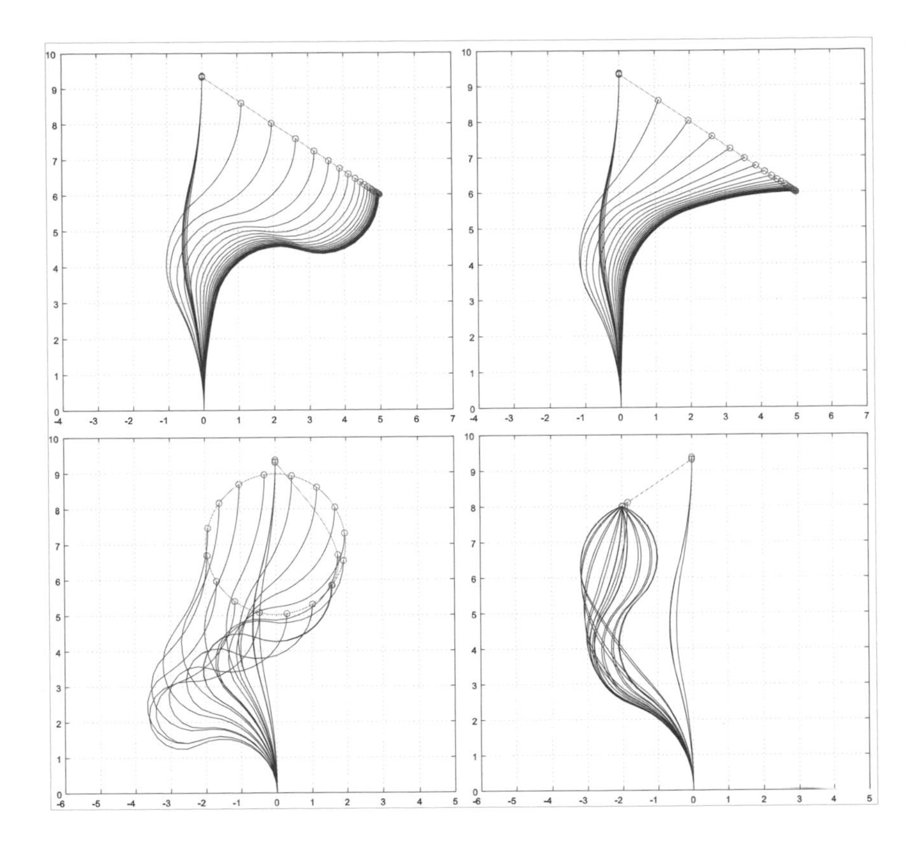

**Figure 7.9.** Results obtained by inverse Jacobian control (upper left: hand is kept in upper direction, upper right: its direction is changed to right, bottom left: tracking a round trajectory, bottom right: move and swing)

All results show that each segment of an IPMC membrane is properly bent to realize the given trajectories of positions and orientations of the end points. By the inverse Jacobian method, trajectories in Cartesian coordinates are straightforward and exponentially converged to the objective, where they are not by the transposed Jacobian. However, an inverse Jacobian is not stable near a singular position, in this case, at the upright position, whereas a transposed Jacobian is defined and stable at any point. From these results, it is confirmed that we can control all three DOF of a 2-D manipulator by the proposed methods.

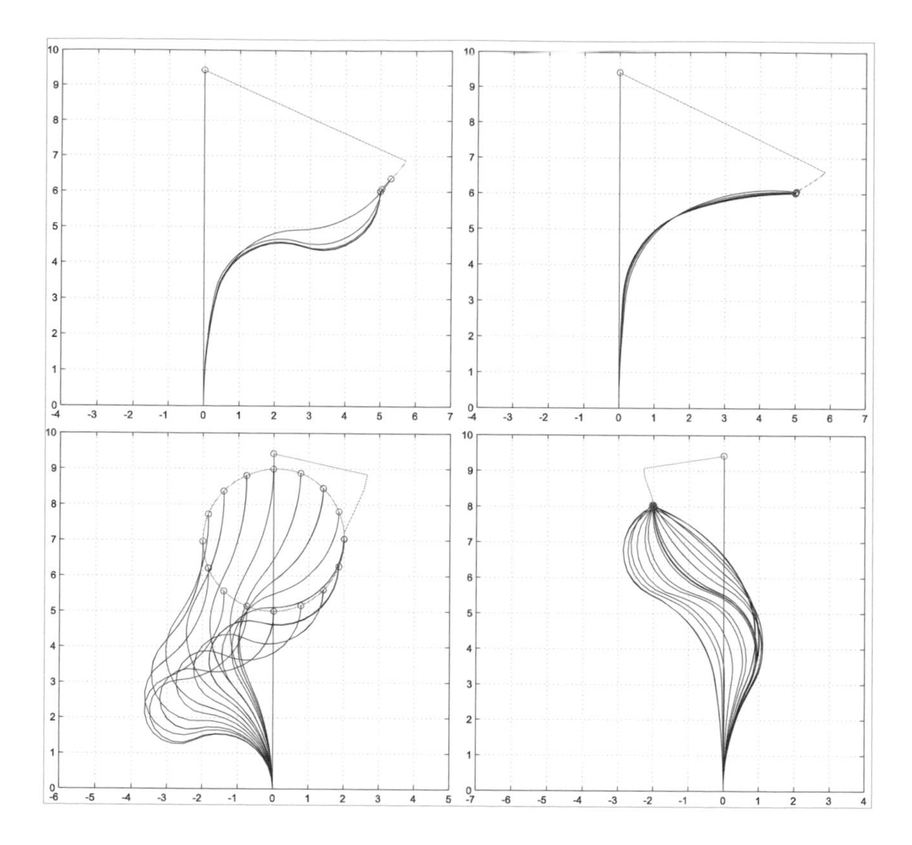

**Figure 7.10.** Results obtained by transposed Jacobian control (upper left: hand is kept in upper direction, upper right: its direction is changed to right, bottom left: tracking a round trajectory, bottom right: move and swing)

## 7.5 Visual Sensing and Control System of a Multi-DOF Manipulator

### 7.5.1 Visual Sensing System

To realize the proposed control method, we need to determine the error of the end point $\boldsymbol{P}_e$ from an objective and the curvature $\boldsymbol{\theta}$ to calculate the Jacobian $\boldsymbol{J} = \boldsymbol{J}(\boldsymbol{\theta})$ in real time.

Because the artificial muscle is a soft actuator and the membrane itself bends, limited types of sensor can be used. We propose the use of a vision sensor that can measure the overall shape changes of multi-DOF motions by contact-free sensing. However, conventional CCD cameras do not have sufficient sampling speed due to their limited video transfer rate. Thus, we used a 1 ms high-speed CPV system [12], which can capture an image and execute a processing algorithm in a cycle time of 1 ms.

Light colored beads on each joint of the manipulator are used as references for the coordinate systems. The curvatures of the links are calculated using (see Figure 7.6)

$$\theta_i = 2\arctan\left( -\frac{{}^{i}P_{i+1x}}{{}^{i}P_{i+1y}} \right), \quad (i = \{0, 1, 2\}) \tag{7.7}$$

where ${}^{i}\boldsymbol{P}_{i+1} = \left({}^{i}P_{i+1x}, {}^{i}P_{i+1y}\right)^t$ is the position vector from the origin of $\Sigma_i$ to $\Sigma_{i+1}$ in the coordinate system of $\Sigma_i$.

### 7.5.2 Control System

The proposed feedback control system is shown in Figure 7.11. It includes a visual sensing system, a Jacobian and error estimation component, a control component, and a multi-DOF patterned artificial muscle (IPMC). Figure 7.12 shows the camera head of our visual sensing system and an IPMC manipulator. The manipulator is hung in water above the electric connector.

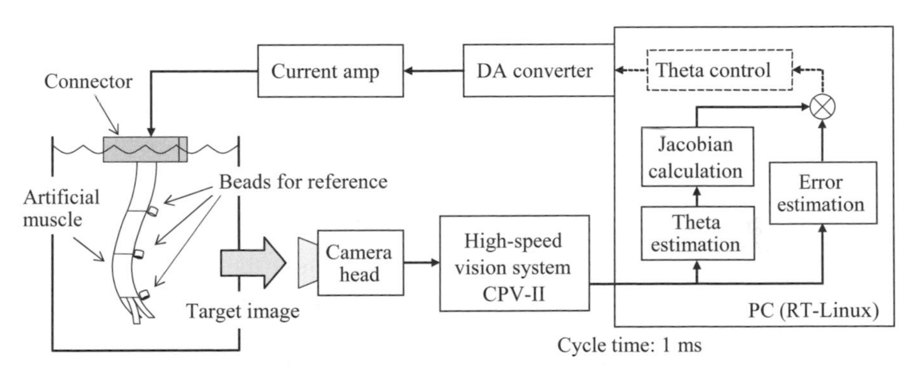


**Figure 7.11.** Block diagram of visual sensing and control system

**Figure 7.12.** Camera head of vision system and artificial muscle

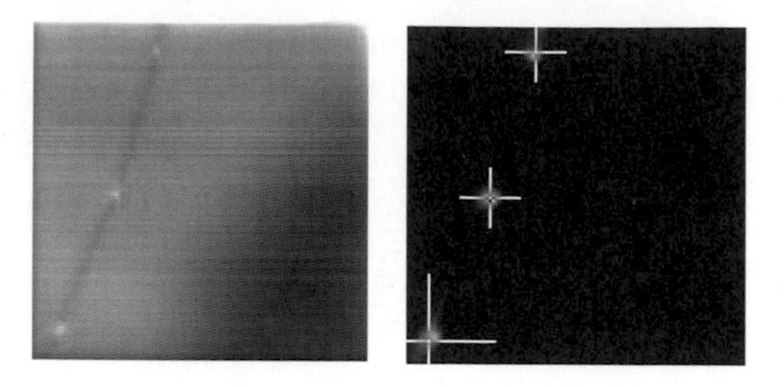

**Figure 7.13.** Image obtained by vision system (left), and result image of recognition of reference beads (right)

## 7.6 Experimental Results of Multi-DOF Manipulator

In this study, we have realized all parts of the control system shown in Figure 7.11 except the theta control stage which is enclosed by the dotted line in the figure.

First, we tested the multi-DOF bending motion of the IPMC. A problem encountered was the voltage drop through interconnection lines on the IPMC. We found that more than 3V is needed at the connector area to provide sufficient voltage at each link segment (up to 2V). With such a high voltage, the electrolysis of water occurs and this heat causes breakdown of the thin interconnections. In subsequent experiments, we wired electrodes directly to each arm segment.

Next, we conducted experiments using the visual sensing system. We applied sine waves with maximum voltages of 2V and cycle times of 1s to the arm segments. First, their phases are synchronized so that a swing motion of the arm was made. An image obtained using the high-speed vision system and a result of recognition of the reference beads are shown in Figure 7.13. The center points of three reference beads are identified accurately, as shown by three crosses.

The curvatures estimated from the images by online calculation using Eq. (7.7) are shown in Figure 7.14. The reconstructed bending motion of the IPMC using the curvature data and the proposed kinematic model is shown in Figure 7.15.

In the next experiment, we applied the same sine waves with a cycle time of 1.5s; their phases are shifted 60° from each successive segment.

A snakelike motion has been realized by a phase-shifted sinusoidal input. Online estimated curvatures of a bending motion caused by the shifted sine waves with a cycle time of 1.5s are shown in Figure 7.16. The bending motions reconstructed from the curvature data and kinematic model are shown in Figure 7.17. The sequential photographs of a real bending motion caused by the shifted sine waves with a cycle time of 3s are shown in Figure 7.18.

From the experimental results of both estimation and reconstruction, we can see that the estimated parameters match the real motions of the segmented IPMC. Our vision system and algorithms work properly for sensing a multi-DOF IPMC manipulator.

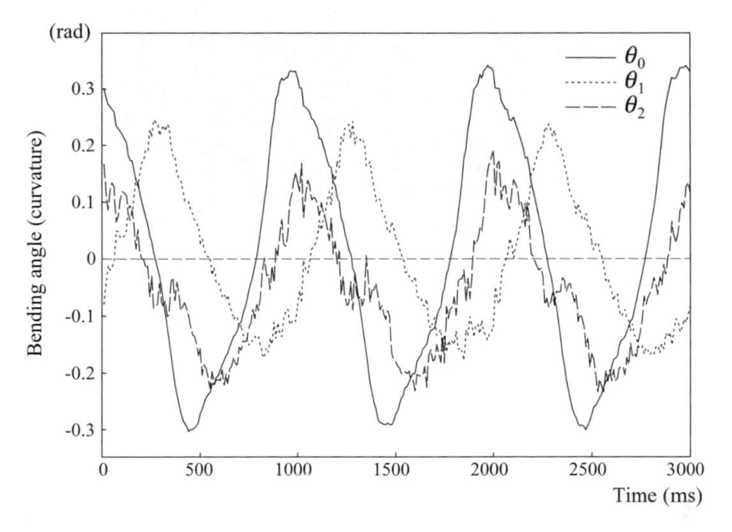


**Figure 7.14.** Results of estimation of curvature in a swing motion

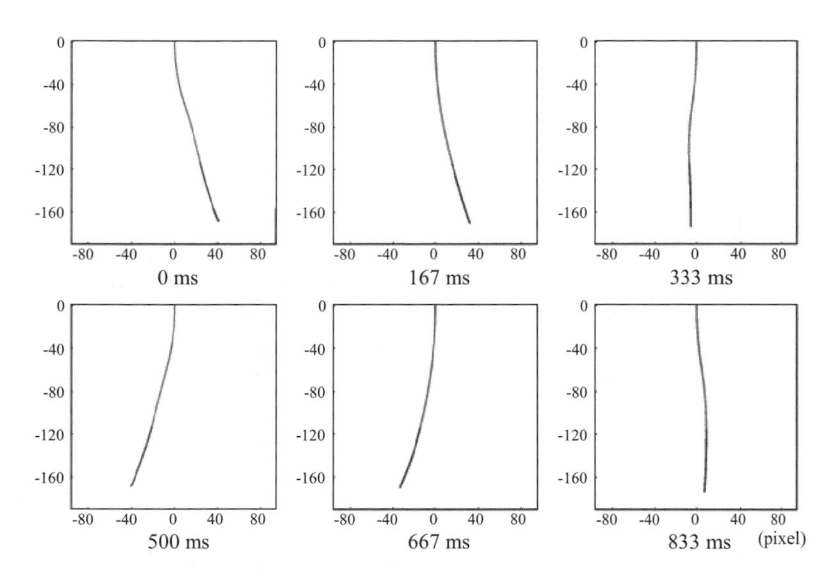


**Figure 7.15.** Link motions reconstructed by curvature data and the kinematic model in a swing motion (from left to right and top to bottom)

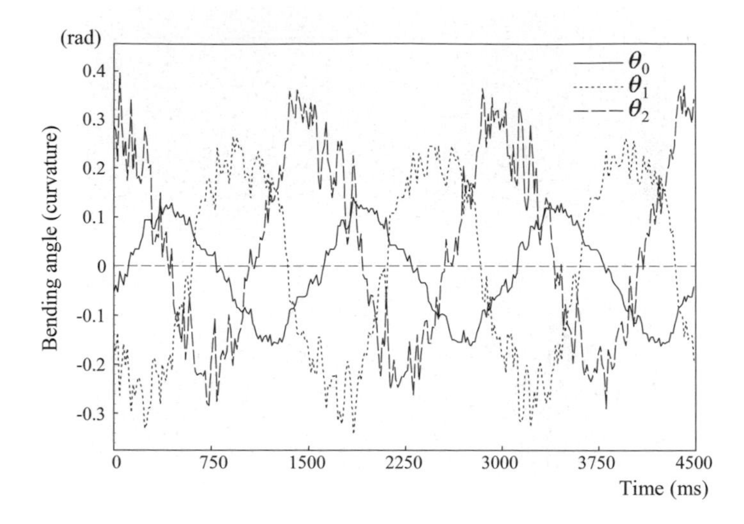


**Figure 7.16.** Results of estimating curvature in a snake-like motion

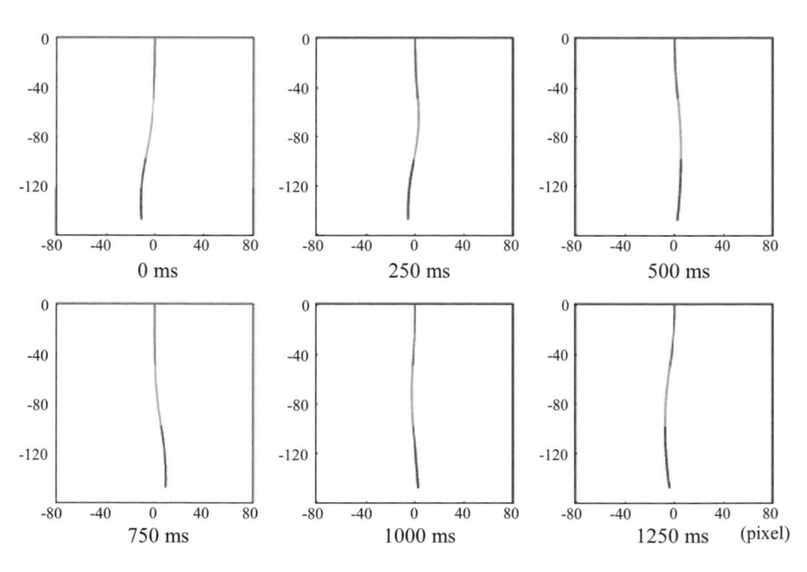


**Figure 7.17.** Link motions reconstructed by curvature data and the kinematic model in a snake-like motion (from left to right and top to bottom)

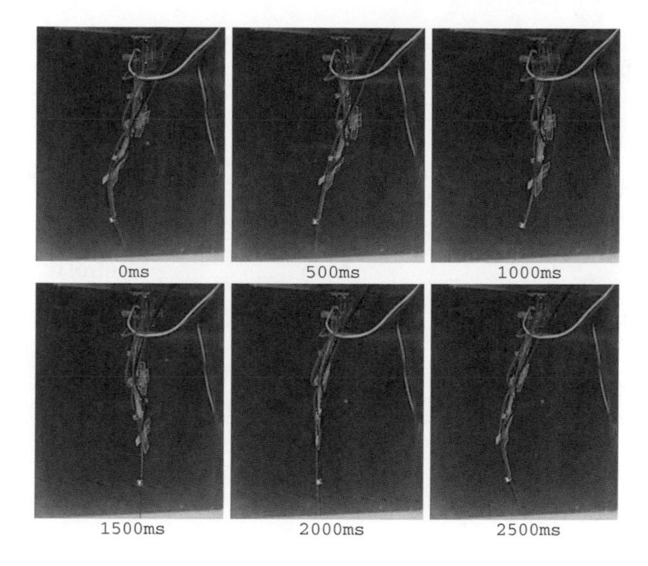


**Figure 7.18.** Snakelike bending motions caused by phase-shifted sine waves

# 7.7 Snake-Like Swimming Robot

## 7.7.1 Concept of a Snake-Like Swimming Robot Using IPMC

The underwater microrobot is also a novel and potential application of the IPMC. There are several studies on underwater propulsion robots using the IPMC. For example, Mojarrad and Shahinpoor developed a propulsion robot using the IPMC as a fin to generate a forward impelling force [11]. Guo *et al.* developed a fish-like microrobot that uses two actuators for right and left turning [5]. However, in these studies, a strip-shaped IPMC is used, which generates only a simple bending motion, thus a backward movement could not be achieved, although forward propulsion and a directional change have been realized. Not only that, such a simple bending motion has a lower efficiency than a snake-like wavy motion in propulsion, which has been found by the biological analysis of the swimming mechanism of fish or other living creatures. To realize more complex motions, such as multi-DOF motions, a patterned IPMC, each segment of which could be bent individually, is required.

Now, we are aiming at realizing a swimming robot with a patterned IPMC that bends like a snake. A snake-like motion sweeps a smaller area than a simple bending motion. Thus, it is suitable for future swimming robots in thin tubes, such as blood vessels, as shown in Figure 7.19. We input voltages as phase-shifted sinusoidal waves to each segment of the patterned IPMC, so that a progressive wave is generated as its bending motion. The system can control its impelling force and swimming directions left, right, forward, and backward. We can change amplitudes, frequencies, and phases of input waves to control the speed and direction of the robot propulsion. In this study, we describe the patterned IPMC and its propulsion by experiments.

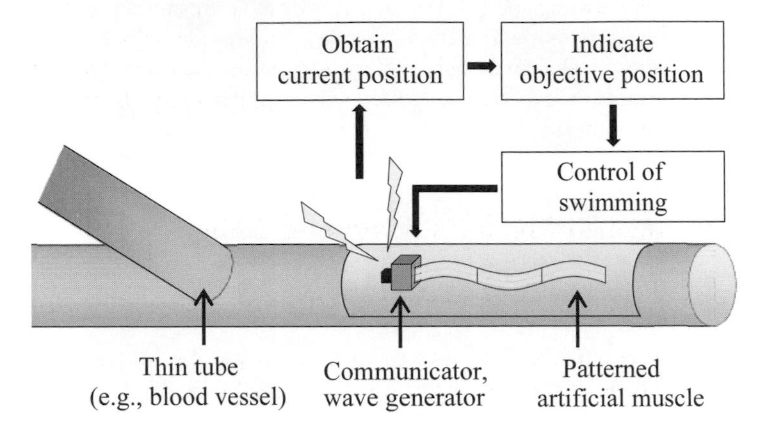


**Figure 7.19.** Concept of a snake-like swimming robot

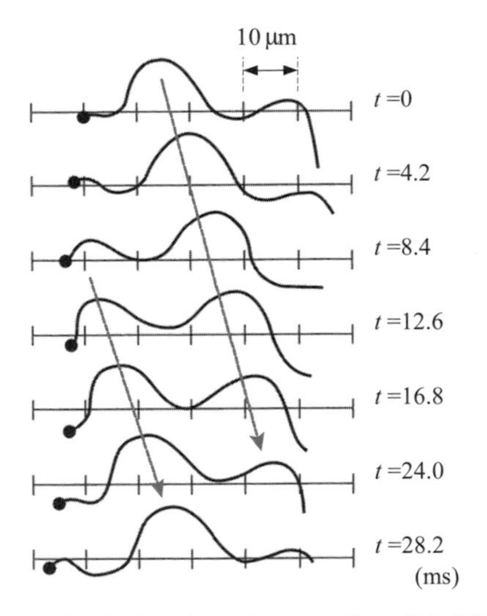


**Figure 7.20.** Swimming of sperm of starfish [7]

### 7.7.2 Natural Creatures Swimming in Water

There are many creatures swimming in water, among which eels, morays, and sea serpents can control their swimming directions forward and backward. Their bodies, called anguilliform, are slender or long and thin like a ribbon. The heights of their bodies are almost constant from head to tail. When they move forward, they generate a forward impelling force by sending progressive waves toward the rear along their bodies [20, 3]. Similarly, to move backward, they send progressive waves toward the front along their bodies. As an example of swimming by progressive waves, the swimming of the sperm of starfish is shown in Figure 7.20 [7].

The snake-like swimming of the anguilliform is most efficient in a high-Reynolds-number environment such as that on a microscale. It is also efficient for slender body fishes in a normal-scale underwater environment, including carps and gibels. A wavy motion is better at reducing an angular recoil from water than a simple bending motion [8].

## 7.8 Patterning of IPMC for a Swimming Robot

### 7.8.1 Comparing Single- and Multi-DOF Motions

Although many natural creatures use a wavy or snake-like motion, on the other hand, a single artificial muscle in the form of an IPMC cannot generate various motions on its own. The static form of an IPMC depends uniquely on an input voltage, indicating that it has only a single DOF. Figures 7.21 (a) and (b) show simulations of a simple swing motion of an IPMC and a multi-DOF motion of the IPMC controlling its three segments independently. These simulations are based on the kinematic model in which a constant curvature for each segment of the IPMC is assumed [14]. A sinusoidal input is given to the IPMC in (a) and 60° phase-shifted sinusoidal waves are given to each segment in (b). A wavy motion is realized in (b).

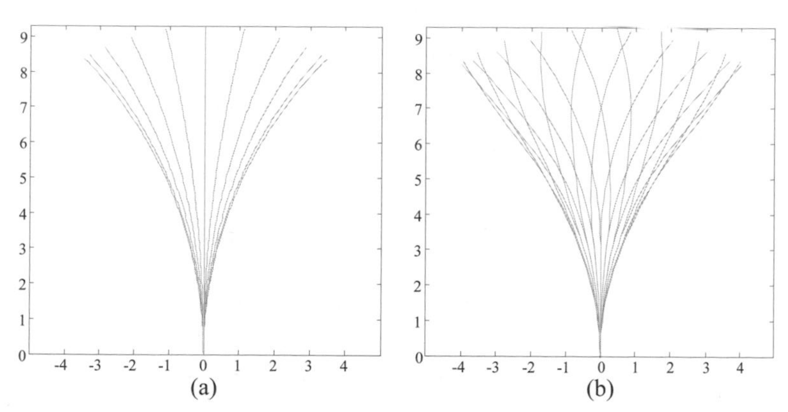


**Figure 7.21.** Simulations of bending motion: (a) 1-DOF simple bend and (b) 3-DOF snake-like bend

### 7.8.2 Patterning of IPMC for Snake-Like Swimming

In our previous study, we have developed a laser patterning method that enables electric insulation at a minimum gap width of 50 μm. However, in this study, to create a simple pattern such as that shown in Figure 7.22, a tentative method, which is the carving of the IPMC surfaces with a small hand chisel, is sufficient. We verified the electric insulation between segments.

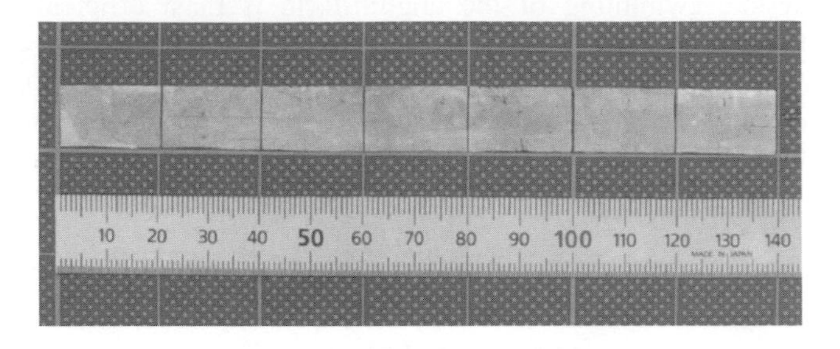


**Figure 7.22.** Patterned artificial muscle

The IPMC used in this study was a Nafion 117 membrane (by DuPont) five times chemically plated with gold. The thickness of this IPMC was 200 μm. This IPMC was electrically isolated to seven segments. To input voltages, we used seven connectors and electric wires touching each segment directly. They were small and sufficiently thin not to disturb the movement of this IPMC. We also used floats to prevent this IPMC from sinking (Figure 7.23).

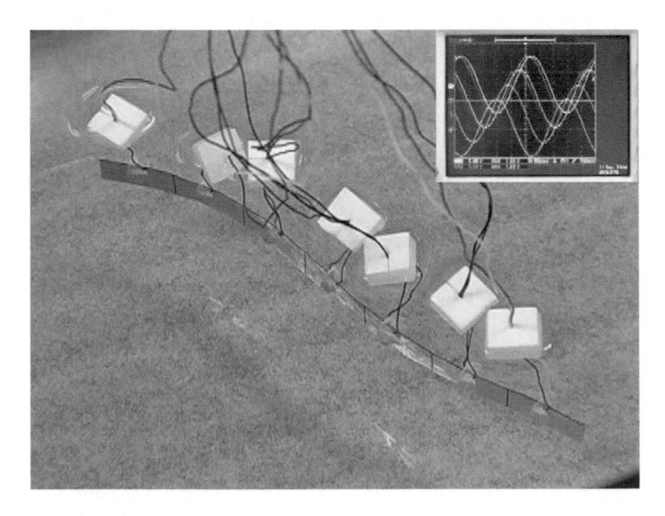

**Figure 7.23.** Snake-like swimming robot with patterned artificial muscle

## 7.9 Control System of Snake-Like Swimming Robot

A block diagram of the control system of the swimming robot is shown in Figure 7.24. Input waves are calculated by a PC and converted to voltage signals by a DA converter. Then, currents of signals are amplified and sent to corresponding segments of the IPMC. Control signals for latter experiments are composed of sinusoidal waves, whose amplitudes $A$V and frequencies $f$ Hz are constant, but their phases are delayed (0, $\theta$, $2\theta$,...)deg from segment 1 to segment 7, as shown in Figure 7.25.

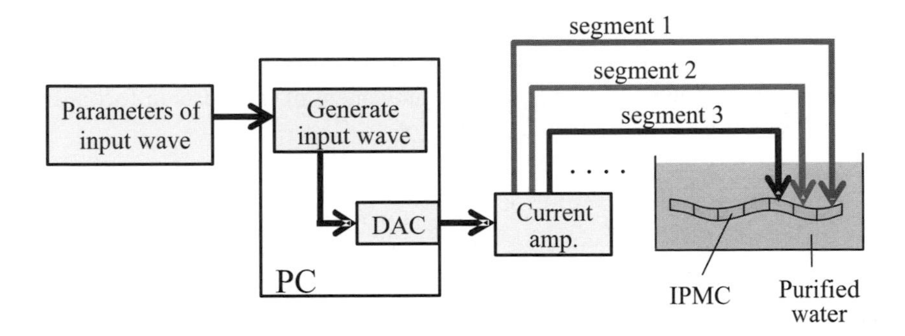


**Figure 7.24.** Block diagram of snake-like swimming robot

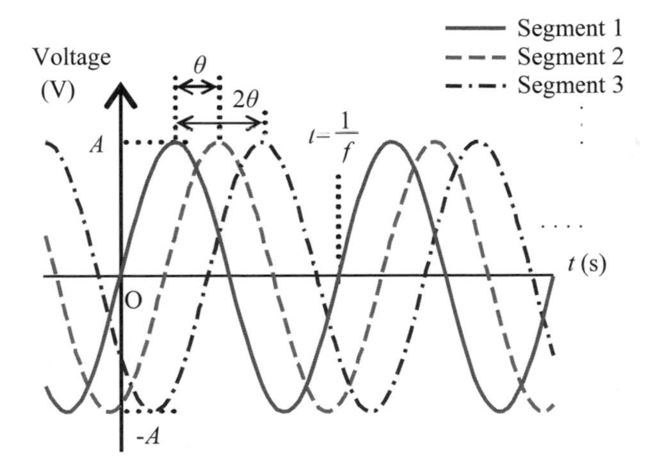


**Figure 7.25.** Input wave of snake-like bending motion

## 7.10 Experiments of Snake-Like Swimming Robot

### 7.10.1 Forward and Backward Propulsions

We first investigated the forward and backward propulsions by the snake-like bending motion of the patterned IPMC. An experimental system for measuring speed is shown in Figure 7.26. A long string and a long balancing rod cancel the obstruction of electric wires and allow free movement of the floats in forward and backward directions.

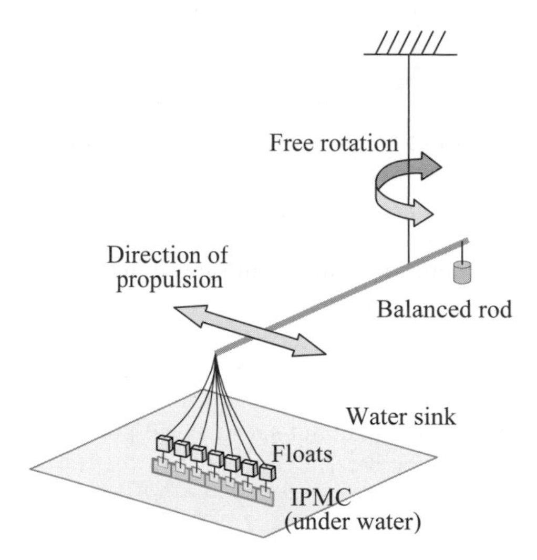


**Figure 7.26.** Experimental setup for measuring propelling speed

We searched for an optimal condition for propulsion by changing the frequency *f* and phase shift $\theta$. In our experiments, an amplitude of input waves is fixed at $A = 2\,\mathrm{V}$. Although a high voltage generates a large bending motion and a large impelling force, a voltage of more than 2V induces the electrolysis of water and this heat harms the IPMC.

Figure 7.27 shows the change in propelling speed with input frequency *f*, where error bars show standard deviations of the trials. Sine waves with the amplitude $A = 2\,\mathrm{V}$ and phase delay $\theta = 60°$ are input to each segment of the patterned IPMC from frequencies $f = 1$ to 10Hz. Speeds were calculated by measuring the forward or backward distance traveled in 10s. We obtained the maximum speed at the frequency *f* of 2Hz.

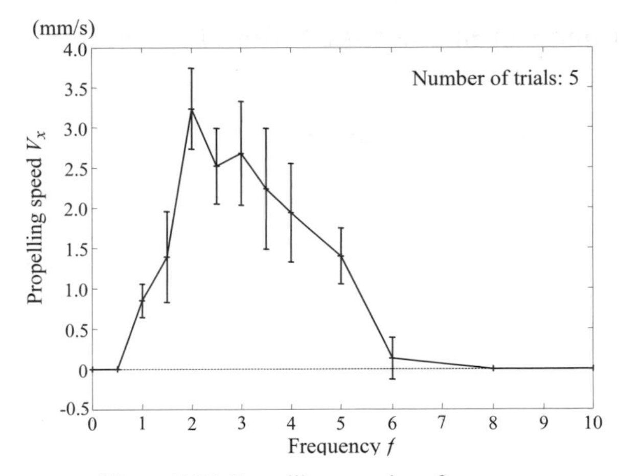


**Figure 7.27.** Propelling speed vs. frequency

We also investigated the change in propelling speed with phase shift. The phase delays are $\theta = -180°$ to $180°$, and the amplitude $A = 2$V and frequency $f = 2$Hz with which our IPMC produced the maximum speed in the first experiment.

Figure 7.28 shows the results of this experiment, where error bars show standard deviations of the trials. We successfully controlled the propelling direction of the IPMC forward or backward by changing the direction and speed of propagation of waves along the body by advancing and delaying the phases of the sine waves. We obtained the maximum speed at the phase shift $\theta$ of $\pm 60°$. Figure 7.29 shows the input voltage and current response of the IPMC. Figure 7.30 shows the bending motions of forward and backward propulsions.

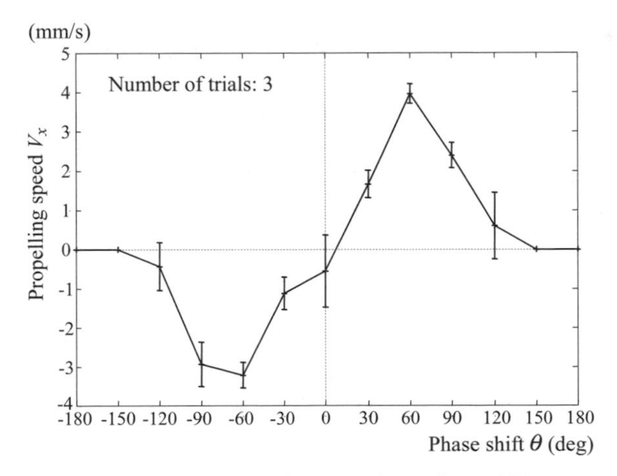


**Figure 7.28.** Propelling speed vs. phase shift

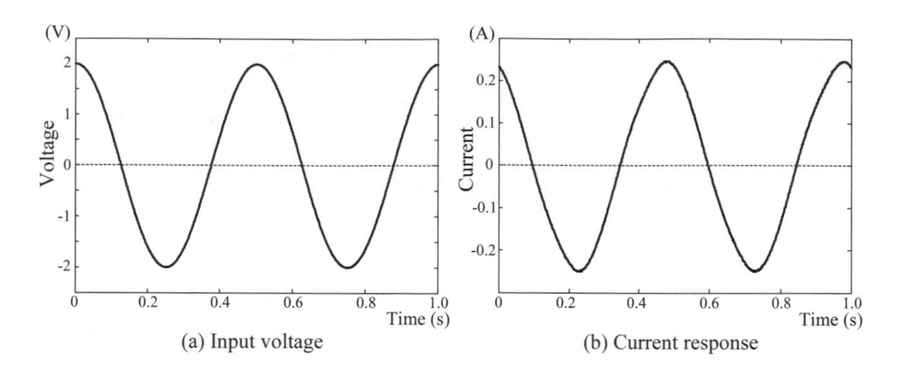


**Figure 7.29.** Input voltage and current response of forward propulsion

### 7.10.2 Right Turn and Left Turn

We next attempted the right turn and left turn of the patterned IPMC. In generating right and left directed biased forces for turning the robot, simply biased sine waves are not effective because nonlinear dynamics of bending responses of an IPMC cancels averagely biased input voltages in a few seconds. Instead of using the biased sine waves, we propose the use of a sawtooth-like waveform, whose rising and falling times of input voltage are not uniform, as shown in Figure 31. In this case, the frequencies during the rising and falling of the input waves are different. The first experiment suggests that the change in frequency induces the change in impelling force; thus, the direction of propulsion may be biased to the right or left by the proposed sawtooth-like waveform.

An experimental setup for measuring turning speed is shown in Figure 7.32. A long string allows free rotation of the robot in the right and left directions.

To evaluate the proposed sawtooth-like waveform, we conducted an experiment to investigate the turning speed of the IPMC by changing the ratio of the rising time to the falling time of the input waves ($T_1/T_2$) from 0.026 to 38.8 using a piecewise sinusoidal waveform, as shown in Figure 31. The amplitude, phase delay, and total frequency of the oscillating waves are fixed at $A$ = 2V, $\theta$ = 60° and $f$ = 2Hz with which the IPMC produced the maximum speed in the previous two experiments. The turning speeds are calculated by measuring the angle of turning right or left in 10s.

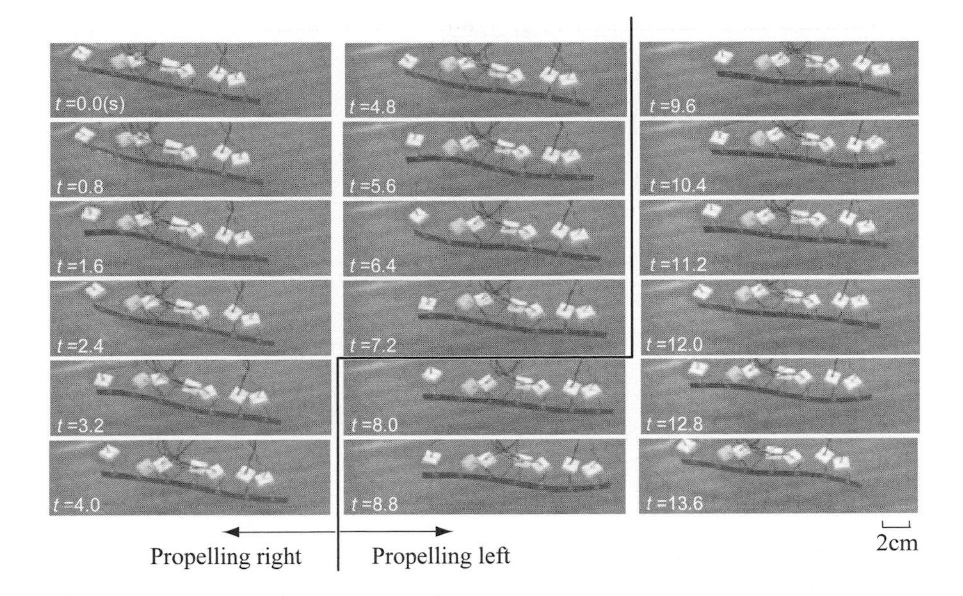


**Figure 7.30.** Forward and backward propulsions

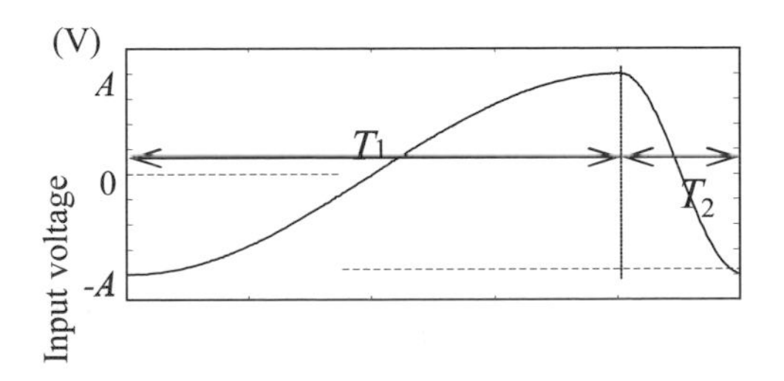


**Figure 7.31.** Sawtooth-like input waveform for turning

The results are shown in Figure 7.35, where error bars show standard deviations of the trials. From these results, we found that the robot can be turned right and left by changing the ratio of the rising time to the falling time of the proposed piecewise sinusoidal waveform. Figures 7.33 and 7.34 show the input voltages and current responses of the IPMC that induce the maximum left-turn and right-turn speeds of the robot. Figure 7.36 shows the bending motion of the IPMC for the right turn and left turn.

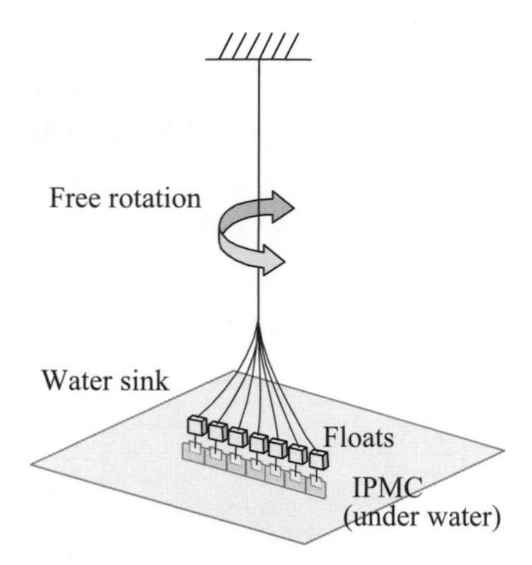


**Figure 7.32.** Experimental setup for measuring turning speed

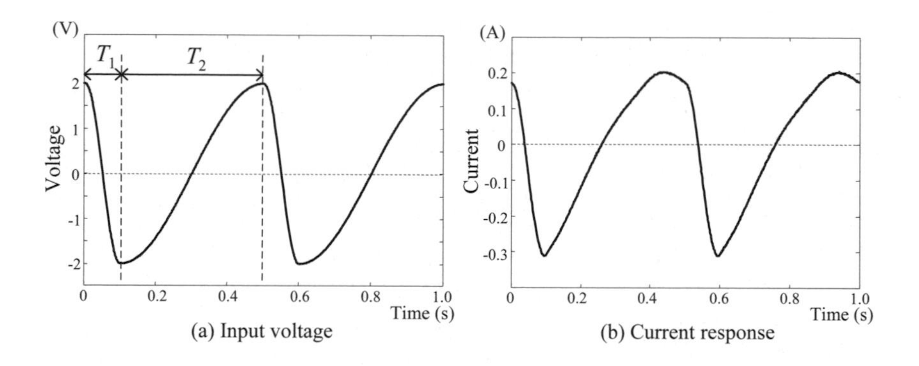


**Figure 7.33.** Input voltage and current response of left turn

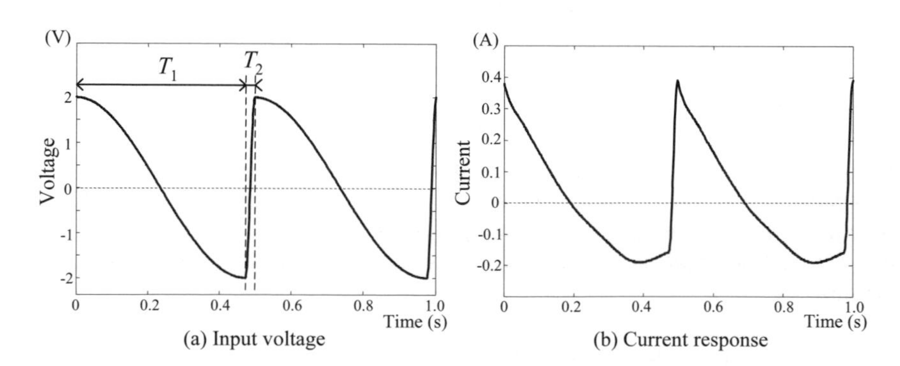


**Figure 7.34.** Input voltage and current response of right turn

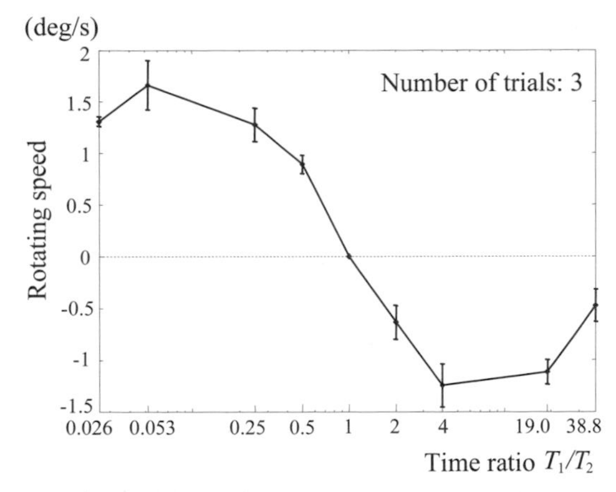


**Figure 7.35.** Rotating speed vs. time ratio

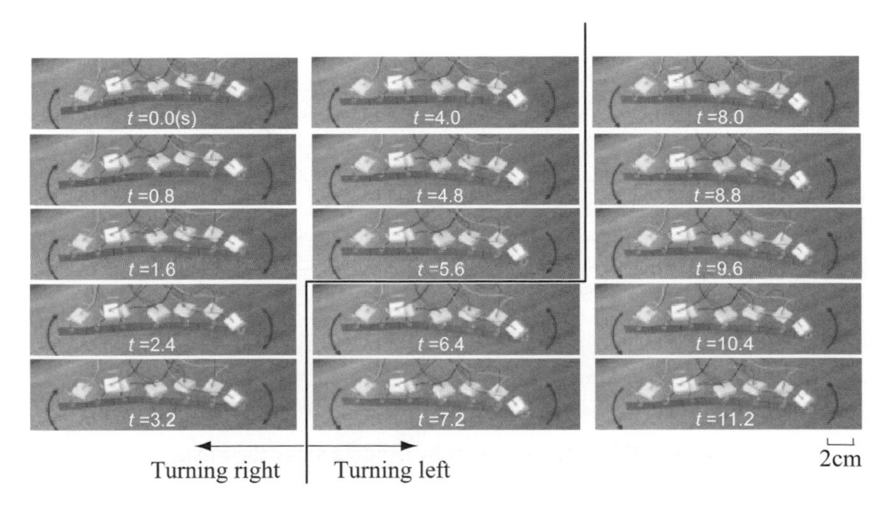


**Figure 7.36.** Right turn and left turn

# 7.11 Modeling of Propulsion and Turning

In this section, we present a propulsion model of the snake-like swimming motion to explain the experimental results of the propulsion and turning of the IPMC. We first consider the forward and backward propulsions and then the right turn and left turn based on a propulsion model for microorganisms [4] that swim with a snake-like motion.

## 7.11.1 Modeling of Forward and Backward Propulsions

We first estimate the propulsion force generated by the bending motion of the IPMC. The $x$ axis is aligned along the opposite direction of the forward propulsion of the IPMC, and the $z$ axis indicates the amplitude of the bending motion.

As shown in Figure 7.37, when a small part of an IPMC $\delta s$ moves along the $z$ direction with its speed of $V_z$, $\delta s$ receives a drag force $\delta L_z$ whose direction is tangent to $\delta s$, and a resistance force $\delta N_z$ whose direction is normal to $\delta s$ from a fluid (in this case, water). If $|\delta s| << 1$ and $V_z << 1$, we can assume that the drag and normal resistance forces are proportional to speeds relative to each direction. They can be calculated by

$$\delta L_z = C_L V_z \sin\theta\, \delta s$$
$$\delta N_z = C_N V_z \cos\theta\, \delta s$$

where $C_L$ and $C_N$ are the drag and normal resistance coefficients and $\theta$ is the tangent angle of $\delta s$.

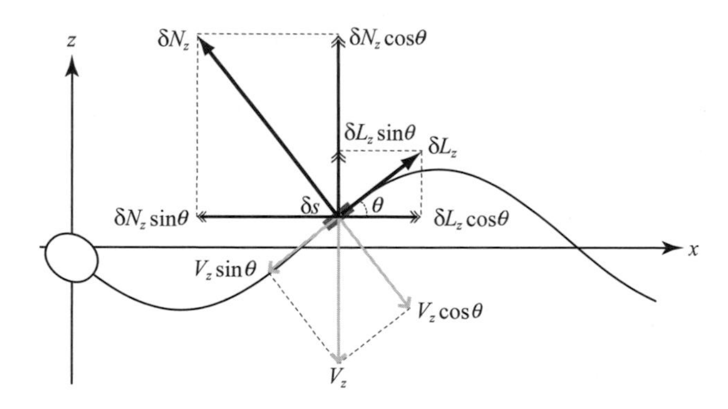


**Figure 7.37.** Force generated by bending along the $z$ axis

Similarly, the forces induced by the $x$ directed motion $V_x$ of $\delta s$ can be calculated by

$$\delta L_x = C_L V_x \cos\theta\, \delta s$$
$$\delta N_x = C_N V_x \sin\theta\, \delta s$$

as shown in Figure 7.38.

Thus, the total drag force $\delta L$ and resistance force $\delta N$ induced by $V_x$ and $V_z$ are calculated by

$$\begin{aligned} \delta L &= \delta L_z + \delta L_x \\ &= C_L (V_z \sin\theta + V_x \cos\theta) \delta s \end{aligned} \tag{7.8}$$

$$\begin{aligned}\delta N &= \delta N_z + \delta N_x \\ &= C_N\left(V_z \cos\theta - V_x \sin\theta\right)\delta s\end{aligned} \tag{7.9}$$

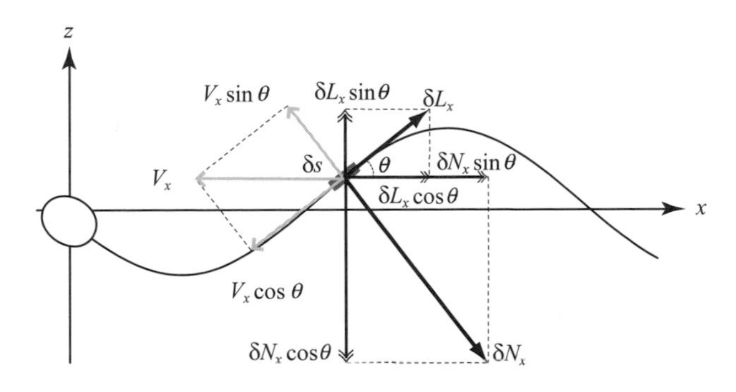


**Figure 7.38.** Force generated by bending along the $x$ axis

The $x$ directed force $\delta F_x$ applied by the fluid can be calculated by summarizing each $x$ element of $\delta L$ and $\delta N$ as

$$\begin{aligned}\delta F_x &= \delta N \sin\theta - \delta L \cos\theta \\ &= \left[\left(C_N - C_L\right)V_z \sin\theta\cos\theta - V_x\left(C_N \sin^2\theta + C_L \cos^2\theta\right)\right]\delta s \\ &= \frac{\left(C_N - C_L\right)V_z \tan\theta - V_x\left(C_L + C_N \tan^2\theta\right)}{1+\tan^2\theta}\delta s\end{aligned} \tag{7.10}$$

From this equation, we can see that the direction of propulsion is forward when $C_N > C_L$ at $V_x = 0$.

Now, let $z = z(x,t)$ be the position of the IPMC along the $x$ axis thus, $V_z = dz/dt$ and $\tan\,\theta = dz/dx$. Equation (7.10) can be rewritten as

$$\frac{dF_x}{ds} = \left\{\left(C_N - C_L\right)\frac{dz}{dt}\frac{dz}{dx} - V_x\left[C_L + C_N\left(\frac{dz}{dx}\right)^2\right]\right\} \Big/ \left[1+\left(\frac{dz}{dx}\right)^2\right] \tag{7.11}$$

When $\theta << 1$, $(dz/dx)^2 \cong 0$ and $ds \cong dx$; thus, (7.11) can be rewritten simply as

$$dF_x = \left[\left(C_N - C_L\right)\frac{dz}{dt}\frac{dz}{dx} - C_L\overline{V}_x\right]dx \tag{7.12}$$

where $\overline{V_x}$ is the average speed of the IPMC propulsion.

Finally, the total propelling force $F_x$ of the IPMC is calculated by

$$F_x = \int_{x=0}^{x=l} dF_x \tag{7.13}$$

where $l$ is the length of the IPMC.

### 7.11.2 Modeling of Right Turn and Left Turn

We now determine the moment of force generated by the bending motion of the IPMC. From Eqs. (7.8) and (7.9), the $z$ directed force $\delta F_z$ applied to $\delta s$ by the fluid is calculated by summarizing each $z$ element of $\delta L$ and $\delta N$ as

$$\begin{aligned}\delta F_z &= \delta N \cos\theta - \delta L \sin\theta \\ &= \left[V_z\left(C_N \cos^2\theta + C_L \sin^2\theta\right) - \left(C_N - C_L\right)V_x \sin\theta\cos\theta\right]\delta s \\ &= \frac{V_z\left(C_N + C_L \tan^2\theta\right) - \left(C_N - C_L\right)V_x \tan\theta}{1+\tan^2\theta}\delta s\end{aligned} \tag{7.14}$$

By substituting $V_z = dz/dt$ and tan $\theta = dz/dx$ to (7.14), the following equation is obtained.

$$\frac{dF_z}{ds} = \left\{\frac{dz}{dt}\left[C_N + C_L\left(\frac{dz}{dx}\right)^2\right] - V_x\left(C_N - C_L\right)\frac{dz}{dx}\right\} \Big/ \left[1+\left(\frac{dz}{dx}\right)^2\right] \tag{7.15}$$

By assuming $\theta << 1$, Eq. (7.15) can be rewritten simply as

$$dF_x = \left[C_N \frac{dz}{dt} - \overline{V_x}\left(C_N - C_L\right)\frac{dz}{dx}\right]dx \tag{7.16}$$

Now the force $F_z$, which is calculated by

$$F_z = \int_{x=0}^{x=l} dF_z \tag{7.17}$$

is not the moment of force but the propelling force directed to the $z$ axis. To calculate the moment of force $M$, the distance $r(x)$ from the $x$ position of $\delta s$ and the center of mass of the IPMC should be multiplied by $\delta F_z$ and integrated as

$$M = \int_{x=0}^{x=l} r(x) dF_z \tag{7.18}$$

## 7.12 Applying Propulsion Model

In the previous section, we presented the propulsion model of the swimming motion in which the position $z(x,t)$ of each point of the IPMC should be determined to calculate the propelling and turning forces of the robot. In this study, we extracted $z(x,t)$ from the real bending motion of the experiments and applied it to the propulsion model to calculate the propelling force.

Moreover, the drag resistance coefficient $C_L$ and normal resistance coefficient $C_N$ should be given in the model. However, we did not estimate them by ourselves and used only the values described in a previous study [9], in which the coefficients of drag resistance of a cylinder $C_L = 0.82$ and of normal resistance of a circular plate $C_N = 1.17$ were obtained. These values may not be so different from the true values, although precise measurements for real IPMC devices should be an important future work.

### 7.12.1 Extracting Bending Motion

To extract the position $z(x,t)$ of each point of the IPMC, we captured a sequence of images of the bending motion in our experiments and extracted the bending curve, as shown in Figure 7.39. We sampled 15 points with equal spaces on the extracted curved line, as shown in Figure 7.39(d).

These sets of points extracted from the image sequence are the sampling points of $z(x,t)$ and were used to calculate the differential $dz/dx$ and $dz/dt$ in the propulsion model. Figure 7.40 shows an example of the extracted $z(x,t)$ and calculated differential $dz/dx$ and $dz/dt$ from the corresponding $z(x,t)$.

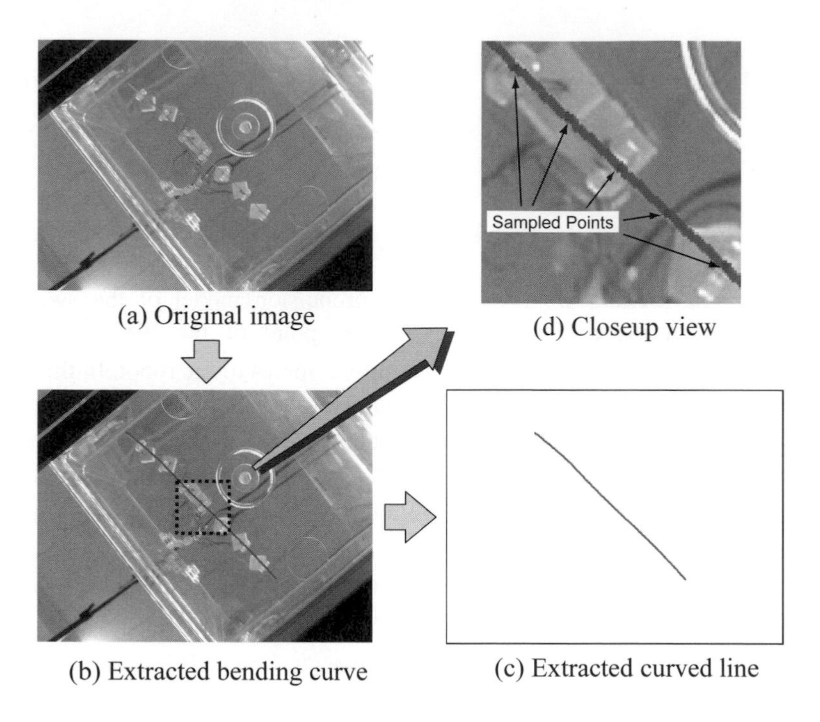


**Figure 7.39.** Extracted curved line from image

### 7.12.2 Applying Model on Forward and Backward Propulsions

We first calculated the propelling force $F_x$ by (7.13) using $z(x,t)$ extracted from the experiments of forward and backward propulsions. This $F_x$ varies with a change in average propelling speed $\overline{V}_x$, as shown in Figure 7.41.

Generally, when a robot is propelling at a constant speed, the propelling force $F_x$ is zero. In this case, if $F_x = 0$ in the result of the analysis shown in Figure 41, the forward and backward constant speeds are $V_x = -5.3\times10^{-4}$m/s and $V_x = 9.8 \times 10^{-4}$m/s, respectively. On the other hand, the true forward and backward speeds measured in the previous experiment are $V_x = -4.2 \times 10^{-3}$m/s and $V_x = 3.5 \times 10^{-3}$m/s, respectively. These differences between the speeds of the real IPMC and the speeds analyzed from the model can be induced by the force that is not proportional to the speed and not considered in this model.

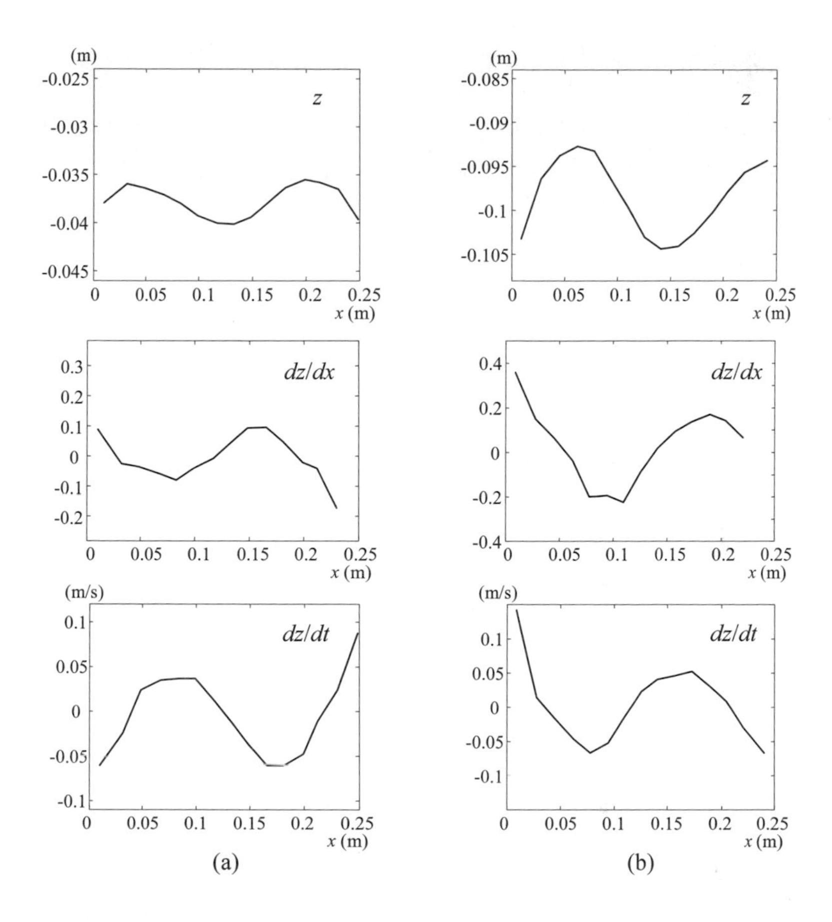


**Figure 7.40.** Examples of extracted plots of (a) forward and (b) backward propulsions

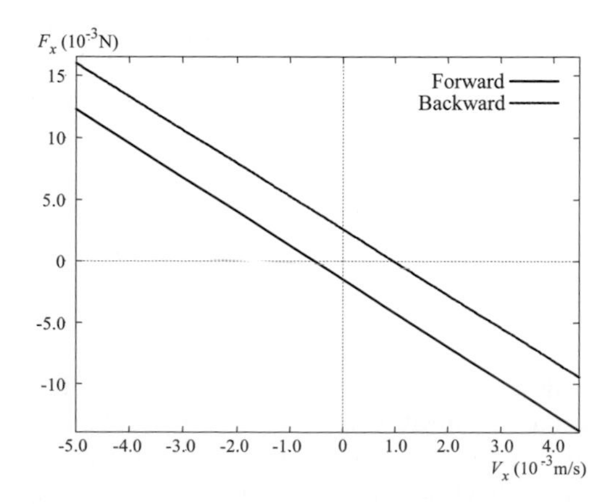


**Figure 7.41.** Propulsion force vs. propelling speed

### 7.12.3 Applying Model to Right Turn and Left Turn

We next calculated the moment of force $M$ by (7.18) using $z(x,t)$ extracted from the experiments of the right turn and left turn of the IPMC. This $M$ varies with a change in average propelling speed $\overline{V}_x$, as shown in Figure 7.42.

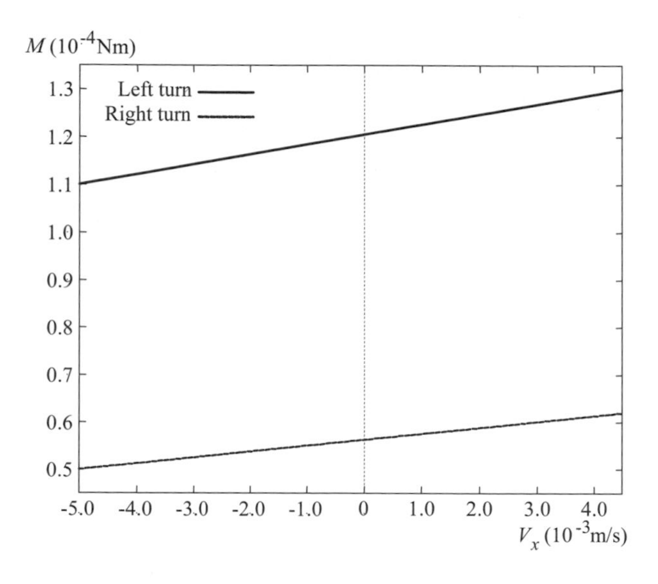


**Figure 7.42.** Moment of force vs. propelling speed

By fixing $V_x$ as zero, the moments of force $M = 1.2 \times 10^{-4}$ Nm for the left turn and $M = 5.6 \times 10^{-5}$ Nm for the right turn are obtained from Figure 7.42. These values indicate the moments of force for the left turn on both. Although this is not true, comparing the relative differences between the two values obtained, the moment of force for the left turn is larger than that for the right turn, which is considered a reasonable result. These biased results can be induced by an error in the estimation of the moment center of the IPMC. In the analysis, we calculated only the average position of the 15 sampled points as the moment center, which may not be accurate. Moreover, force that is not proportional to speed can be the other reason for this error.

## 7.13 Conclusions

In this study, we described our new type of multi-DOF manipulator using a patterned artificial muscle (IPMC), and proposed the kinematic modeling of multi-link motions for the control of the manipulator using an inverse Jacobian or a transposed Jacobian with visual sensing. Simulations and experiments verify our approach. Our future work should address the problem of the low conductivity of segment interconnections. We can increase the width of the interconnections and/or increase the amount of gold plating, so that their conductive layers become thicker. Another issue is the dynamic aspect of the bending motions of the IPMC membrane. Although bending velocity generally follows a given voltage signal, nonlinear characteristics, such as swing back motion, also contribute to a dynamic response. Extensive work has been carried out to identify and model the dynamics. In the future, we will examine these models and realize dynamic control of an IPMC manipulator.

We have also developed a biomimetic snake-like swimming robot using an IPMC. We demonstrated that the swimming speed and direction can be controlled using a patterned IPMC. The IPMC has segments, each of which can be controlled individually. By inputting sine waves with different phases, we can make progressive waves move along the body of the IPMC, and these waves induce an impelling force. We also proposed an analytical model for the robot propulsion considering the drag and resistance forces induced by the bending motion of the IPMC. We applied the results of the experiments to the proposed model and estimated the propelling forces of the robot propulsion. The following are our future plans: (1) make an IPMC that can generate a stronger impelling force and swim faster and (2) mount a battery and a controlling circuit on the IPMC, instead of connecting them to a host PC.

## 7.14 References


[1] K. Asaka and K. Oguro (2000) Bending of polyelectrolyte membrane platinum composites by electric stimuli Part II. Response kinetics. J. of Electroanalytical Chemistry, 480:186–198.

[2] Y. Bar-Cohen (2002) Electro-active polymers: current capabilities and challenges. Proc. of SPIE Int. Symp. on Smart Structures and Materials, EAPAD

[3] J. Gray (1957) The movement of the spermatozoa of the Bull. J. of Experimental Biology, 35(1):97–111.

[4] J. Gray and G. J. Hancock (1955) The propulsion of sea-urchin spermatozoa. J. of Experimental Biology, 32:802–814.

[5] S. Guo, T. Fukuda, and K. Asaka (2003) A New Type of Fish-Like Underwater Microrobot. IEEE/ASME Trans. on Mechatronics, 8(1):136–141.

[6] S. Guo, T. Fukuda, K. Kousuge, F. Arai, K. Oguro, and M. Negoro (1995) Micro catheter system with active guide wire. Proc. IEEE Int. Conf. on Robotics and Automation, 79–84.

[7] Y. Hiramoto (1979) Flagellar movements. J. of the Japan Society of Mechanical Engineers, 82(732):1003–1007. (in Japanese)

[8] M.J. Lighthill (1960) Note on the swimming of slender fish. J. Fluid Mechanics, 9:305–317.

[9] M. Makino (1991) Fluid resistance and streamline - design of vehicle shape from viewpoint of hydrodynamics -. Sangyo-tosho press. (in Japanese)

[10] K. Mallavarapu and D. J. Leo (2001) Feedback control of the bending response of ionic polymer actuators. J. of Intelligent Material Systems and Structures, 12:143–155.

[11] M. Mojarrad and M. Shahinpoor (1997) Biomimetic robotic propulsion using polymeric artificial muscles. Proc. IEEE Int. Conf. on Robotics and Automation, 2152–2157.

[12] Y. Nakabo, M. Ishikawa, H. Toyoda, and S. Mizuno (2000) 1 ms column parallel vision system and its application of high speed target tracking. Proc. IEEE Int. Conf. on Robotics and Automation, 650–655.
[13] Y. Nakabo, T. Mukai, and K. Asaka (2004) A multi-DOF robot manipulator with a patterned artificial muscle. The 2nd Conf. on Artificial Muscles. Osaka
[14] Y. Nakabo, T. Mukai, and K. Asaka (2005) Kinematic modeling and visual sensing of multi-DOF robot manipulator with patterned artificial muscle. Proc. IEEE Int. Conf. Robotics and Automation, 4326–4331.
[15] Y. Nakabo, T. Mukai, K. Ogawa, N. Ohnishi, and K. Asaka (2004) Biomimetic soft robot using artificial muscle. IEEE/RSJ Int. Conf. Intelligent Robots and Systems, in tutorial, WTP3 Electro-Active Polymer for Use in Robotics.
[16] K. Ogawa, Y. Nakabo, T. Mukai, K. Asaka, and N. Ohnishi (2004) A snake-like swimming artificial muscle. The 2nd Conf. on Artificial Muscles. Osaka
[17] K. Ogawa, Y. Nakabo, T. Mukai, K. Asaka, and N. Ohnishi (2005) Snakelike swimming artificial muscle. Video Proc. IEEE Int. Conf. Robotics and Automation.
[18] K. Oguro, Y. Kawami, and H. Takenaka (1992) Bending of an ion-conducting polymer film-electrode composite by an electric stimulus at low voltage. J. of Micromachine Society, 5:27–30. (in Japanese)
[19] S. Tadokoro, S. Yamagami, M. Ozawa, T. Kimura, and T. Takamori (1999) Multi-DOF device for soft micromanipulation consisting of soft gel actuator elements. Proc. of IEEE Int. Conf. on Robotics and Automation, 2177–2182.
[20] T.Y. Wu (1961) Swimming of waving plate. J. of Fluid Mechanics, 10:321–344.

# 8

# Robotic Application of IPMC Actuators with Redoping Capability

M. Yamakita[1, 2], N. Kamamichi[2], Z. W. Luo[3, 2], K. Asaka[4, 2]

[1] Department of Mechanical and Control Engineering, Tokyo Institute of Technology
2-12-1 Oh-okayama, Meguro-ku, Tokyo, 152-8552, Japan
yamakita@ctrl.titech.ac.jp

[2] Bio-Mimetic Control Research Center, RIKEN
2271-130 Anagahora, Shimoshidami, Moriyama-ku, Nagoya, 463-0003, Japan
nkama@bmc.riken.jp

[3] Department of Computer and Systems Engineering, Kobe University
1-1 Rokkodai, Nada, Kobe, 657-8501, Japan
luo@gold.kobe-u.ac.jp

[4] Research Institute for Cell Engineering, AIST
1-8-31 Midorigaoka, Ikeda, Osaka, 563-8577, Japan
asaka-kinji@aist.go.jp

## 8.1 Introduction

Machines and robots have big impacts on our life and industry to realize high-speed, high-power, and high-precision motion; however, recently other factors are demanded, *e.g.*, miniaturization or flexibility. For robots working in ordinary human life, it is desired to use safe and soft actuators, which are sometimes called artificial muscle. A high polymer gel actuator is one of the candidates for artificial muscle actuators due to their softness and miniaturizability.

For several decades, electroactive polymers (EAP) 0, which respond to electric stimuli with shape change, received little attention because of their actuating limitations. During the last ten years, development of EAP materials with large displacement and quick response changed the potential capability, and EAP received much attention from engineers and researchers in many disciplines, *e.g.*, robotics, medical service, and the toy industry.

The ionic polymer-metal composite (IPMC) is one of the most promising EAP actuators for applications. IPMC is produced by chemically plating gold or platinum on a perfluorosulfonic acid membrane which is known as an ion-exchange membrane. When an input voltage is applied to the metal layers of both surfaces, they bend at high speed. The phenomenon of this motion was discovered by Oguro *et al.* in 1992 0. The characteristics of an IPMC are as follows:

Driving voltage is low (1~2 V).
Speed of response is fast (> 100 Hz).

It is durable and stable chemically. (It is possible to bend more than $10^6$ times.)
It is a flexible material.
It moves in water and in wet conditions.
Miniaturization and weight saving is possible.
It is silent.
It can be used as a sensor.

By exploiting the characteristics, IPMC actuators have been applied to robotic applications such as an active catheter [3,4], a fish-type underwater robot [4~10], a wiper for a nanorover 0, a micropump 0, a micromanipulator 0, and a distributed actuation device 0.

It, however, also has disadvantages that the actuation force is still small and that the input voltage is restricted to the range where electrolysis of the ionic polymer does not occur. To improve performance, development of the ionic polymer membrane and plating method are required.

IPMC actuators also have another noteworthy property; the characteristics of bending motion depend highly on counterions. In application to mechanical systems such as robots, the possibilities exist to change the properties of the dynamics adequately according to the environment or purpose. We have called this property the “doping effect”, and verified the effect on robotic applications.

The goal of our study is applying an artificial muscle actuator to robotic applications especially to a bipedal walking robot, and we developed a linear actuator using IPMC. The structure of our proposed actuator is very simple, and the actuator transforms bending motion into linear motion. We assume that elementary units are connected in parallel and series to realize the desired displacement and force. In this paper, we describe the structure of the actuator and identify an empirical model of the actuator. Numerical simulations of bipedal walking are demonstrated, and the doping effect is investigated by the walking simulation and an experiment using a snakelike robot. Finally, control of doping speed by exercise is also considered.

## 8.2 Proposed IPMC Linear Actuator

In this section, the structure and basic properties of the proposed IPMC linear actuator will be explained 0.

### 8.2.1 Structure of IPMC Linear Actuator

The proposed linear actuator is composed of many basic units connected in parallel and series so that enough force and displacement can be obtained. The structure of the elementary unit is shown in Figure 8.1. This elementary unit consists of four IPMC films. One side of the unit is formed from a pair of films that are connected by a flexible material or the same thin film. When an input voltage is applied to electrodes on the surface with the anode outside, each membrane bends outside, then the actuator is constricted. The actuation force and displacement of each unit are small; however, the elementary units can be connected in parallel and series as in Figure 8.2, so the actuator can realize the desired force and displacement. By shifting the series of elementary units by a half pitch to avoid interference as in Figure 8.2, the total actuator is made compact, and high power/volume and miniaturization are realized.

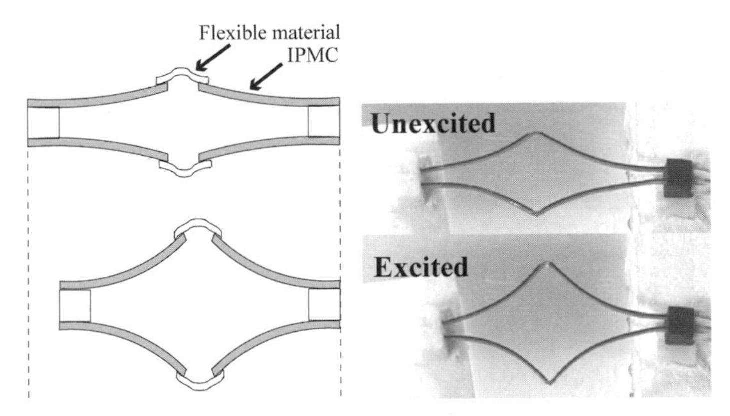


**Figure 8.1.** Structure of IPMC actuator

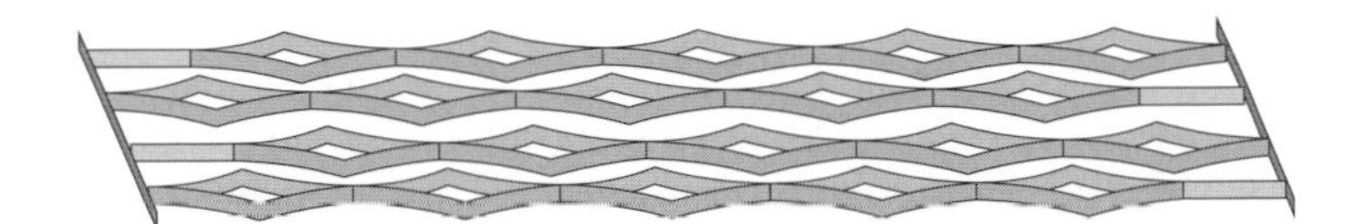

**Figure 8.2.** Basic concept of IPMC linear actuator

### 8.2.2 Properties of the Elementary Unit

To check the characteristics of the actuator, we carried out fundamental experiments. Figure 8.3 shows the experimental setup. In this experiment, one edge of the actuator is fixed on a board floating on water to reduce the effects of the weight of electrodes and ties. Displacement of the linear actuator was measured by a laser displacement meter.

*8.2.2.1 Response in Step Voltage*

Figure 8.4 shows the response in a step voltage without loads, where step input voltages of 1.5, 2.0, and 2.5 V were applied at 0 s. The IPMC film which we used in this experiment is Nafion®117 (DuPont) plated with gold. A counterion doped in the film is $Na^+$. Though the response of the actuator varies depending on its condition, it was confirmed that the unit whose total length is 40 mm is constricted by 10 mm with a step input voltage of 2.5 V in average. As the applied voltage is increased, the peak value of the displacement is also increased.

In this experiment, it is observed that when a step voltage is applied, the IPMC membrane bends toward the anode side quickly and bends back gradually. The characteristic varies according to the counterion, as mentioned below. It was also observed that the current increased sharply at the moment when the input voltage was applied, and then it decreased exponentially.

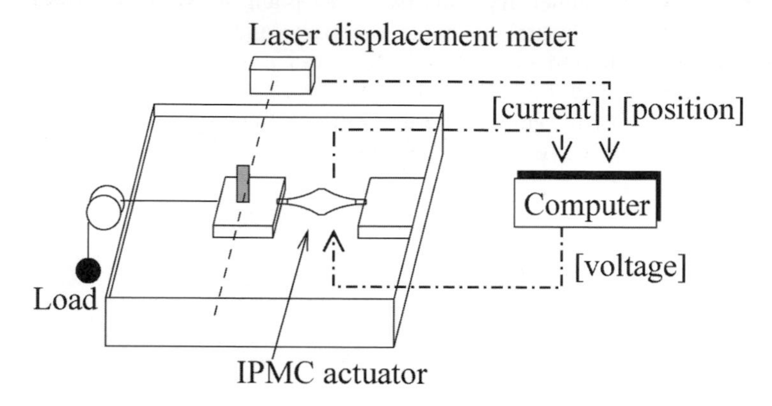


**Figure 8.3.** Experimental setup

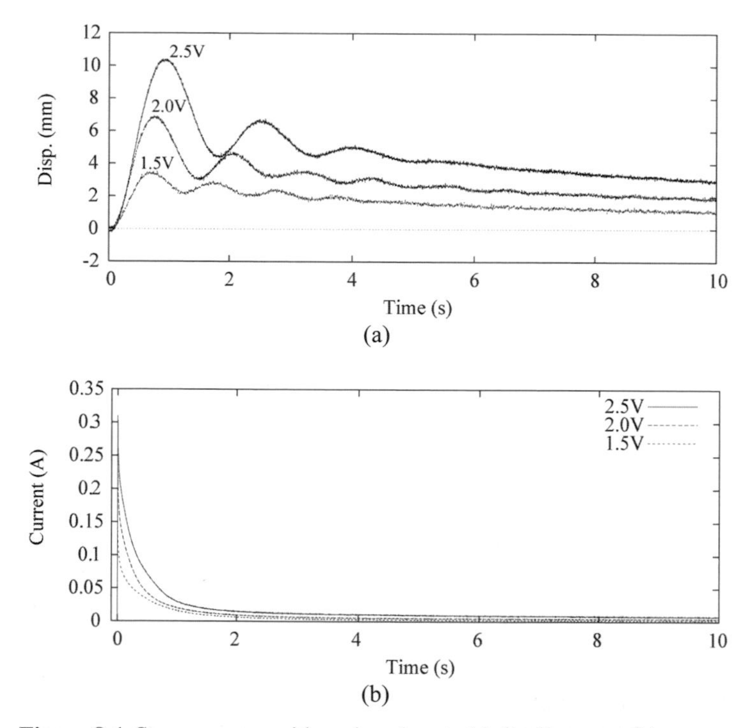


**Figure 8.4.** Step response with various inputs (a) displacement (b) current

*8.2.2.2 Response with Loads*

Figure 8.5 shows the response in step voltage with loads; a step input voltage of 2.5 V was applied in loading. As the load is increased, displacement becomes small but the current almost does not change.

Note that to avoid damage of IPMC actuators by electrolysis, the control input voltage is limited to about 3.0 V. So it is not so effective to change the dynamic properties by changing the control voltage. In the following, changing the dynamic properties chemically is considered by changing the doped ions in the film.

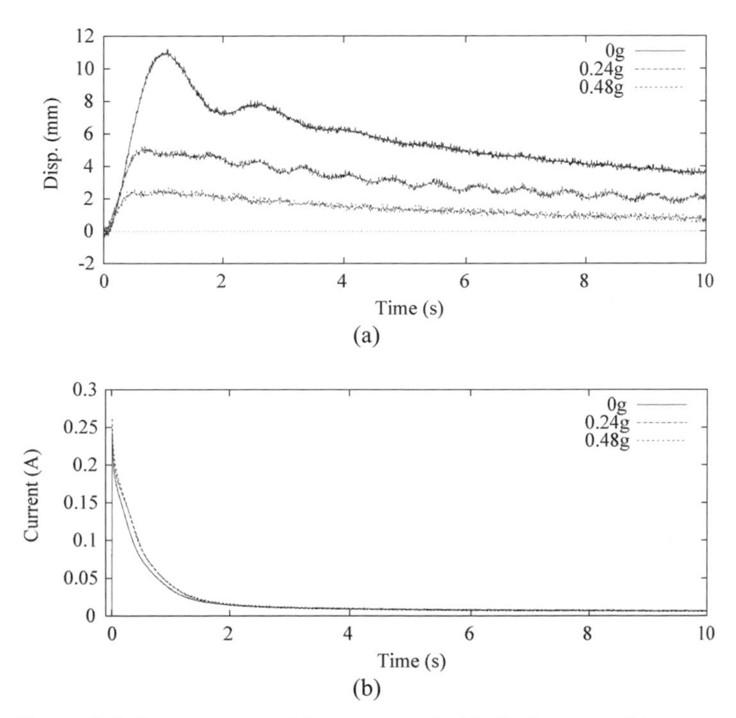


**Figure 8.5.** Step response with various loads (a) displacement (b) current

*8.2.2.3 Responses with Different Doped Counterions*

It is known that IPMC changes bending characteristics with respect to doped counterions 0. Figure 8.6 shows the responses of actuators for the same input voltage of 2.5 V, which are doped with sodium ($Na^+$), cesium ($Cs^+$), and tetraethylammonium ($TEA^+$) as the counterion, respectively. From the figure, it is observed that the raising time of the unit with $Na^+$ is shorter than that with $Cs^+$ and the rising time of the unit with $TEA^+$ is largest. On the other hand, the tendency of the response to decay is large for the unit with $Na^+$ or $Cs^+$, but it is very small for the unit with $TEA^+$. The doping of the counterion is easily done by just putting the unit in a solution containing the target counterion, and higher condensed counterions are doped into IPMC films. Also, the change of the doped ion is reversible. The property suggests that the characteristics of the actuator can be changed for specific purposes. In Section 8.4.3, we will show the effect of doping in the walking pattern and efficiency of a small bipedal robot, and in Section 8.5 we will show experimental results of doping on a snakelike robot. In 0, it was discussed how to change equivalent characteristics of the actuator mechanically.

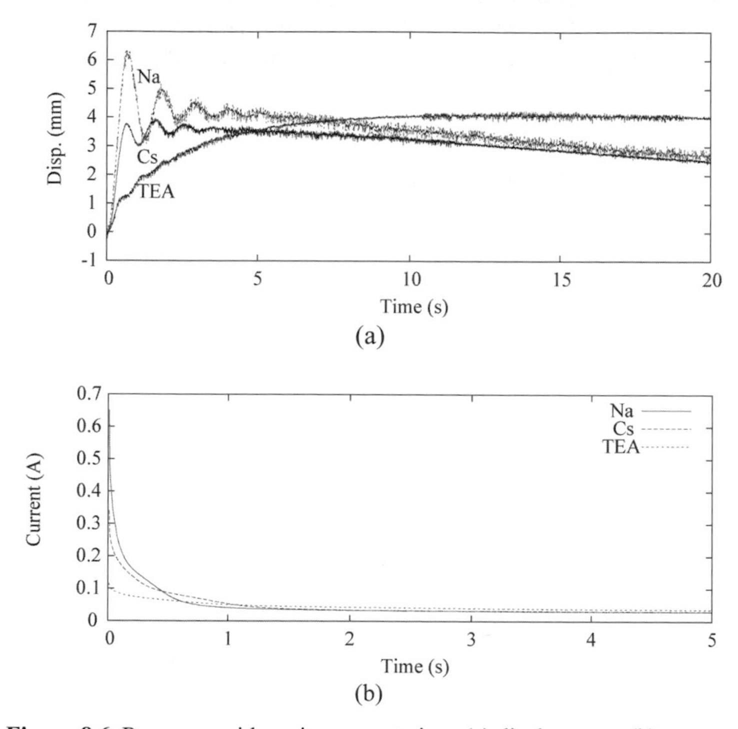


**Figure 8.6.** Response with various counterions (a) displacement (b) current

## 8.3 Model Identification

To know the capability of the linear actuator for a robotic system, we identify the linear actuator as a mathematical model. IPMC actuators have been modeled in various ways as a black or gray box model 0, and a detailed model based the physical and chemical phenomena [19~21]; however, it is difficult to represent these models by systems of ordinary differential equations because of their complexity. In this paper, in consideration of model-based control, we use a gray box model which has two inputs, *i.e.*, control voltage and external force. We identify the actuator as a linear time-invariant model with static nonlinearity from input-output data using a subspace identification algorithm [15,22].

### 8.3.1 Identification Method

First, the model of the actuator is assumed to be represented by the system in Figure 8.7. This model has two inputs and one output, and it consists of two subsystems, $P_1(s)$ and $P_2(s)$, which are connected in series. $P_1(s)$ is a system with one input $v$, input voltage, and one output $f_1$ which is a force generated by electric stimuli. $P_2(s)$ is a system with one input force $f_2$ which is exerted on the actuator, and one output y, displacement of actuator. $f_2$ is assumed to be the difference between $f_1$ and $f_l$. $P(s)$ is defined as $P_2(s)\ P_1(s)$.

For the moment, it is assumed that the actuator is driven in a small operating range, and the dynamics is identified as an LTI model as follows:

Identification of $P(s)$: Measure a response from input voltage $v$ to displacement $y$; then compute the system $P(s)=P_2(s)P_1(s)$ from input-output data using a subspace identification algorithm.

Identification of $P_2(s)$: Measure a response from load $f_l$ to displacement $y$; then compute the system $P_2(s)$ from input-output data using a subspace identification algorithm.

Computation of $P_1(s)$: Compute the system $P_1(s)$ as $P_2(s)^{-1}P(s)$.

In procedures 1 and 2, we performed the system identification using the N4SID function in MATLAB, and from the discrete time model obtained, the corresponding continuous model was determined.

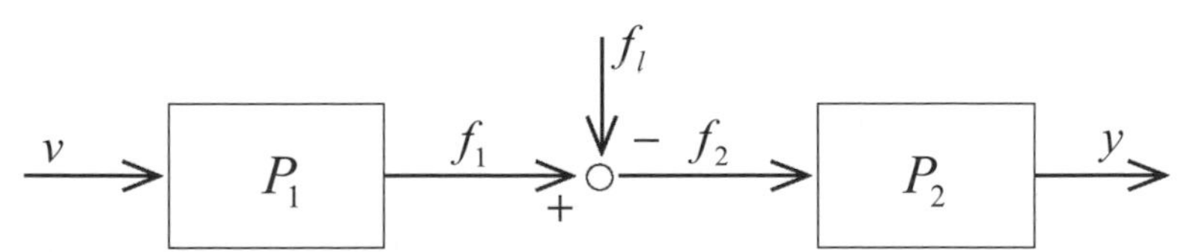


**Figure 8.7.** Block diagram

### 8.3.2 Identification Results

We obtained estimated transfer functions of the systems as

$$
\begin{aligned}
P_1(s) &= -\frac{1.50\times10^{-3}s^2+1.09\times10^{-2}s+3.93\times10^{-2}}{s^3+6.13s^2+3.23\times10s+7.12\times10} \\
P_2(s) &= -\frac{3.49\times10^3s^2+1.23\times10^6s+3.81\times10^6}{s^4+7.19s^3+6.49\times10^4s^2+4.14\times10^5s+1.33\times10^6}
\end{aligned}
\tag{8.1}
$$

If the relative degree of $P_2(s)$ is 2, then it is easy to simulate an impulsive effect of collision. In procedure 2, the relative degree of $P_2(s)$ was estimated as 1, but the coefficient of the highest order term in the numerator of $P_2(s)$ was much smaller than the other. Thus we eliminated it, and obtained the transfer function whose relative degree is 2. In procedure 3, small coefficients of $P_1(s)$ ware also eliminated.

Figure 8.8 shows a comparison between the experimental result and the simulation result using the identification model. It is shown that a quick transient of simulation result is nearly equal to the experimental one. But in part of the slow decay, there is a little error between them. As the results of identification with a linear approximate model, the characteristics of the actuator are captured.

When the control voltage and external load are given simultaneously, as in Figure 8.9, we can observe the large error between them. The reason for the error can be inferred as nonlinear effects due to the large deformation of the structure of the unit. So we should consider other models to deal with such a large deformation and the nonlinearity.

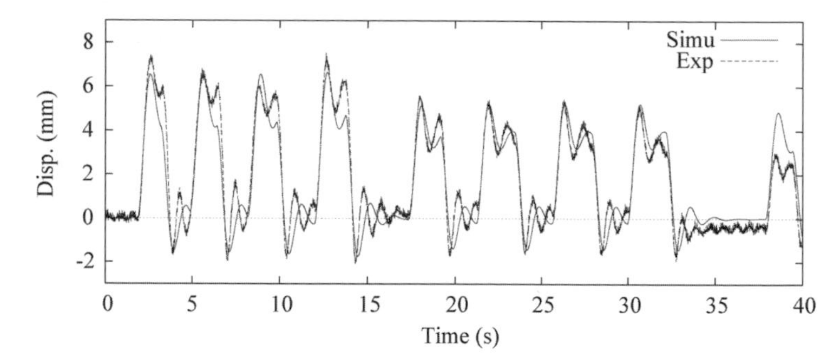


**Figure 8.8.** Identification result 1

### 8.3.3 Introduction of Nonlinear Effect

To reduce the error due to nonlinearity, a nonlinear compensation term is introduced. First, the system is identified as an LTI model and static nonlinearity which is represented as a polynomial or weighted sum of Gaussian functions is introduced, *i.e.*, we use a Hammerstein model for the $P_2$ part. Weights for the basis functions are determined to minimize the mean square error. Figure 8.9(b) shows the result with the nonlinear terms. To the inputs of $P_2(s)$, $f_1$ and $f_l$, nonlinear elements which are represented as a weighted sum of Gaussian functions and polynomials are introduced. Compared to the result of Figure 8.9 (a), it can be seen that the mean square error is reduced by 40% and the model is valid for a large operating range.

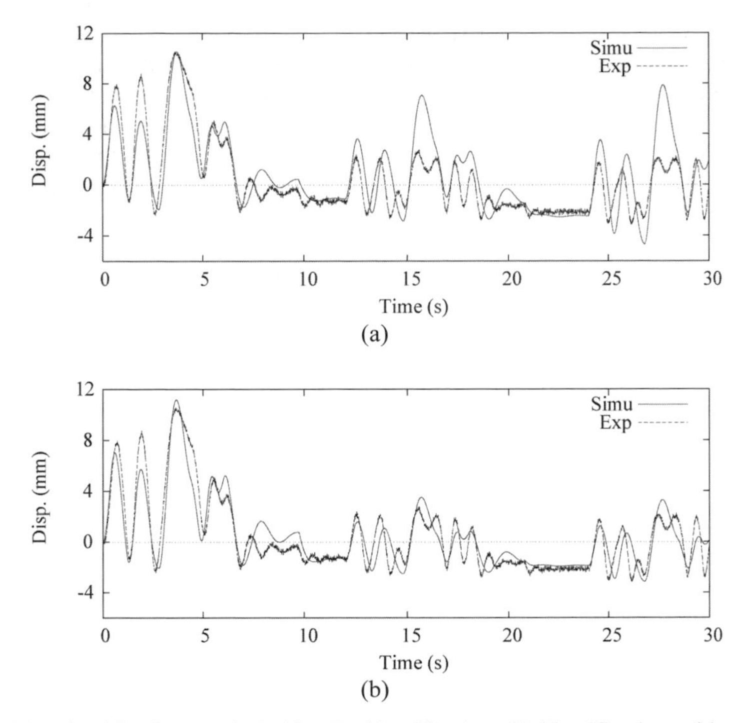


**Figure 8.9.** Identification result 2 (a) LTI identification (b) identification with nonlinear terms

### 8.3.4 Experiment of Feedback Control

To check the validity of the identified model, a feedback control experiment using an LQ servocontroller was conducted.

A position controller was designed based on the Hammerstein model and an LQ servocontroller was designed for a compensated Hammerstein model. Because the Hammerstein model contains static nonlinearity in front of the LTI part, the nonlinearity is compensated by the inverse system of the nonlinear function, and a linear controller was designed for the LTI part. Figure 8.10 shows a whole system composed of the Hammerstein model and the designed controller. The controller is designed as follows:

(1) Design a linear controller $K$ for LTI part $P$,

(2) Put $N^{-1}K$ in front of the Hammerstein model $P\,N$.

Please notice that in general the nonlinear block N might not be invertible. There, however, exists $N^{-1}$ for our model during the considered operating range.

Figure 8.11 shows the control results. The state of the system required for an LQ servecontroller was estimated by a linear observer. The actuator used for the experiment was composed of IPMC films doped with $Na^{+}$. From the figure, it is observed that the desired position control is achieved, though some oscillation is observed. In figure (b), the control voltage of the experiment and simulation are shown and we can see some deviation. Especially, the actual control voltage gradually increased due to integral operation. The reason inferred for the deviation is that such a slow mode was not identified by our Hammerstein model. We consider, however, that our model is valid for a periodic motion with a short period considered below.

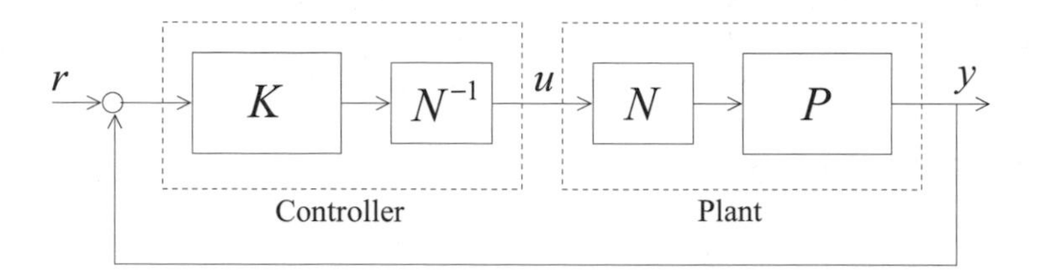


**Figure 8.10.** Design of controller

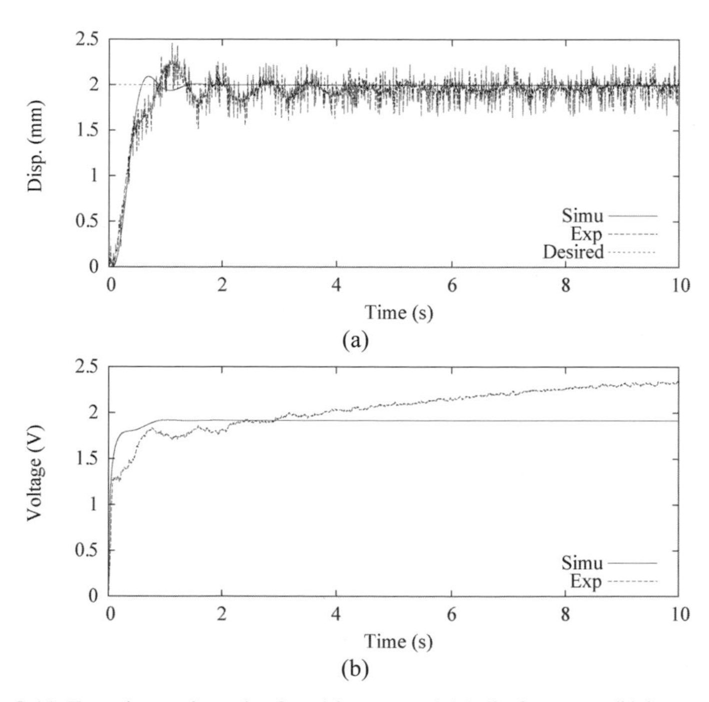


**Figure 8.11.** Experimental result of position control (a) displacement (b) input voltage

## 8.4 Application to Walking Robot

This section addresses an application of the linear actuator to a small-sized bipedal walking robot shown in Figure 8.12, and realization of walking with the proposed actuator is investigated by numerical simulations. The parameters of the robot are set as $m_l$=5 g, $m_h$=10 g, $a$=50 mm, $b$=50 mm, $l$=100 mm, $r_h$=4 mm, $g$=9.81 m/s$^2$, and $r_f$=0 mm. This small bipedal robot can exhibit passive dynamic walking 0 without any actuator on a gentle slope. In the following simulation, we assume that actuators are attached between legs, as in the right side of Figure 8.12. In the simulations, we assume that contact between a leg and ground is pin contact and collision of the swing leg with the ground is perfectly inelastic.

### 8.4.1 Simulation Results

Figure 8.13 shows the simulation results of walking on level ground. The number of units connected in parallel and series is set as 4 and 3, respectively. In this simulation, we applied a square pulse as input voltage whose cycle was 0.48 s and whose amplitude was 2.5 V. From the results, it can be seen that a one-period walking gait is generated and the walking cycle synchronizes with the cycle of the input signal.

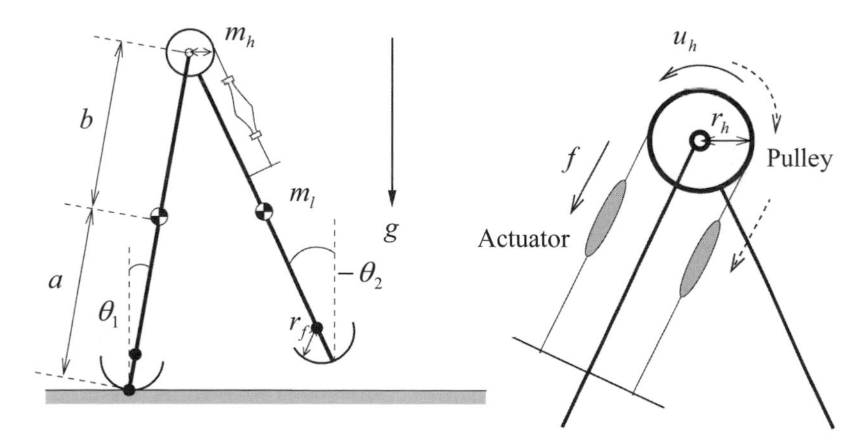


**Figure 8.12.** Model of bipedal walking robot

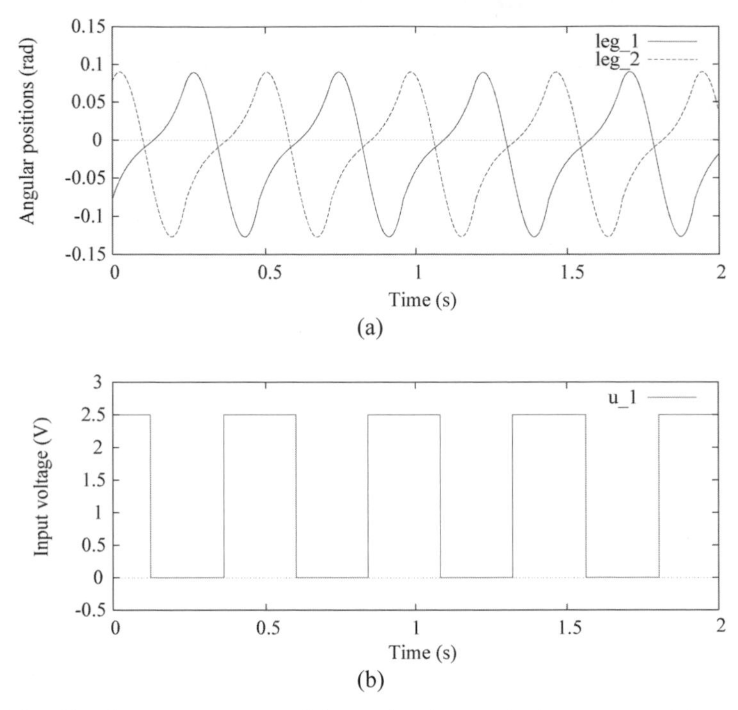


**Figure 8.13.** Simulation results of walking on level ground (a) angular positions (b) input voltage

## 8.4.2 Walking Control with Synchronization

### *8.4.2.1 Open-Loop Control*

In the previous section, we applied time-variant signals as open-loop inputs, *i.e.*, constant periodic square pulses, and it was shown that the bipedal walking robot with actuators can walk on level ground with the period synchronized with the period of the input signal. In this section, we consider time-invariant input signals, that is, switch the input signals in response to the state of the robot, *e.g.*, the angle of a stance leg.

Figure 8.14 shows the simulation results of open-loop control. In this simulation, the parameters of the robot are set as $m_l$= 9.6 g, $m_h$ =32.1 g, $a$=42.9 mm, $b$=56.9 mm, $l$=99.8 mm, $r_h$ =11.3 mm, $g$=9.81 m/s$^2$, and $r_f$=0 mm due to an experimental system, and the input signals are switched by the angle of a stance leg. From the results, it is observed that a one-period walking gait is generated. In comparison with time-variant input, there is not much difference in the convergence to steady state or the basin of attraction; however, we are able to simplify the experimental setup and to adjust easily the timing of inputs and the start of walking. Moreover, the systems with the open-loop inputs are autonomous, then we can apply a simple feedback control method, as described below.

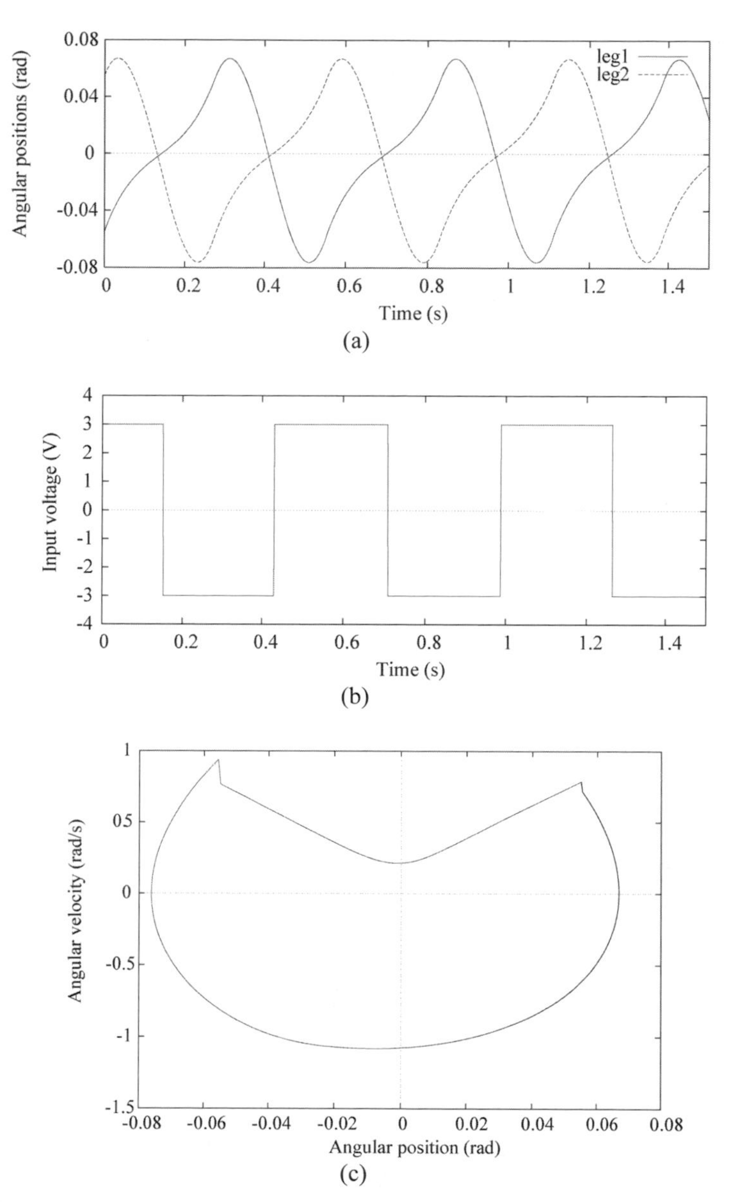


**Figure 8.14.** Simulation results of open-loop control (a) angular positions (b) input voltage (c) phase plane

*8.4.2.2 Feedback Control*

Robots with the actuator can walk on level ground with open-loop control; however, the walking gait is not robust, and the basin of attraction of the limit cycle is limited; then it is difficult to continue walking under the disturbance. To realize robust walking and to enhance the basin of attraction, we apply a feedback control based on the linear approximate model of trajectories on a limit cycle 0.

Figure 8.15 shows a conceptual diagram of trajectories. $\gamma$ is a periodic orbit through the point $q_0$, and $\Sigma$ is a hyperplane perpendicular to $\gamma$ at $q_0$; then for any point $q \in \Sigma$ sufficiently near $q_0$, the orbit will cross $\Sigma$ again at a point $P(q)$; $\Sigma$ is called a Poincaré section. It is considered a discrete time nonlinear system on $\Sigma$, called a Poincaré map, as follows:

$$q_{k+1} = P(q_k, u_k) \tag{8.2}$$

$q^*$, $u^*$ are the equilibrium state and input where the equation $q^*=P(q^*, u^*)$ holds. Let $\delta q_k, \delta u_k$ be small perturbations from the equilibrium state and input, *i.e.*, $\delta q_k = q_k - q^*, \delta u_k = u_k - u^*$. Linearizing Eq. (8.2) around $q^*$ and $u^*$ results in the linear system

$$\begin{aligned} \delta q_{k+1} &= \left.\frac{\partial P}{\partial q}\right|_{\substack{q=q^* \\ u=u^*}} \delta q_k + \left.\frac{\partial P}{\partial u}\right|_{\substack{q=q^* \\ u=u^*}} \delta u_k \\ &= \Phi \delta q_k + \Gamma \delta u_k \end{aligned} \tag{8.3}$$

Assume the cycle is stable and the pair $(\Phi, \Gamma)$ is controllable. To stabilize the walking motion and enhance a basin of attraction, we consider the regulator problem based on the discrete time linear system of Eq. (8.3). In this paper, a performance index is defined as

$$J := \sum_{k=0}^{\infty} (\delta q_k^T Q \delta q_k + r \delta u_k^2) \tag{8.4}$$

and we solve the LQR optimal control problem to determine a feedback control input $\delta u_k = F \delta x_k$ where $F$ is an optimal feedback gain matrix determined from $J$.

*8.4.2.3 Numerical Simulation*

We demonstrate numerical simulations, and the parameters of the robot are the same as in the previous simulation. In this simulation, we define the Poincaré section $\Sigma$ in the state just after the heel strike collision, and the state vector $q_k$ is defined as $q_k = [q_{1k}, q_{2k}, q_{3k}]^T$, where $q_{1k}$ is the angle of the hip joint, $q_{2k}$ is the angular velocity of the stance leg, and $q_{3k}$, is the angular velocity of the swing leg.

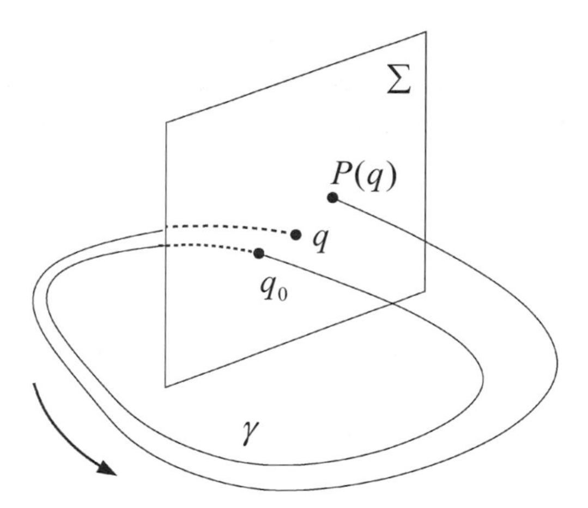


**Figure 8.15.** Poincaré map

Because $\Phi, \Gamma$ can not be obtained analytically, we computed them numerically by computer simulation as

$$\Phi = \begin{bmatrix} 2.82\times10^{0} & -3.94\times10^{-1} & 3.73\times10^{-2} \\ 1.30\times10^{1} & -1.59\times10^{0} & 1.55\times10^{-1} \\ -1.44\times10^{0} & 7.10\times10^{-1} & -2.29\times10^{-2} \end{bmatrix} \quad (8.5)$$

$$\Gamma = \begin{bmatrix} 9.25\times10^{-3} \\ 5.42\times10^{-2} \\ 4.26\times10^{-2} \end{bmatrix} \quad (8.6)$$

The weighting matrices $Q$, $r$ are determined as follows:

1. Check the limit of stability; let $q_{1f}$, $q_{2f}$, $q_{3f}$ be the quantity of state in the stability limit, respectively, and check them by numerical simulation, that is, we search the maximum perturbation that the robot does not even fall down.

2. Determine $Q$; $Q$ is set as $Q = \mathrm{diag}(1/q_{1f}^2, 1/q_{2f}^2, 1/q_{3f}^2)$.

3. Determine $r$; $r$ is adjusted manually to obtain a suitable input.

Figure 8.16 shows the simulation results of feedback control; deviations are included in initial conditions. $Q$, $r$, and feedback vector $F$ are

$Q = \mathrm{diag}(3.42\times10^{5}, 9.61\times10^{3}, 8.73\times10^{1})$

$r = 1.0$

$F = [7.76\times10^{1}, -1.02\times10^{2}, 9.80\times10^{-1}]$

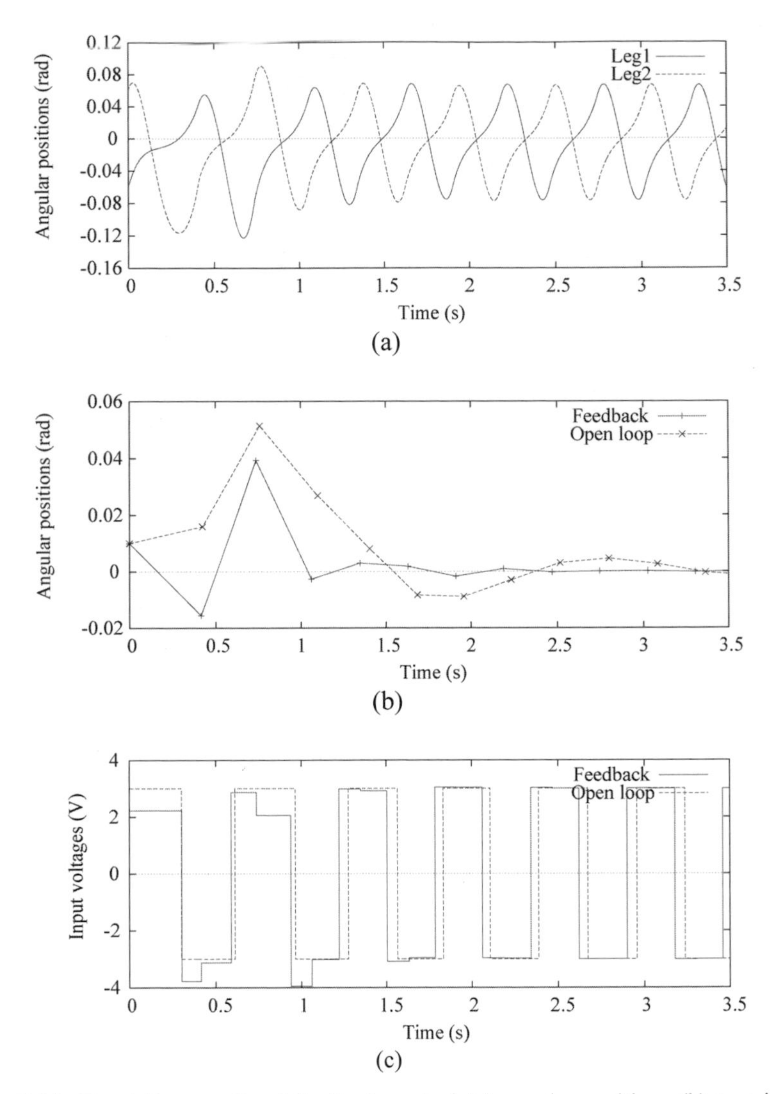


**Figure 8.16.** Simulation results of feedback control (a) angular positions (b) transition of $\delta q_1$ (c) input voltage

Figure 8.16(a) shows angular positions, figure (b) shows the transition of $\delta q_1$ on Poincaré section $\Sigma$, and figure (c) shows the input voltage to the actuator, the total of the open-loop signal and feedback signal. From the results, it is observed that the convergence to steady state becomes fast in comparison to open-loop control. The validity of this feedback control was investigated, but more detailed analysis of the basin of attraction and the robustness of the control is left for future work.

### 8.4.3 Doping Effect on Walking

As shown in the previous section, the bending characteristics of IPMC film are highly affected by the doped counterion. There exist possibilities to change the properties of the actuator according to the environment or purpose. If we consider walking application, we can change the property so that the actuator is suitable for slow walking with low energy consumption or fast walking with high energy consumption, or possibly running. We investigate the possibility of adaptation with doping of the actuator for walking control by numerical simulations 0. Recall that the doped ion can be exchanged as many times as required.

We compare walking speeds and walking efficiencies with actuators composed of IPMC films doped with $Na^+$ and $Cs^+$ for the same input voltage. The input voltage is rectangular, its amplitude is 2.5 V, and it is applied to the system in an open-loop fashion. The parameters of the robot are set as $m_l$=5.0 g, $m_h$=10.0 g, $a$=50.0 mm, $b$=50.0 mm, $l$=100.0 mm, $r_h$=4.0 mm, $r_f$ = 0.0 mm, and $g$=9.81 m/s$^2$. We assume also that in the simulation the number of units connected in parallel and series is set as 4 and 3, respectively.

Figure 8.17(a) shows a plot of average walking speed vs. the applied frequency of the input where the solid line shows the plot for the actuator with $Na^+$ and dotted line for that with $Cs^+$. From the figure, it can be seen that if the same control frequency input is applied to the robot, faster walking is realized by the actuator doped with $Na^+$ rather than by that with $Cs^+$. The maximum speed of the robot doped with $Na^+$ is higher than that with $Cs^+$. Note here that this kind of property may not exist if the parameters of the robot are not designed properly. So the design of the robot is important for the doping to be effective for walking. Figure 8.17(b) shows a plot of walking speed vs. the average consumed power. Because the input current for the actuator is almost irrelevant to the walking pattern, the peak value of the injected current of the actuator doped with $Na^+$ is large, and the corresponding consumed power is large.

From the observation, it can be suggested that if the input voltage is the same, the actuator doped with $Na^+$ realizes high-speed walking with high energy consumption, and the one doped with $Cs^+$ can generate a slow walking pattern with low energy consumption when the mass is rather heavy, *i.e.*, $m$=5 g. On the other hand, when $m$=1 g, the actuator with $Cs^+$ can realize a wide range of walking speeds with low energy consumption. Note here that even if the average input power is increased in the case of $Cs^+$, the walking speed is not increased because the walking pattern is not proper and the energy dissipated in a collision is increased.

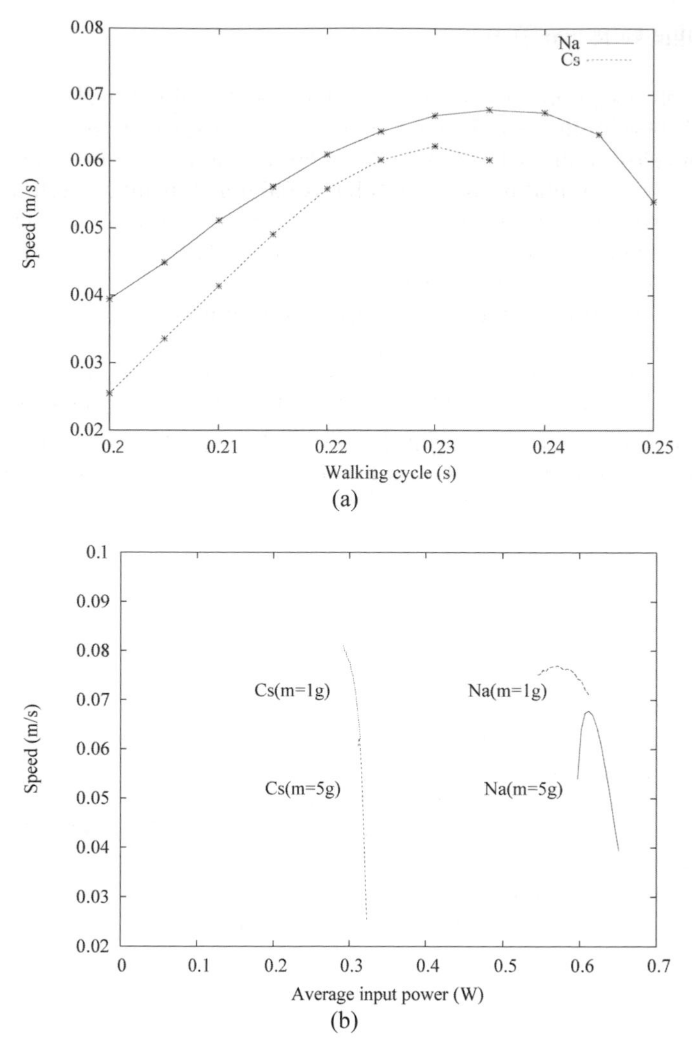


**Figure 8.17.** Simulation results of the doping effect on bipedal walking (a) average speed vs. walking cycle (b) average speed vs. average input power

## 8.5 Application to Snakelike Robot

In the last section, it was shown that the efficiency of walking with different walking speeds was confirmed by numerical simulation. In this section, the effect is checked by a snakelike robot swimming in water experimentally.

### 8.5.1 Snakelike Robot

Figure 8.18 shows an experimental machine, a three-link snakelike swimming robot with IPMC actuators. The frame of the robot is made of styrene foam. Thin fins are attached to the bottom of the body frame, and each frame is connected by an IPMC film. The total mass of the robot is 0.6 g and its total length is 120 mm. The IPMC film which we used in this experiment is Nafion®117 (by DuPont) plated with gold; the thickness of this film is about 200 μm in a wet condition, and it was cut into a ribbon with a width of 2 mm and length of 20 mm.

To check the performance of the robot, we also performed experiments using the snakelike robot as shown in Figure 8.18.

Figure 8.19 shows the experimental results with input signals whose cycle is 2 s, amplitude is 2.5 V, phase shift is 90 °, and the kind of counterion is sodium ($Na^+$). From figures (a) and (b), it can be confirmed that the robot performs an undulating motion and moves forward. Figure 8.20 shows sequential photographs of the experiment. For more details of the experimental setup and the properties of the motions, refer to 0.

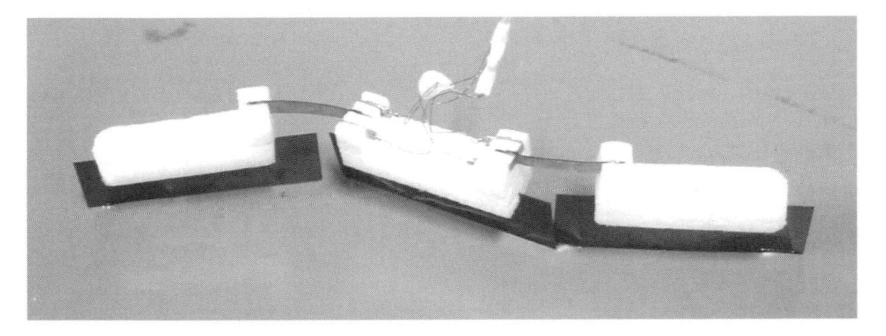

**Figure 8.18.** Snakelike robot using IPMC

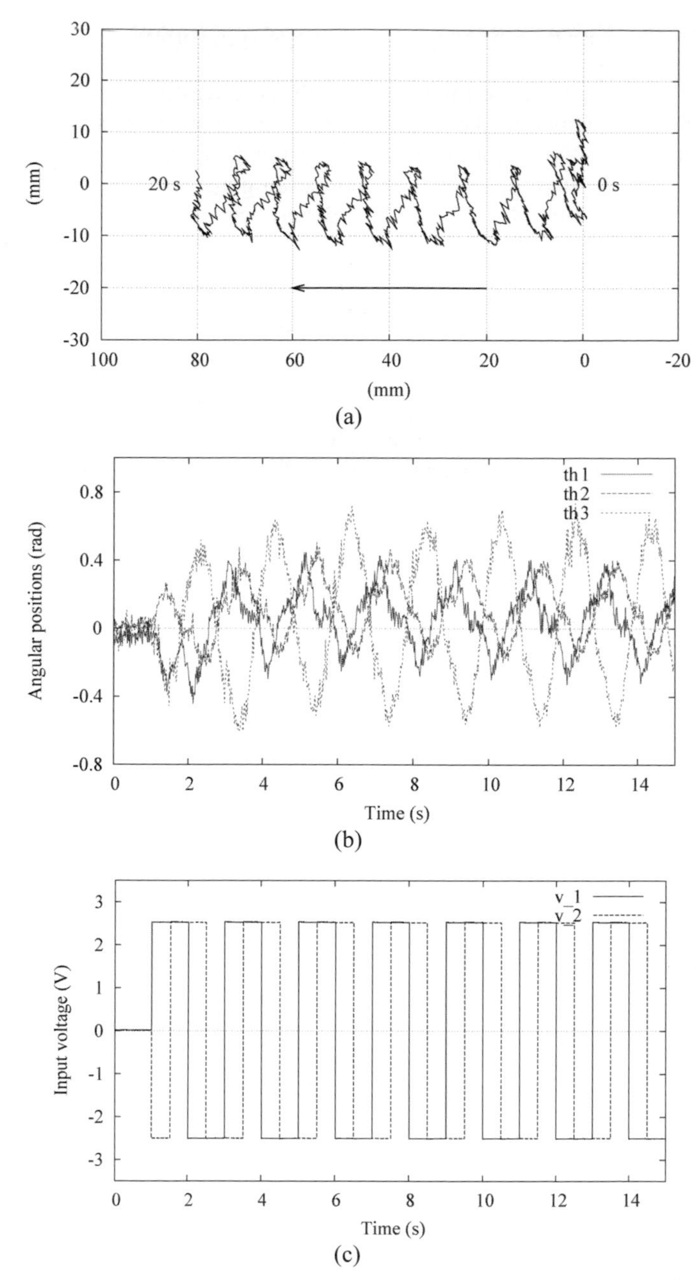


**Figure 8.19.** Experimental results (a) trajectory of head position (b) angular positions (c) input voltages

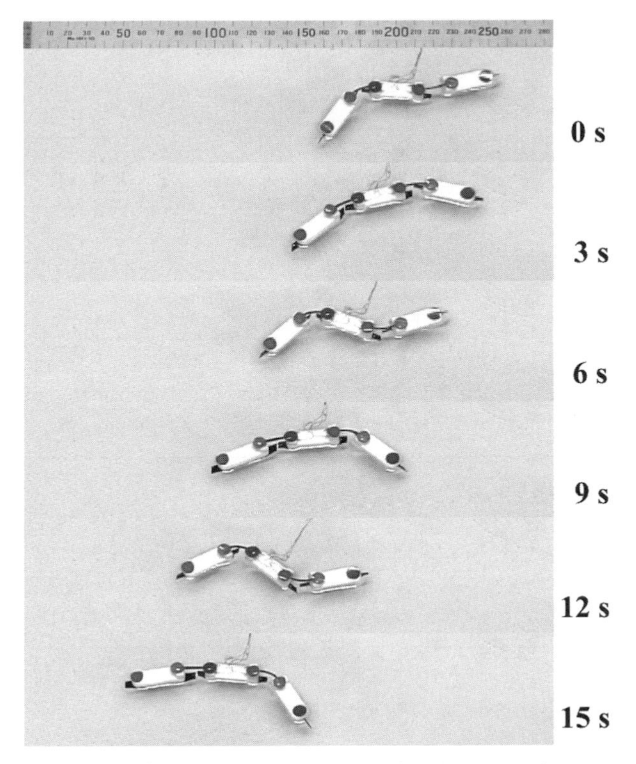


**Figure 8.20.** Sequencial photographs of the experiment

### 8.5.2 Doping Effect

To verify the doping effect, we performed experiments on IPMC actuators which were doped with $Na^+$, $Cs^+$ and $TEA^+$ as counterions. We compare propulsive speed and efficiencies of the actuators doped with each ion for the same input voltage. The inputs voltages were square pulses whose amplitude was 2.5 V and phase shift was 90 °, and we repeated measurements at various input frequencies.

In Figure 8.21 (a), the average propulsive speed vs. consumed power is plotted. The snakelike robot doped with $Na^+$ can move faster; however, consumed power is large. If it need not move at high speed, we should use the actuators doped with other counterions that can be driven by low power. Figure 8.21(b) shows the average propulsion speed vs. power consumed per distance. If there is no limit to the capacity of a power source, it can be considered that the actuators doped with $Na^+$ are effective because the robot can move for a short time; however, there is a region of low consumed power achieved only by the robot doped with $TEA^+$.

From the observation, it can be summarized that if the input voltage is the same, the actuator doped with $Na^+$ realizes a high-speed swimming motion with high energy consumption, the one doped with $TEA^+$ can generate slow swimming speed with low energy consumption, and the one doped with $Cs^+$ has characteristics between those of $Na^+$ and $TEA^+$. Note that the actuators can be adjusted to various characteristics by selecting an appropriate counterion or by mixing several ions in appropriate proportions.

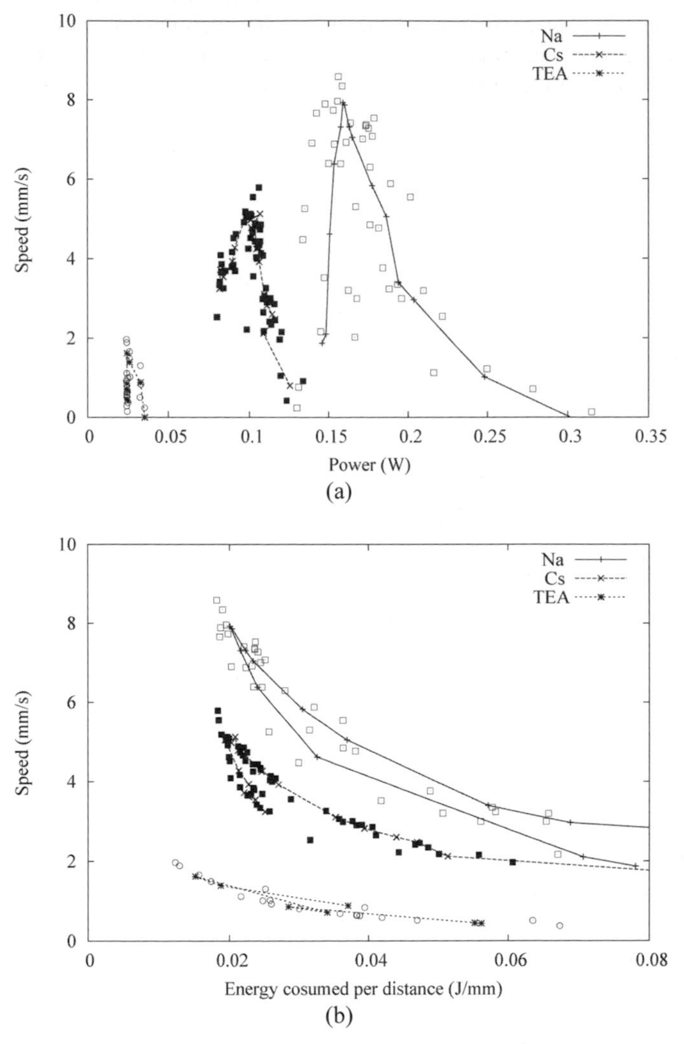


**Figure 8.21.** Experimental results of doping effect (a) consumed power vs. average speed (b) consumed energy per distance vs. average speed

## 8.6 Control of Partial Doping Effect by Exercise

The doping effect is caused by exchanging counterions and a higher condensed counterion is doped into IPMC films. The doping of the counterions is easily done just putting the actuators in a solution containing the target counterion just as the robots take a bath containing a nutritional supplement. When the robots cannot take a bath, liquid containing the counterion can be delivered to the actuators through tubes like blood vessels. Figure 8.22(a) illustrates these doping processes. If the speed of changing the ion can be controlled by exercises, *i.e.*, bending IPMC films, the property of particular actuators can be changed by such motions. This phenomenon can be considered similar to muscles in a human body that can be trained by exercise for a particular purpose, as in Figure 8.22 (b).

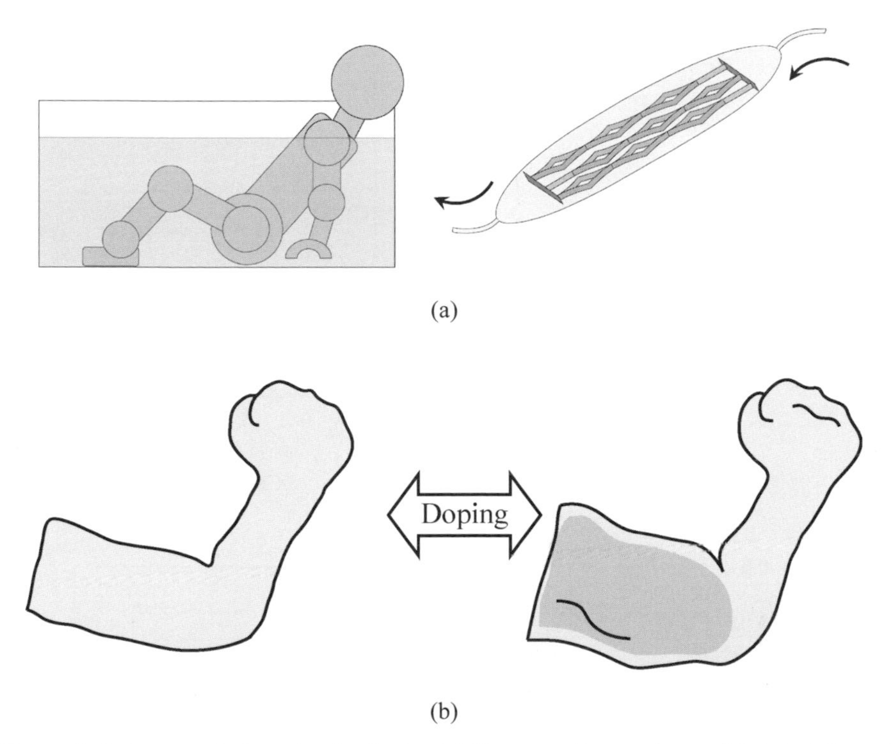


**Figure 8.22.** Image of adaptation by doping (a) Process of ion-exchange (b) Adaptation of partial elements by doping

### 8.6.1 Experiment

To investigate the possibility of the effect in IPMC actuators, we conducted an experiment as follows. Two linear actuators doped with $TEA^+$ were prepared, and one of the actuators was just immersed in the $Na_2SO_4$ solution with $Na^+$. On the other hand, another actuator was actuated in the same solution so that the bending motion was caused frequently.

At every interval, the characteristics of the two actuators were measured. In our experiment, step responses for a constant voltage input are stored.

The length, width, and thickness of the films were 25 mm, 2 mm, and 200 μm, respectively, and they were immersed in the liquid by 15 mm. For the activated film, a rectangular input whose levels were $-1 \leftrightarrow 1$ V and whose frequency was 0.5 s was injected. The step responses of the films were measured at 0, 10, 30, 60, 120, and 180 minutes where the input voltage was 2.0 V.

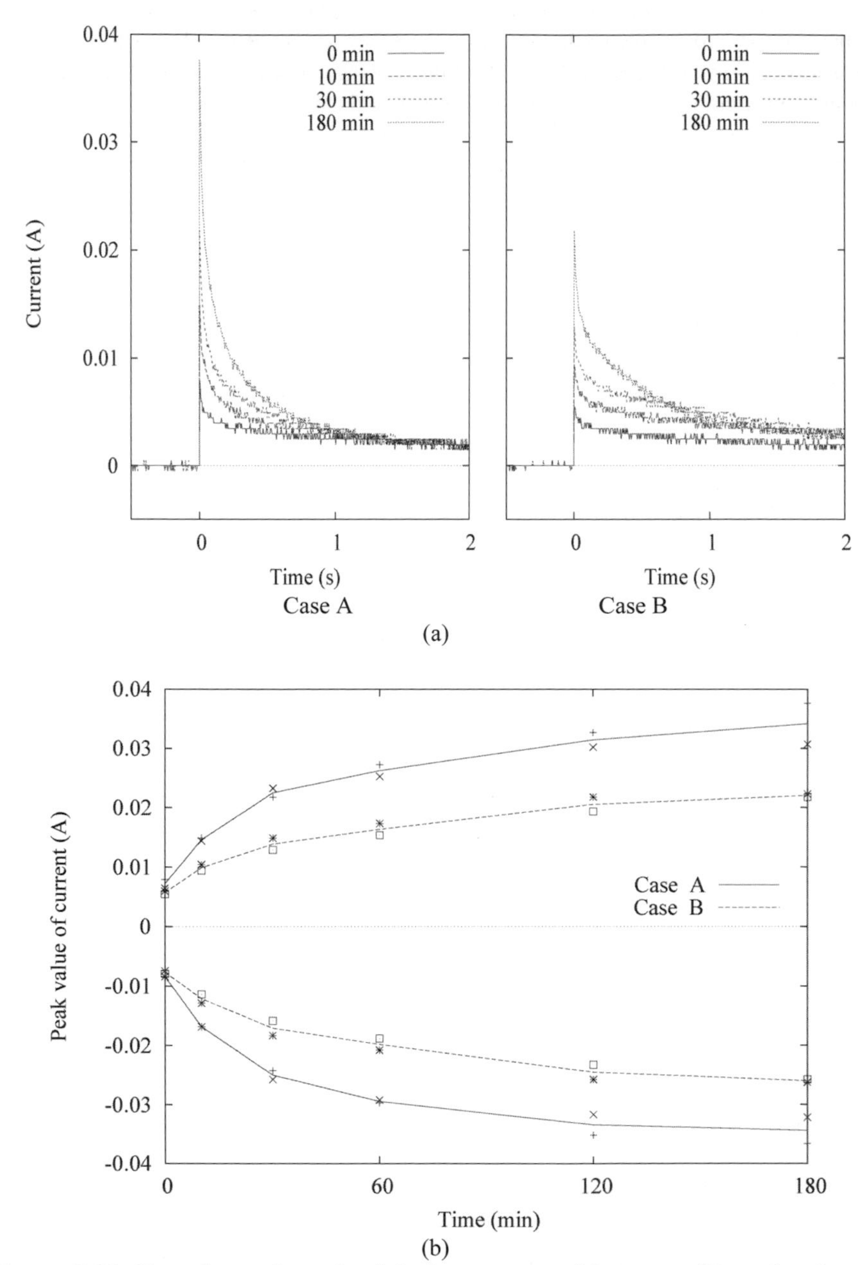


**Figure 8.23.** Experimental result of doping progress (a) current (b) peak value of current

When the step response was measured, the actuators were immersed in pure water for 10 minutes to avoid changes due to the mechanical effects of motion.

Figure 8.23 shows the experimental results. In the figure, (a) shows the changes in the current profile for each step input, and (b) shows peak values of the current, with respect to the intervals. From the figure, it can be seen that the peak values of the current increased according to the increase of the interval, and the property was changed from the property of $TEA^+$ to that of $Na^+$. Note that the property of the actuator immersed with motion changed more quickly than that without motion. Actually, the property of the film without motion at 180 minutes was achieved by the film with motion at 30 minutes.

## 8.7 Conclusions

We have discussed the development of a linear actuator using IPMC materials and its applications to a walking robot and a snakelike robot. In this monograph, the doping effects on motion were focused on especially, and it was shown by numerical simulations of walking control and by an experiment of a swimming control of the snakelike robot that the properties of the actuator can be adjusted according to particular motions, *i.e.*, slow speed motion with low energy consumption or high speed motion with high energy consumption. Also, a possibility that some actuators distributed in a system can be partially doped with a desired ion by moving the actuators mechanically was shown by a preliminary experiment. The authors consider that the developed IPMC linear actuator can be used for biomimetic control systems where the properties of the system can be adapted to an environment using doping effects.

To apply the artificial muscle actuator to a general robotic system, there exist a lot of problems such as limitation of output force; however, we think the mutual evolution of improvement of actuator technology and design of control system is important for further applications.

## 8.8 Acknowledgments

The contents of the paper are collections of works in the last few years by co-workers in the development of the IPMC linear actuator. The authors give their special thanks to Mr. Kaneda, Mr. Kozuki, and Mr. Sera at Tokyo Tech.

## 8.9 References

Y. Bar-Cohen, Electroactive Polymer (EAP) Actuators as Artificial Muscles: Reality, Potential, and Challenges, SPIE Press, 2001.

K. Oguro, Y. Kawami and H. Takenaka, ``Bending of an ion-conducting polymer film-electrode composite by an electric stimulus at low voltage," Journal of Micromachine Society, 5, 27-30, 1992. (in Japanese)

S. Guo, T. Fukuda, K. Kosuge, F. Arai, K. Oguro and M. Negoro, ``Micro catheter system with active guide wire," Proc. of IEEE Int. Conf. on Robotics and Automation, pp.79-84, 1995.

EAMEX Corporation, http://www.eamex.co.jp/

M. Mojarrad and M. Shahinpoor, ``Biomimetic robotic propulsion using polymeric artificial muscles," Proc. of IEEE Int. Conf. on Robotics and Automation, pp.2152-2157, 1997.

S. Guo, T. Fukuda and K. Asaka, ``A new type of fish-like underwater microrobot," IEEE/ASME Trans. on Mechatronics, Vol. 8, No. 1, pp.136-141, 2003.

J. Jung, B. Kim, Y. Tak and J. O. Park, ``Undulatory tadpole robot (TadRob) using ionic polymer metal composite (IPMC) actuator," Proc. of IEEE/RSJ Int. Conf. on Intelligent Robots and Systems, pp.2133-2138, 2003.

J. W. Paquette, K. J. Kim and W. Yim, ``Aquatic robotic propulsor using ionic polymer-metal composite artificial muscle," Proc. of IEEE/RSJ Int. Conf. on Intelligent Robots and Systems, pp.1269-1274, 2004.

A. Punning M. Anton, M. Kruusmaa and A. Aabloo, ``A biologically inspired ray-like underwater robot with electroactive polymer pectoral fins," Proc. of IEEE/ Int. Conf. on Mechatronics and Robotics, Vol. 2, pp.241-245, 2004.

Y. Nakabo, T. Mukai, K. Ogawa, N. Ohnishi and K. Asaka, ``Biomimetic soft robot using artificial muscle," in tutorial ``Electro-Active Polymer for Use in Robotics", IEEE/RSJ Int. Conf. on Intelligent Robots and Systems, 2004.

Y. Bar-Cohen, S. Leary, A. Yavrouian, K. Oguro, S. Tadokoro, J. Harrison, J. Smith and J. Su, ``Challenges to the application of IPMC as actuators of planetary mechanisms," Proc. of SPIE Int. Symp. on Smart Structures and Materials, EAPAD, Vol. 3987, 2000.

S. Guo, S. Hata, K. Sugumoto, T. Fukuda and K. Oguro, ``Development of a new type of capsule micropump," Proc. of IEEE Int. Conf. on Robotics and Automation, pp.2171-2176, 1999.

S. Tadokoro, S. Yamagami, M. Ozawa, T. Kimura and T. Takamori, ``Multi-DOF device for soft micromanipulation consisting of soft gel actuator elements," Proc. of IEEE Int. Conf. on Robotics and Automation, pp.2177-2182, 1999.

S. Tadokoro, S. Fuji, M. Fushimi, R. Kanno, T. Kimura and T. Takamori, ``Development of a distributed actuation device consisting of soft gel actuator elements," Proc. of IEEE Int. Conf. on Robotics and Automation, pp.2155-2160, 1998.

M. Yamakita, N. Kamamichi, Y. Kaneda, K. Asaka and Z. W. Luo, ``Development of an artificial muscle linear actuator using ionic polymer-metal composites," Advanced Robotics, Vol. 18, No. 4, pp.383-399, 2004.

K. Onishi, S. Sewa, K. Asaka, N. Fujiwara and K. Oguro, ``The effects of counter ions on characterization and performance of a solid polymer electrolyte actuator," Electrochemica Acta, Vol. 46, No. 8, pp.1233-1241, 2001.

Y. Kaneda, N. Kamamichi, M. Yamakita, K. Asaka and Z. W. Luo, ``Development of linear artificial muscle actuator using ionic polymer -introduce nonlinear characteristics to attain a higher steady gain-," Proc. of the Annual Conf. of RSJ, 2003. (in Japanese)

S. Tadokoro and T. Takamori, ``Modeling IPMC for design of actuation mechanisms," Electroactive Polymer (EAP) Actuators as Artificial Muscles, Reality, Potential, and Challenges, Ed. Y. Bar-Cohen, SPIE Press, pp.331-366, 2001.

K. Asaka and K. Oguro, ``Bending of polyelectrolyte membrane platinum composites by electric stimuli Part II. Response kinetics," Journal of Electroanalytical Chemistry, 480, pp.186-198, 2000.

S. Tadokoro, S. Yamagami and T. Takamori, ``An actuator model of ICPF for robotic applications on the basis of physicochemical hypotheses," Proc. of IEEE Int. Conf. on Robotics and Automation (ICRA), pp. 1340-1346, 2000.

S. Tadokoro, M. Fukuhara, Y. Maeba, M. Konyo, T. Takamori and K. Oguro, ``A dynamical model of ICPF actuator considering ion-induced lateral strain for molluskan robotics," Proc. of IEEE Int. Conf. on Robotics and Automation, pp. 2010-2017, 2002.

K. Mallavarapu, K. Newbury and D. J. Leo, "Feedback control of the bending response of ionic polymer-metal composite actuators," Proc. of SPIE Int. Symp. on Smart Structures and Materials, EAPAD, Vol. 4329, pp.301-310, 2001.

T. McGeer, ``Passive dynamic walking," The Int. Journal of Robotics Research, Vol. 9, No. 2, pp.62-82, 1990.

M. Yamakita, N. Kamamichi, T. Kozuki, K. Asaka and Z. W. Luo, ``Control of biped walking robot with IPMC linear actuator," Proc. of IEEE/ASME Int. Conf. on Advanced Intelligent Mechatronics, 2005.

M. Yamakita, N. Kamamichi, Y. Kaneda, K. Asaka and Z. W. Luo, ``IPMC linear actuator with re-doping capability and its application to biped walking robot," Proc. of 3rd IFAC Symposium on Mechatronic Systems, pp.359-364, 2004.

M. Yamakita, N. Kamamichi, T. Kozuki, K. Asaka and Z. W. Luo, ``A snake-like swimming robot using IPMC actuator and verification of doping effect," Proc. of IEEE/RSJ Int. Conf. on Intelligent Robots and Systems, 2005.

# 9

# Applications of Ionic Polymer-Metal Composites: Multiple-DOF Devices Using Soft Actuators and Sensors

M. Konyo[1], S. Tadokoro[2], K. Asaka[3]

[1] Graduate School of Information Science, Tohoku University,
6-6-01 Aramaki Aza Aoba, Aoba-ku, Sendai 980-8579, Japan
konyo@rm.is.tohoku.ac.jp
[2] Graduate School of Information Science, Tohoku University
tadokoro@rm.is.tohoku.ac.jp
[3] Research Institute for Cell Engineering, National Institute of AIST,
1-8-31 Midorigaoka, Ikeda, Osaka, 563-8577, Japan
asaka-kinji@aist.go.jp

## 9.1 Introduction

The ionic polymer-metal composite (IPMC, which is also known as ICPF* ) [1, 2] is one of the electroactive polymers that have shown potential for practical applications. IPMC is an electroless plated electroactive polymer (EAP) material that bends when subjected to a voltage across its thickness (see Figure 9.1). IPMC has several attractive EAP characteristics that include:

(1) Low drive voltage is 1.0 – 5.0 V).
(2) Relatively high response (up to several tens of Hertz).
(3) Soft material ($E = 2.2 \times 10^8$ Pa).
(4) Possible to miniaturize (< 1 mm).
(5) Durability to many bending cycles (> $1 \times 10^6$ bending cycles).
(6) Can be activated in water or in a wet condition.
(7) Exhibits distributed actuation allowing production of mechanisms with multiple degrees of freedom.

The IPMC generates a relatively small force where a cantilever-shaped actuator (2 × 10 × 0.18 mm) can generate about 0.6 mN, and therefore its applications need to be scoped accordingly. Some of the applications that were investigated for IPMC include an active catheter system [3, 4], a distributed actuation device [5–7],

* Kanno and Tadokoro named the Nafion-Pt composite ICPF (Ionic Conducting Polymer gel Film) in 1992. In the field of robotics, most researchers use the name ICPF, and it is well recognized.

an underwater robot [8], micromanipulators [9, 10], a micropump [11], a face-type actuator [7], a wiper of an asteroid rover [12, 13], and a tactile haptic display for virtual reality [14–17]. The actual number of applications that were considered is still small, but the list is expected to grow in the coming years with the emergence of requirements that account for the limitations while taking advantage of the unique capabilities.

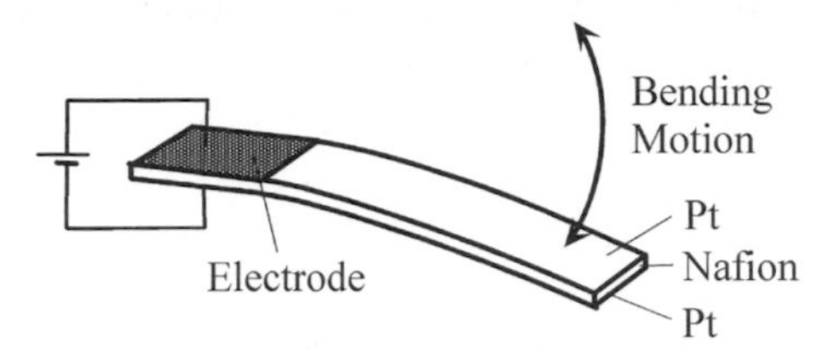


**Figure 9.1.** Ionic polymer metal composite (IPMC) actuator shown for Pt/Nafion composite EAP

Many investigators have studied models for IPMC, with the largest number addressing Nafion-Pt composite EAP [18–27].

A soft sensing system is also important for advanced applications of IPMC actuators, because conventional solid sensors may cancel the flexibility of an IPMC. One possible sensor would be an IPMC itself. An IPMC can also be used as a sensor, because an electric potential will be generated across the composite when the strip is bent suddenly. The authors showed that the velocity of deformation of an IPMC strip was in proportion to the sensor output voltage and two kinds of velocity-sensing systems were proposed [28]. One is a 3-DOF tactile sensor that has four IPMC sensor modules combined in a cross shape and can detect both the velocity and the direction of the motion of the center tip. Another is a patterned IPMC strip that has both actuator and sensor functions. This strip can sense the velocity of bending motion made by the actuator part.

In this chapter, we describe several robotic applications developed using IPMC materials, which the authors have developed as attractive soft actuators and sensors. In Sections 9.2 to 9.4, several applications of IPMC actuators which have soft actuation mechanisms are described. We introduce several unique applications as follows:

(1) Haptic interface for a virtual tactile display
(2) Distributed actuation device
(3) Soft micromanipulation device with three degrees of freedom

In Section 9.5, we focus on aspects of the sensor function of IPMC materials. The following applications are described:

(1) 3-DOF tactile sensor
(2) Patterned sensor on an IPMC film

## 9.2. Haptic Interface for Virtual Tactile Display

### 9.2.1 Background

A novel technology to display to humans more realistic tactile sensation including qualitative information will realize advanced telecommunication directly connected to human physical skills and human mental sensibilities. A cutaneous display in addition to a force display helps human dexterous telemanipulation for use in medicine, space, and other extreme environments. For virtual reality applications, a tactile display is also effective to produce human emotional responses such as a rich texture feel, comfort of touch, and high presence of virtual objects.

A number of tactile displays have been proposed for evoking the cutaneous sense accepted by subcutaneous receptors for rough or frictional feeling on the surface of an object [29]. Conventional mechanical stimulation displays are equipped with a dumbbell-shaped vibration pin, a linear motor, and a pneumatic device. Consequently, it is difficult for the subject to perform contact motion freely in a 3-d space with this type of display due to the weight and size of its actuator .

EAP materials have many attractive characteristics as a soft and light actuator for such a stimulation device. The authors have developed a tactile display using IPMC actuators [14–17]. In our research, the target of tactile information is quite different from conventional ones. Our display can produce a delicate touch including even qualitative information such as a haptic impression or material feel when we stroke the surface of cloth.

The most characteristic feature of tactile sensation is a diversity of perceptual content. This variety is reflected in physical factors of target materials such as rigidity, elasticity, viscosity, friction, and surface shapes. It is interesting that tactile receptors in human skin cannot sense the physical factors directly. They can detect only the inner skin deformations caused by contacting to the objects. This suggests that the reproduction of the same physical factors of materials is not necessary for representing the virtual touch of materials. Virtual touch needs only the reproduction of internal deformations in the skin. Furthermore, a tactile illusion can even be provided by reproduction of nervous activities of tactile receptors, regardless of the inner deformations.

Based on this standpoint, several researchers proposed tactile display methods that make a selective stimulation on each tactile receptor using a magnetic oscillator and air pressure [30] and electrocutaneous stimulation [31]. However, selectivity of stimuli for all kinds of receptors was not enough to reproduce various tactile sensations. By using IMPC actuators, the authors proposed a tactile synthesis method that could control three physical characteristics, which are roughness, softness, and friction, as tunable parameters of textures. This method realizes selective stimulations on each kind of tactile receptors based on its temporal response characteristics [14–17].

In addition, an active perceptual process based on contact motion is very important for human tactile perception. To confirm the feel against hands (haptic impression) people use hand movements consciously or actively to clarify the properties of an object. Such an active touch in connection with contact motion excels passive sensory perception qualitatively and quantitatively. We successfully developed a wearable tactile display presenting mechanical stimuli on a finger in response to hand movements by using a small interface [16]. Almost no studies had realized a wearable tactile display that could make a multi degree-of-freedom mechanical stimulation on the skin.

In this chapter, haptic interfaces using IPMC actuators are described. Our display can realize a selective stimulation on human skin. We also describe a tactile synthesis method that can control three physical characteristics, which consists of roughness, softness, and friction, as tunable parameters of textures.

### 9.2.2 Wearable Tactile Display Using ICPF Actuators

Haptic interfaces for presenting human tactile feel were developed using IPMC actuators [14–17]. To express delicate tactile feel including even qualitative information such as tactile impression or material feel, we need to control the sensory fusion of elementary sensations that are generated by different sensory receptors.

Conventional tactile displays could hardly control such delicate sensation because it was difficult to make fine distributed stimuli on a human skin under the limitation of their actuators such as magnetic oscillators, piezoelectric actuators, shapememory alloy actuators, pneumatic devices, and so on. EAP materials have many attractive characteristics as a soft and light actuators for such a stimulation device. IPMC is suitable for the following reasons:

(1) *High spatial resolution*: The required spatial resolution for stimulating sensory receptors, especially Meissner's corpuscle in the finger tip, is less than 2 mm. IPMC films are easy to shape, and their simple operating mechanism allows miniaturizing a stimulator to make a high-density distributed structure. Conventional actuators can hardly control such minute force because of their heavy identical mass and high mechanical impedance. IPMC has enough softness that special control methods are not required to use the passive material property.

(2) *Wide frequency range*: Tactile display can stimulate several tactile receptors selectively by changing frequency ranges because each tactile receptor has different time response characteristics for vibratory stimulation [17]. The required frequency range is from 5 Hz to 200 Hz to stimulate all kinds of tactile receptors. The response speed of IPMC is fast enough to make a vibratory stimulation on a skin higher than 200 Hz. This means that IPMC can stimulate all receptors selectively.

(3) *Stimuli in multiple directions*: Each of the tactile receptors has selectivity for the direction of mechanical stimuli. Meissner's corpuscle detects especially the shearing stress toward the skin surface. Figure 9.2 shows that bending motions of an IPMC, which contacts with a surface of skin in a tilted position, make a stress in both the normal direction and shearing direction.

(4) *Wearability*: In human tactual perception, an active perceptual process based on hand contact motion is very important. To generate the virtual reality of tactile feel, we should move our hand actively and freely, and

receive appropriate stimuli in response to the hand movements. For conventional mechanical stimulation devices of tactile display, it is difficult to attach the device to a finger, so that the subjects cannot perform contact motion freely in a 3-D space. An IPMC based wearable display was successfully developed, which was made so smaller in size and weight that there was no interference with hand movements [16].

(5) *Safety*: The low driving voltage (less than 5 V) is safe enough to touch with a human finger directly.

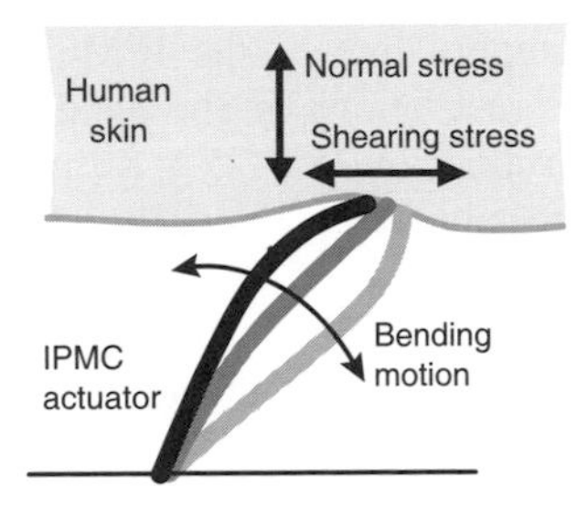


**Figure 9.2.** Multidirectional stimulation of a human skin using an IPMC actuator

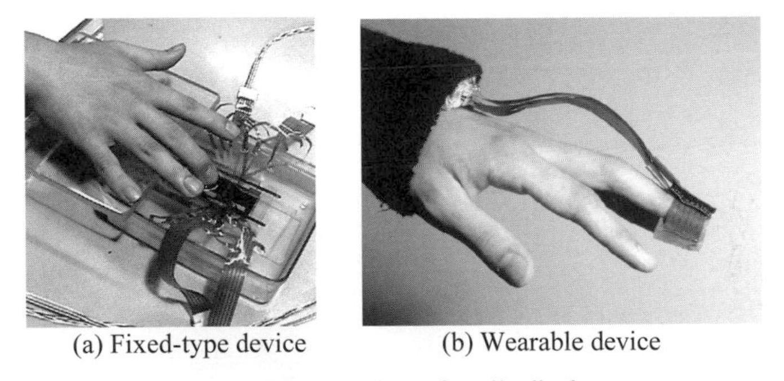

(a) Fixed-type device (b) Wearable device

**Figure 9.3.** Overview of tactile displays

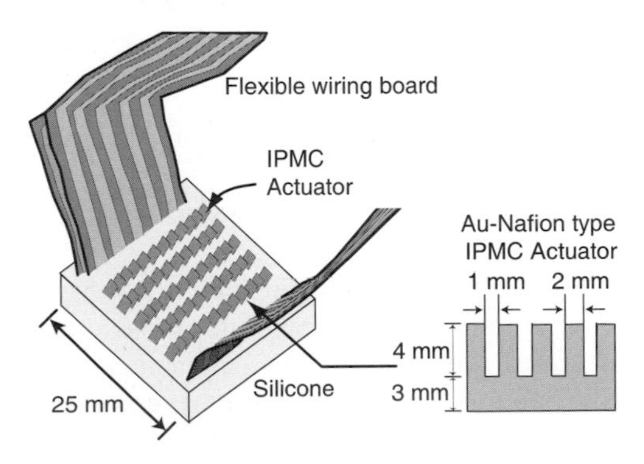


**Figure 9.4.** Structure of ciliary device using IPMC actuators

In an early prototype [14], tactile feel has been presented as shown in Figure 9.3a. In that case, subjects obtained only passive tactual perception because they could not perform contact motion. As shown in Figure 9.3b, the new wearable device [16] can be attached to the tip of a finger.

The structure of the wearable stimulation device is shown in Figure 9.4. The ciliary part is provided with Nafion-Au composite actuators, where each cilium is 3 mm long and 2 mm wide, in 12 rows leaving 1 mm gaps horizontally and 1.5 mm gaps vertically. All cilia are tilted 45° to transmit mechanical stimuli both in the normal and the tangential directions to the surface of the skin efficiently as shown in Figure 9.2. The power supply line of the IPMC is provided with a flexible wiring board in to minimize restrictions on the hand, so the fingertip can be bent flexibly. The use of silicon rubber of 25 × 25 × 8 [mm] applied to the base of the ciliary part has made it possible to lighten the device to approximately 8 g including the flexible wiring board.

An IPMC needs to be kept moistened because its actuators are operated by ionic migration. Even in a little wet condition in the air, however, the device can provide stimuli sufficiently for several minutes.

Figure 9.5 shows the total display system. The stimulation device is attached to the middle finger tip. The system is designed to read positional information of the hand using Polhemus' FASTRAK, which can read information according to a magnetic field.

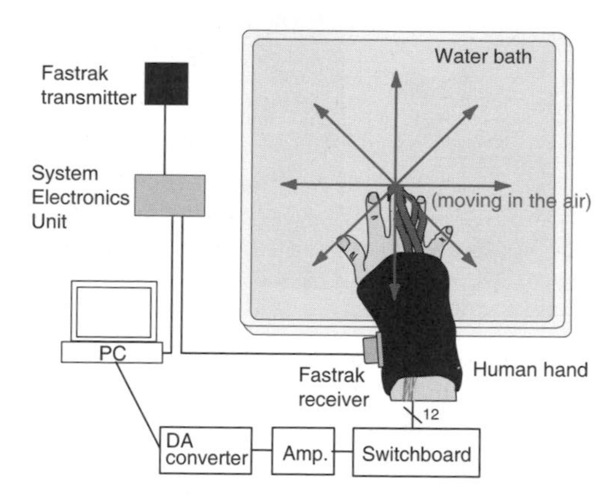


**Figure 9.5.** Wearable tactile display system in response to virtual contact motion

### 9.2.3. Concept of the Selective Stimulation Method

In human skin, tactile receptors generate elementary sensations such as touch, pressure, vibratory sensation, pain, temperature sense, and so on. A Tactile impression is an integrated sensation of these elementary sensations. To present tactile feel arbitrarily, stimuli applied to these receptors should be controlled selectively and quantitatively. As mentioned previously, tactile receptors cannot sense the physical factors of environments directly. They detect only the skin deformation caused by contacting objects. A tactile illusion can be provided by reproduction of activities of tactile receptors, regardless of the inner deformations.

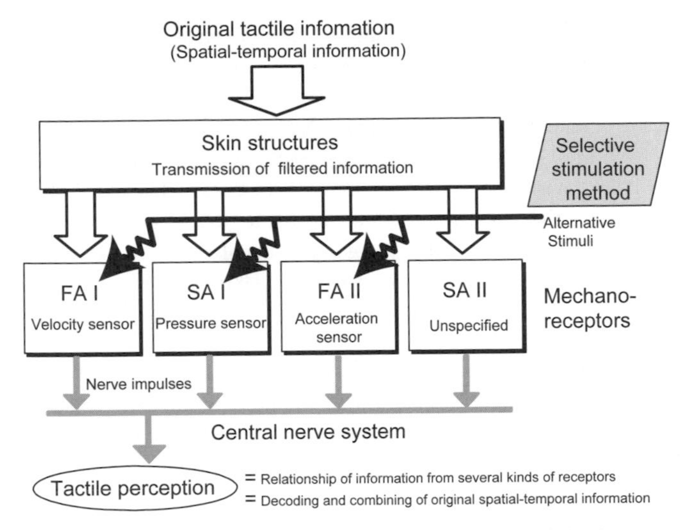


**Figure 9.6.** Concept of selective stimulation method

Figure 9.6 illustrates the concepts of the selective stimulation method. There are four types of mechanoreceptors embedded in human fingers, FA I type (Meissner's corpuscle), SA I type (Merkel corpuscle), FA II type (Pacinian corpuscle), and SA II type (Ruffini endings) [32]. It is known that each receptor has temporal response characteristics for mechanical stimulation and causes subjective sensation corresponding to its responsive deformation. For example, SA I detects static deformations of skin and produces static pressure sensation, and FA I detects the velocity of the deformation and produces the sense of fluttering vibration. Tactile impression is an integrated sensation of these elementary sensations. To present tactile feel arbitrarily, stimuli applied to these receptors should be controlled selectively.

The first problem is how to stimulate each receptor selectively. We have focused on the frequency response characteristics of tactile receptors. Figure 9.7 [33] illustrates the human detection threshold against vibratory stimuli, which represents the sensibility of each receptor to frequency variation. A smaller amplitude threshold means higher sensibility. This figure shows that there are three frequency ranges in which the most sensitive receptor changes. In the lowest frequency range, SA I is most sensitive relatively. The best becomes FA I in the middle range and FA II in the highest range, respectively. This suggests that the selective stimulation can be realized using these frequency characteristics, and arbitral tactile feels can be produced by synthesizing several frequency components.

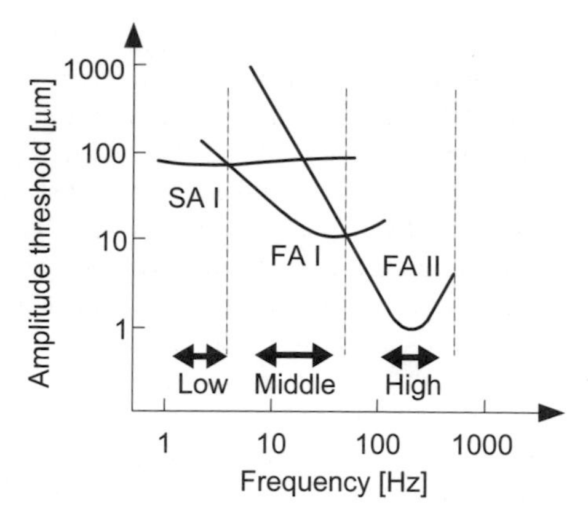


**Figure 9.7.** Thresholds of tactile receptors for vibratory stimulus and selective stimulation ranges (revised from Maeno [33], which was originally based on Talbot and Johnsson[34] and Freeman *et al.* [35]).

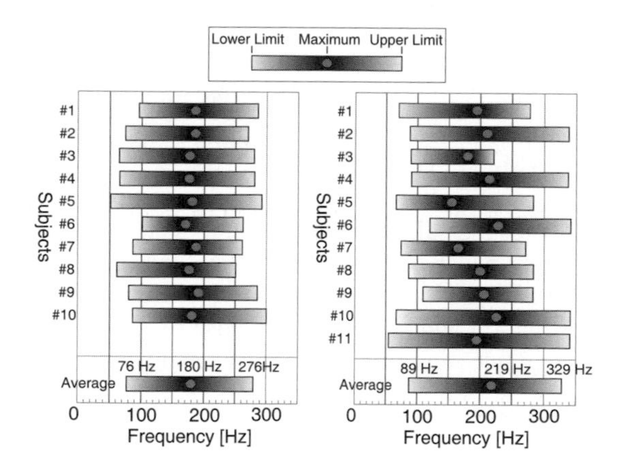


**Figure 9.8.** Perceptual range of simple vibratory sensation

For the IPMC tactile display, selective stimulation is realized by changing drive frequencies, utilizing the receptors' response characteristics. It was confirmed by subject's introspection that the contents of sensation vary with the change of drive frequency as follows:

(1) *Less than 5 Hz*: static pressure sensation (SA I).
(2) *10 – 100 Hz*: periodical pressing or fluttering sensation, as if the surface of a finger is wiped with some rough material (FA I).
(3) *More than 100 Hz*: simple vibratory sensation (FA II).

Figure 9.8 shows the experimental results of the perceptual range of simple vibratory sensations for (a) fixed-type display and (b) wearable display. It considered that the subjects begin to feel simple vibratory sensation when the information from FA II exceeds that from FA I. Figure 9.7 shows that the detection threshold of FA II exceeds that of FA I in the vicinity of a frequency from 50 to 100 Hz. This agrees with the results of the perceptual range of vibratory sensation. To create integrated sensations, a stimulating method using composite waves of several frequencies was proposed. Composite waves can stimulate the different kind of tactile receptors at the same time based on the selective stimulation method. In the earlier experiment using the fixed-type IPMC display [14], composite waves of high and low frequencies that present both pressure sensation and vibratory sensation at the same time were applied. The result clearly shows that over 80 % of the ten subjects sensed some special tactile feeling, which is clearly different from a simple vibratory sensation. The authors confirmed that the composite stimulations of two frequency components selected from both the middle and high frequency range illustrated in Figure 9.8 could produce the various qualitative tactile feelings like cloth such as a towel and denim fabric [14].

### 9.2.4 Texture Synthesis Method

We focused on the following three sensations to produce total textural feeling related to the physical properties of materials: (1) roughness sensation, (2) softness sensation, and (3) friction sensation. These sensations are fundamental to express the textural feel of cloth like materials. The three sensations are produced by the following parameters based on the proposed method described later:

(1) *Roughness sensation:* changes in the frequency and the amplitude caused by the relationship of the wavelength of the desired surface and the hand velocities (Section 9.2.5).
(2) *Softness sensation:* the amount of pressure sensation when the finger contacts the surface (Section 9.2.6).
(3) *Frictional sensation:* changes in the amount of subjective sensation in response to hand accelerations when the finger slides across a surface (Section 9.2.7).

The problem is how to connect the stimulation on each receptor with contact phenomena caused by hand movements and physical properties of objects. We have proposed stimulation methods connected to the relationship between hand movements and the physical properties of objects [17]. For roughness sensation, the frequencies of natural stimuli caused by contacting rough surfaces are changed in response to hand movements. Human beings have the possibility to use those changes of frequencies positively. It is known that the slope of the detection threshold of FA I is –1 in the range of less than 40 Hz, as shown in Figure 9.7. The activities of FA I reflects vibratory frequencies proportionally. This means that FA I can perform as a frequency analyzer in a certain range. Based on this hypothesis, we proposed a frequency modulation method for displaying the roughness sensation in response to hand velocity, as described in the next section.

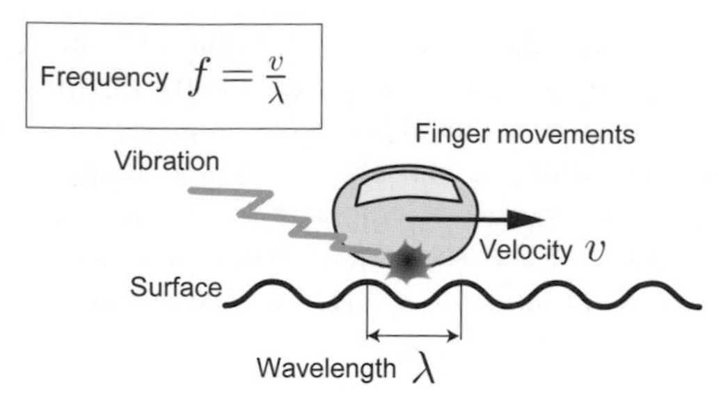


**Figure 9.9.** Definition of surface form using the wavelength

### 9.2.5 Display Method for Roughness Sensation

*9.2.5.1 Method*

As mentioned in Section 9.2.4, we suppose that human beings perceive roughness sensation as the change in frequency detected by FA I in the relationship between their hand movements and the physical properties of the roughness of materials. The roughness of the surface is defined approximately as a sinusoidal surface, which has a given wavelength $\lambda$ as shown in Figure 9.9. When the finger slides on the sinusoidal surface at a given velocity $v$, the frequency of stimuli $f$, which are generated in a finger point, is expressed by a wave equation as follows.

$$f = \frac{v}{\lambda} \tag{9.1}$$

This equation shows that if the hand velocity becomes faster or if the wavelength $\lambda$ becomes shorter, the frequency f increases. We should consider the response characteristics of FA I, which is known as a tactile receptor related to the roughness sensation. It is known that FA I respons to the velocity of mechanical stimuli [32]. Here, when the finger slides across the surface, as shown in Figure 9.9, a displacement of stimulus $y$ at a given time $t$ is defined as a sinusoidal function as follows,

$$y = a\sin(2\pi f t) \tag{9.2}$$

where, $a$ is the amplitude of stimulation. Thus, the velocity of stimulation is expressed by substituting Equation (1) in the following equation.

$$\frac{dy}{dt} = 2\pi a \frac{v}{\lambda} \cos(2\pi \frac{v}{\lambda} t) \tag{9.3}$$

This equation presents the information detected by FA I and shows that both the amplitude $2\pi a v / \lambda$ and the frequency change in response to the velocity $v$. Based on this assumption, the roughness sensation can be presented by changing both the frequency and the amplitude of stimulation in accordance with hand velocity. In this manner, the roughness sensation can be defined by the wavelength $\lambda$. For practical use of this method, we applied phase adjustments to produce smooth outputs in response to changing frequencies with respect to each sampling time.

Note that these frequencies are just in the high responsive range of FA I. Although the proposed frequency-modulation method is not allowed to apply a suitable range of frequency for FA I explicitly, the appropriate frequencies can be generated by human hand movements consequently, when the wavelength is defined of the order of several millimeters.

*9.2.5.2 Evaluations*

As evaluation indexes of roughness sensation, nine kinds of close-set lead balls that had different diameters from 0.5 to 10 mm were used as shown in Figure 9.10. The wearable tactile display system shown in Figure 9.5 was used. The amplitudes of stimulations were fixed at 6.0 V (= the maximum input) and each offset was 0.5 V. The offset was needed to avoid an insensitive zone caused by shortage of amplitudes of the actuators.

The subjects put the device on the right middle finger. They touched the index with their left hand at the same time. There was no restriction on time to explore. The subjects were six males in their twenties.

Figure 9.11 shows the relationship between the defined wavelengths and the mean value of selected indexes with each error bar representing one standard deviation. The results showed that as the defined wavelength became longer, the roughness sensation seemed to increase when the two half groups were considered separately. Especially, as the wavelengths became shorter, the standard deviations became smaller and the roughness sensations were expressed clearly.

From the results, it was confirmed that roughness sensation could be expressed by the parameter of the wavelength in the case of relatively short wavelengths. In addition to the wavelength, it is confirmed that the maximum amplitude of stimulus affects the amount of the subjective sensation of roughness.

**Figure 9.10.** Overview of indexes of roughness

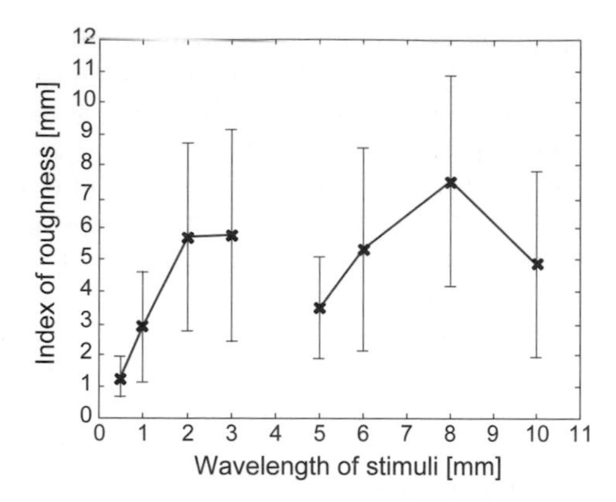


**Figure 9.11.** Wavelength of stimuli vs. average indexes of roughness sensation

### 9.2.6. Display Method for Pressure Sensation

*9.2.6.1 Method*

It is known that SA I detects static deformations of the skin and generates static pressure sensation [32]. Therefore, selective stimulation on SA I can generate pressure sensations. As shown in Figure 9.7, the detection thresholds of SA I hasve flat frequency characteristics in the range of less than 100 Hz. In most of the range of Figure 9.2, FA I is more sensitive than SA I. However, in the range of less than 5 Hz, SA I becomes more sensitive than FA I. This means that the very low frequency vibration can generate pressure sensations relatively larger than the sensation of FA I. The authors confirmed that this assumption was true when the amplitude of simulation was enough small not to sense the vibratory sensation.

*9.2.6.2 Evaluations*

In this experiment, the wearable tactile display system shown in Figure 9.3b was used. The subjects put the device on the right middle finger. They could perform

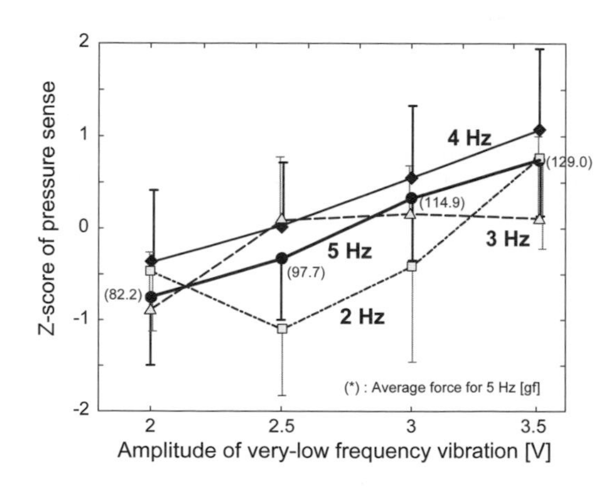


**Figure 9.12.** Pressure force vs. driving voltage of low-frequency stimulation for SA I

stroke motions in the horizontal direction. The stimulation was simple sinusoidal vibrations at a frequency from 2 to 5 Hz. The stimulations were generated only when the hand velocity was higher than 25 mm/s despite the direction of movement. For measuring pressure sensation, the subjects pushed their left middle finger on a sponge that was set on an electric balance, controlling their finger to the same amount of pressure sensation of the artificial pressure sensation for 3 seconds. And then, the amount of the pressure sensation was calculated as the mean of the force for 3 seconds.

Figure 9.12 shows the relationship between the amplitude of vibration and the amount of pressure sensation at each frequency. The amounts of pressure sensation were calculated by a Z-score because the subjects had different sensitivities for the amount of the subjective sensation. The number in the parenthesis shows the mean value of actual forces at the frequency of 5 Hz as a reference. It was confirmed that as the amplitudes increase, the pressure sensations became larger for every frequency component. Utilizing this method, the softness of materials, which we feel instantaneously when the finger touches a surface, can be expressed by the parameter of amplitude for the frequency components of 5 Hz. If the pressure sensation is larger, the contacting object has more stiffness.

### 9.2.7 Display Method for Friction Sensation

To express a cloth-like textural feeling in response to contact motions, synthesis of both the roughness sensation and softness sensation is not enough. In this section, we introduce friction sensation. In this study, the definition of friction sensation is not a usual description based on physical contact conditions. We assumed that the friction sensation can be produced as changes in the amount of subjective sensation in response to hand acceleration when the finger slides across the surface. Especially, the friction sensation is used for expressing the sticking tendency of materials at the beginning of sliding motion.

The authors confirmed that stimulation of high-frequency components corresponding to the acceleration of hand movements could produce a natural sliding feeling [16]. It is known that FA II detects the acceleration of stimuli, and it seems that FA II is related to the detection of hand movements such as by a gyro sensor. Figure 9.13 illustrates the relationship between hand acceleration and amplitudes of the high-frequency component. The high-frequency component is fixed at 200 Hz, in which FA II become most sensitive. Therefore, the parameters of the friction sensation are the maximum and minimum values of the amplitude shown in Figure 9.13.

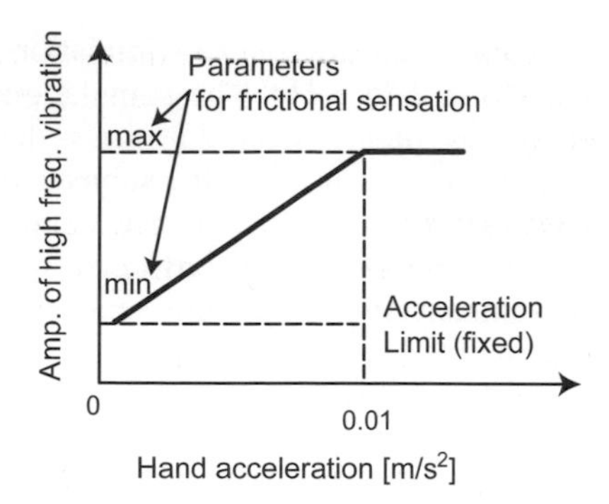


**Figure 9.13.** Relation between the amplitude of high-frequency components for the friction sensation and the acceleration of hand movements

### 9.2.8 Synthesis of Total Textural Feeling

*9.2.8.1 Method*

In this section, syntheses of total textural feeling related to the physical properties of materials based on the three methods described above were evaluated. The voltage inputs generated by the three methods were combined into a signal by a simple superposition. Four materials were selected as targets of the tactile syntheses. The artificial textural feelings were tuned subjectively by changing the parameters of the roughness, softness, and friction sensations. The tunings of textural feelings were extremely easy compared with the author's conventional study because each parameter was related to the physical properties of the materials. The following were the properties of the four materials and the tuned parameters:

(1) *Boa*: shaggy, thick, uneven and very rough surface
($\lambda = 10$, $a = 5.0$, $P = 0.0$, $F_{max} = 2.0$)
(2) *Towel*: rough surface, thick, and soft
($\lambda = 2.0$, $a = 3.0$, $P = 2.0$, $F_{max} = 1.0$)
(3) *Fake leather*: flat surface, thin, hard, and high friction
($\lambda = 8.0$, $a = 1.0$, $P = 4.0$, $F_{max} = 3.0$)
(4) *Fleece*: smooth surface, thin, soft, and low friction
($\lambda =0.5 = 1.0$, $P = 5.0$, $F_{max} = 1.0$)

*9.2.8.2 Evaluations*

As shown in Figure 9.14, four artificial textures, which were tuned as mentioned above, were set in a matrix. The four real materials, which were boa, towel, fleece, and fake leather, were put on the cardboard in the same order as the artificial textural feelings. The wearable tactile display system shown in Figure 9.5 was used. The subjects put the device on the right middle or index finger. They could perform stroke motions with their left hand in the horizontal direction. Before the experiments began, the subjects had experience with the four textural feelings only once. The subjects compared each artificial texture with the corresponding real material. They were to evaluate the similarity of the both feelings at five levels (1: Poor, 2: Fair, 3:Good, 4:Very Good, and 5:Excellent). There was no restriction on time to explore the textures.

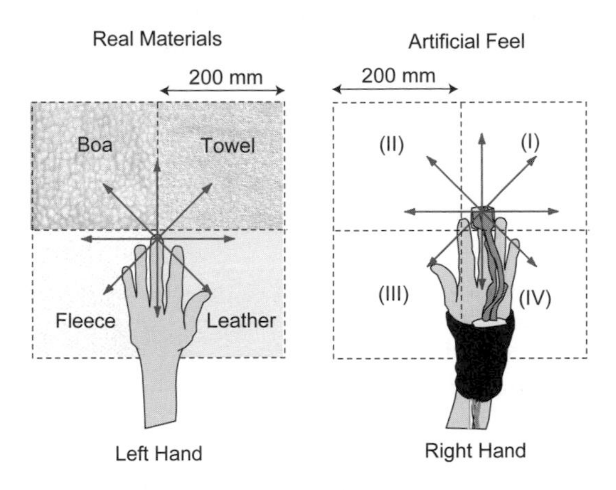


**Figure 9.14.** Comparison between real materials and artificial tactile feelings

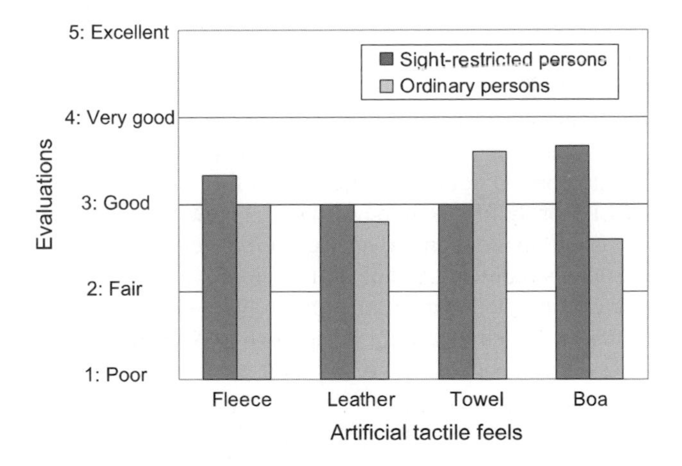


**Figure 9.15.** Evaluations of artificial tactile feeling compared with the real materials

The subjects were divided into two groups: three sight-restricted people (two females in their fifties and one female in her forties) and five ordinary persons (five males in their twenties). The sight-restricted people have more sensitive tactile sensation than ordinary persons. It was expected that the sight-restricted people could evaluate more correctly.

Figure 9.15 shows the evaluation results for the sight-restricted people and the ordinary persons, respectively. Both of the sight-restricted people and the ordinary persons judged more than score of 3, that is "Good", for the almost all artificial textures. These results demonstrated that the proposed methods could synthesize the artificial textural feeling corresponding to the real materials. In addition, the sight-restricted people gave higher evaluations than the ordinary persons so that the synthesized textural feelings had the reasonable reality.

Our tactile synthesis method is based on the physical properties of a material. These parameters of textural feeling can be measured as physical properties. This means that the artificial textural feelings could be synthesized automatically, if the tactile sensors could detect such physical parameters. The authors are also developing the tactile transmission system combining the tactile display and tactile sensors as a master-slave system.

## 9.3. Distributed Actuation Device

The softness of end-effectors is important in manipulation of soft objects like organs, food materials, micro-objects, *etc*. This softness can be actualized using two approaches: (1) drive by hard actuators with soft attachments and (2) direct drive by soft actuators by themselves. The former appears to be a sure method because of present technological development. However, to create micromachines or compact machines like miniature robot hands, the former is limited so it is difficult to find a breakthrough. The problem with the latter is that a readily available soft actuator material does not exist. However, the material revolution currently underway will surely result in the discovery of an appropriate material in the near future. For these reasons, it is meaningful to study methodologies for the effective use of such materials for manipulation with an eye to future applications.

A promising candidate for such a soft actuator material is gel. Many gel materials for actuators have been studied up to the present. The Nafion-platinum composite (IPMC or ICPF) is a new material that is closest to satisfying the requirements for our applications. Because such materials are soft, it is impossible to apply large forces/moments at only a few points on an object, contrary to the case with conventional robot manipulation. At the same time, however, it is an advantage that large pressures cannot be applied actively or passively. So as not to detract from this feature, a number of actuator elements should be distributed for applying the driving force.

The distributed drive is also desirable from the viewpoint of robust manipulation. Even if there are elements that cannot generate appropriate force, in principle, it is possible for the other elements to compensate for them. This signifies insensitivity to environmental fluctuation. In human bodies, for example, excretion of alien substances is performed by a whipping motion of numerous cilia. Paramecia move by paddling their cilia. Centipedes crawl by the cooperative wavy motion of a number of legs. Any of these can robustly accomplish their objectives irrespective of environmental change.

An elliptical friction drive (EFD) element is an actuator element that generates driving force by friction using bending actuators. Figure 9.16 shows an experimental development using the Nafion-Pt composite. It has two actuator parts with platinum plating for actuation and one Nafion part without plating for an elastic connection. The whole structure is fixed to form the shape of an arch.

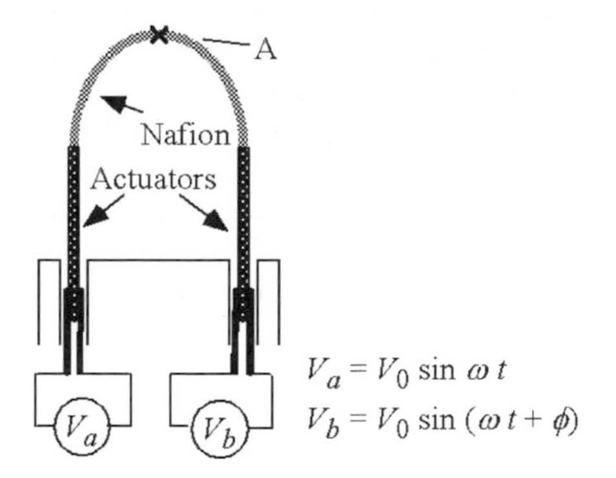


**Figure 9.16.** Structure of EFD actuator element

When sinusoidal voltages with a phase difference are applied to the two actuators, the excited sinusoidal bending motions also have a phase difference. This results in an elliptical motion at the top point (A) of the connecting part. Figure 9.17 shows a developed distributed EFD device. It has 5 × 8 EFD elements on a plate. They cooperatively apply a driving force to an object.

The driving principle is shown in Figure 9.18. Adjacent elements make elliptical motions with a phase difference of $\pi$ (a two-phase drive). On the planar contact face, a frictional force in the $x$ direction is generated alternately by adjacent elements, and then the object is driven.

This element could be applied to a robot hand, for example, as shown in Figure 9.19. The Nafion-Pt composite is produced by a process consisting of surface roughening, adsorption of platinum, reduction, and growth on a Nafion membrane. A masking technique using crepe paper tape with a polyethylene coating can be used to form any arbitrary shape of actuator on the Nafion. This technique is called the pattern plating method. It is an essential technique for creating the various shapes in the gel material required for the actuator. It is also important for supplying electricity efficiently.

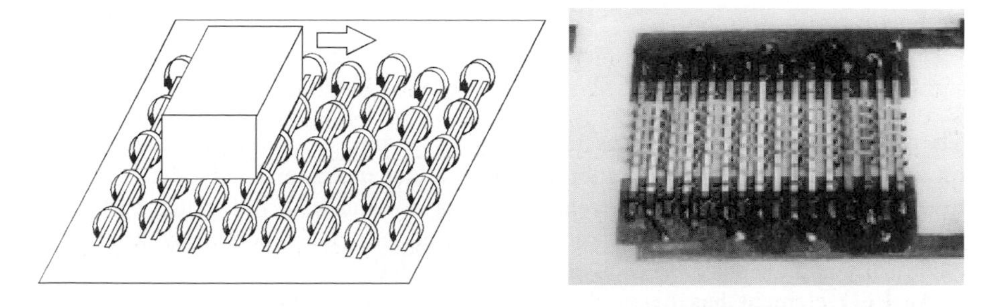

**Figure 9.17.** Distributed actuation device consisting of multiple EFD elements

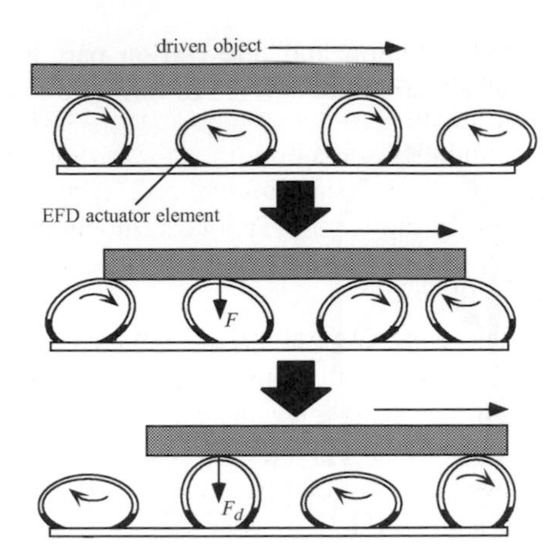


**Figure 9.18.** Principle of distributed drive

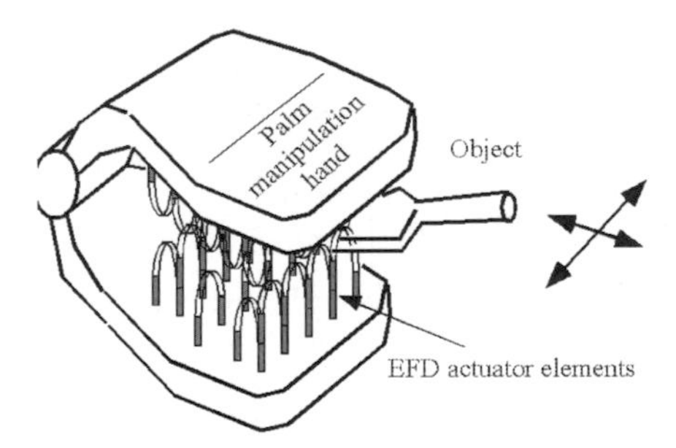


**Figure 9.19.** Application of device developed as a robotic hand (palm manipulation hand)

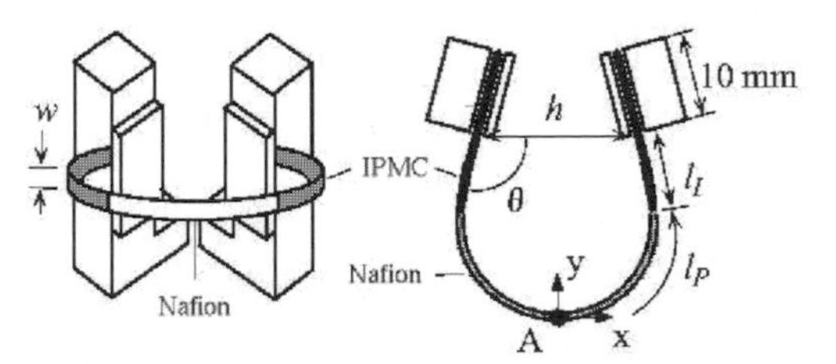


**Figure 9.20.** Design parameters of EFD element

The manufacturing process is as follows. First, composite actuators are produced on the supports of a Nafion membrane cut in a ladder shape. The ladder is formed into rings by rolling the membrane around a column in a hot water bath. The resultant multiple EFDs are fixed together on a plate. Then, they are wired electrically and the shape of each element is adjusted.

An EFD element has many design parameters of mechanism and control, as shown in Figure 9.20. These parameters have effects on the performance of the element. It is difficult for analytical methods to give an optimal design of these parameters because (1) the actuator part is not a point, (2) the motion of each part of the actuator is not uniform (output internal stress and time constant), and (3) function of each part interferes each other.

Investigating changing parameters by trial and error, such as in Figure 9.21, is necessary. Models and simulation tools minimize the number of experiments.

The design parameters of the mechanism and control are determined by simulation and analysis using the Kanno-Tadokoro model and an assumption of a viscous friction driving mechanism, as shown in Figure 9.22. The mechanical parameters are listed in Table 9.1.

Figure 9.23 shows the experimental result obtained by varying the phase difference, $\phi$, between the sinusoidal input voltages for each EFD element. The shape of the elliptical motion changes according to this phase difference. At the same time, the velocity of the plate changes. Consideration of the analytical and experimental results shows that the elliptical motion in the $x$ direction becomes large when the two phases are similar, and that the $y$ displacement increases when the difference between them is close to $\pi$. The speed of the object is determined by these two displacements. The optimal value found in this experiment is $\phi = \pi/2$.

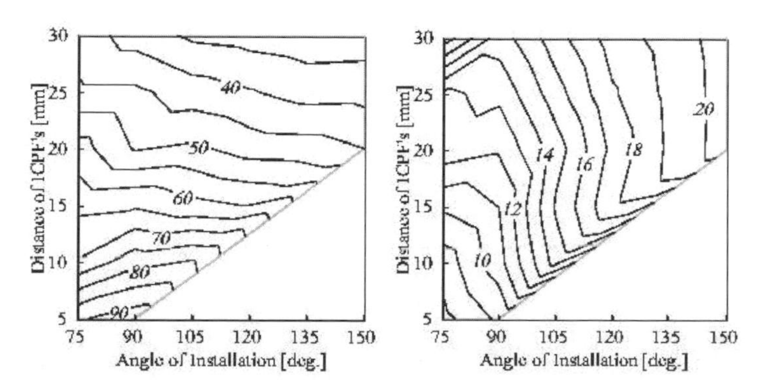


**Figure 9.21.** Motion of EFD tip vs. design parameters (1 = 62 μm)

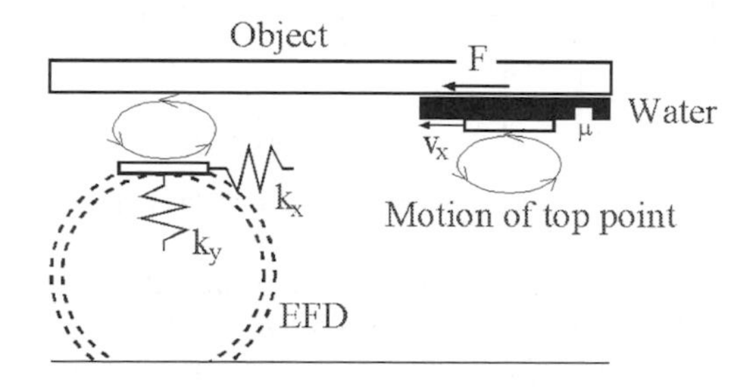


**Figure 9.22.** Mechanism of distributed drive

**Table 9.1.** Mechanical design parameters of a distributed device

| Parameter | Value |
|---|---|
| Number of elements ($N$) | 40 |
| Thickness ($t$) | 180 $\mu$m |
| Width ($w$) | 1 mm |
| Length of ICPF part ($l_I$) | 2 mm |
| Length of PFS part ($l_P$) | 11 mm |
| Distance between supports ($h$) | 7 mm |
| Angle of supports ($\theta$) | 180° |
| Element interval in the x direction ($d_x$) | 14 mm |
| Element interval in the y direction ($d_y$) | 4 mm |

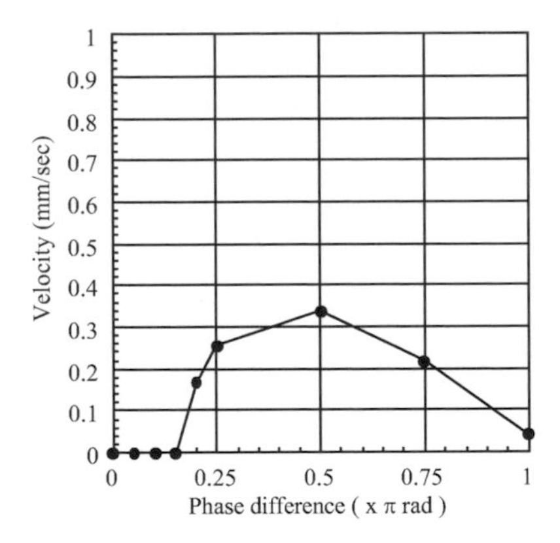


**Figure 9.23.** Effect of phase difference between the input voltage on the velocity of plate transfer ($f$ = 2 Hz, $V_0$ = 1.5 V)

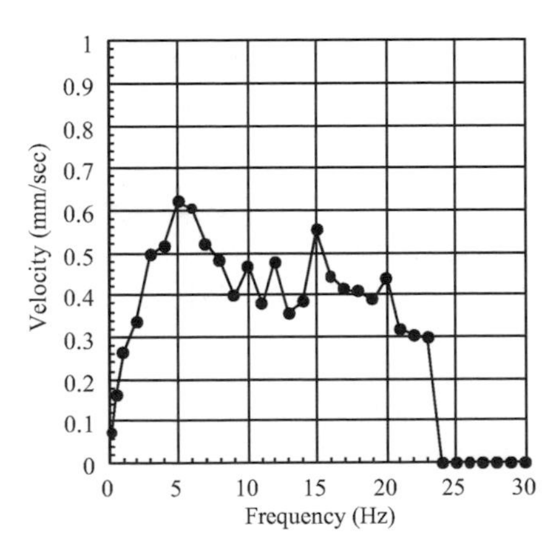


**Figure 9.24.** Effect of input voltage frequency on the velocity of plate transfer ($V_0$ = 1.5 V, $\phi$ = 0.5 π rad)

The results from varying the frequency of the sinusoidal input are shown in Figure 9.24. Analytical results, assuming elasticity of the material, indicate that the resonant frequencies of the element are approximately 5, 10, and 16 Hz. The experimental curve shows good correspondence because the speed is high at these frequencies. The viscous resistance of water is so minor that complex flow near the device can be ignored. The major analysis error results from a modeling error of the contact and the element shape. Under 3 Hz, the experimental result indicates that the velocity increases linearly with frequency. In this range, the speed of the elements is proportional to the frequency because the effect of distributed flexibility is small.

The resonant frequencies vary by 20% depending on the shape of the elements. These frequencies depend on the initial shape and the pressure exerted by the object being manipulated.

## 9.4. Soft Micromanipulation Device with Three Degrees of Freedom

Micromanipulators require the following elements: (1) compact micromechanisms, (2) passive softness, (3) many DOF of motion, and (4) multimodal human feedback. The microactuator is one of the key issues for actualization of such advanced micromanipulators. They are difficult to construct using conventional actuators [36–38].

The important features of the actuators are (1) soft material, (2) force output, (3) ease of miniaturization and machining, and (4) multi-DOF motion ability. A 3-DOF manipulation device is developed by crossing a pair of EFD elements perpendicularly at the end point, as shown in Figure 9.25.

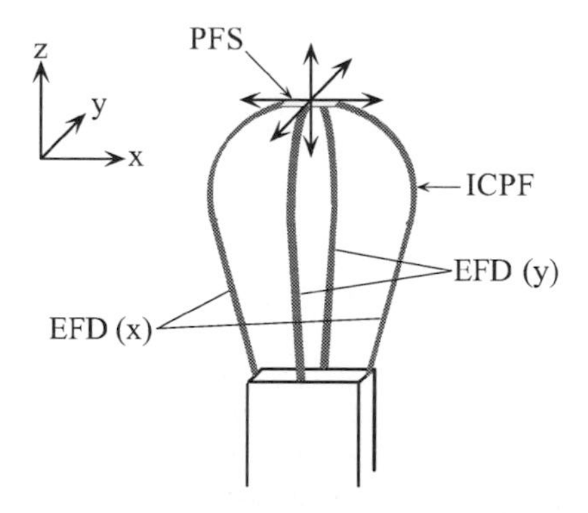


**Figure 9.25.** Structure of the 3-DOF micromotion device developed

The most important factors in its design are: (1) high flexibility and softness, (2) minimum internal force, and (3) large displacement (especially in 2 DOF). According to a characteristic synthesis using the Kanno-Tadokoro model, the width of the actuator is designed to be as thin as possible ($w$ = 0.4 mm), and the pair of actuators is installed on a fixture in parallel. When external force and moment $\boldsymbol{f}_i = [f_x, f_y, f_z, n_x, n_y, n_z]^T$ are exerted at the end of the $i$'th actuator segment, They make a minute translation and rotation $\boldsymbol{x}_i = [x, y, z, a_x, a_y, a_z]^T$. The relation is approximated by

$$x_i = C_i f_i, \tag{9.4}$$

where $\boldsymbol{C}_i$ is a compliance matrix. It is a constant matrix if the actuator can be modeled as an elastic body, as in Kanno *et al.* [18]. Considering the viscoelastic property, it becomes a time-varying matrix depending on the deformation history. Because minute translation and rotation of each actuator are expressed by

$$x_i = D_i x_0 \tag{9.5}$$

using the minute motion of the end-effector $\boldsymbol{x}_o$, the force and moment exerted on the end $\boldsymbol{f}_o$ have a relation of a compliance $\boldsymbol{C}_o$ of the end-effector

$$x_0 = C_0 f_0 \tag{9.6}$$

$$C_0 = \left( \sum_i D_i^T C_i^{-1} D_i \right)^{-1}. \tag{9.7}$$

Using an analysis based on the Kanno-Tadokoro model, it was revealed that the developed device was much softer than conventional micromanipulators because the compliance matrix under a static equilibrium condition was

$$C_0 = \begin{bmatrix} 1.9 & 0 & 0 & 0 & 9.3\times10 & 0 \\ 0 & 1.9 & 0 & -9.3\times10 & 0 & 0 \\ 0 & 0 & 3.5\times10^{-1} & 0 & 0 & 0 \\ 0 & -9.3\times10 & 0 & 2.2\times10^4 & 0 & 0 \\ 9.3\times10 & 0 & 0 & 0 & 2.2\times10^4 & 0 \\ 0 & 0 & 0 & 0 & 0 & 1.6\times10^4 \end{bmatrix} \text{[m/N, rad/Nm]} \tag{9.8}$$

Therefore, this device is safer than a conventional device in manipulation of fragile microstructures. Only the dynamics of the micromanipulator itself were considered here. The actual compliance is affected by surface tension, Van der Waals force, electrostatic force, *etc*. Addressing this issue is beyond the scope of this chapter because these parameters depend on the working environment and cannot be estimated for the manipulator.

Figure 9.26 shows the relation between the magnitude of step input voltage and the resultant maximum displacement. Frequency characteristics under sinusoidal input are shown in Figure 9.27. All experiments in this chapter were performed in water. It was observed that the displacement increases by a quadratic curve according to the voltage. Displacement in the *z* direction is small because the IPMC (or ICPF) actuators are installed in parallel and this characteristic has been predicted at the stage of design. Displacement in the *xy* direction (diagonal motion in a direction 45° to the *x* axis) is larger than the others because of the effect of the section shape and reduction of internal friction caused by water molecule movement in the material. The maximum displacement observed was 2 mm, and the available frequency range was up to 13 Hz. These performances are sufficient for the micromanipulation application. These characteristics can be predicted by computer simulation using models. In this application, rough estimation was performed using the Kanno-Tadokoro model. However, the accuracy was insufficient to determine the final design.

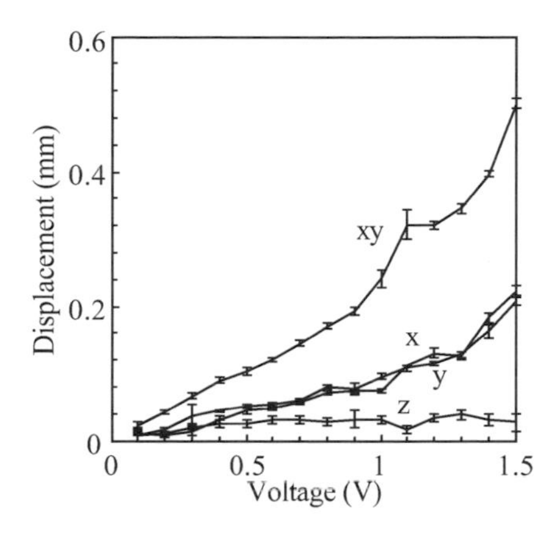


**Figure 9.26.** Effect of the step input voltage on the displacement

Figure 9.28 illustrates experimental Lissajou figures in the *xy* plane. This demonstrates that the developed device has sufficient motion for micro-manipulation.

The feasibility of telemanipulation by the 3-DOF device was proven using the experimental setup shown in Figure 9.29. Motion commands from an operator are given by a joystick and transmitted to the device via a PC. Actual motion is fed back to the operator by a microscopic image on a monitor. Figure 9.30 is a photographic view of the manipulation setup. The following are revealed as a result of various motion tests:

(1) If the joystick is moved without rest, the device responds to high-speed motion commands completely, and the operator can control the motion very easily.

(2) When the joystick stops, the device does not stop and return to the initial position. The period for which the device could stop at arbitrary points was 3 seconds. The latter characteristic exists because the Nafion-Pt composite material is not a position-type actuator but a force-type actuator.

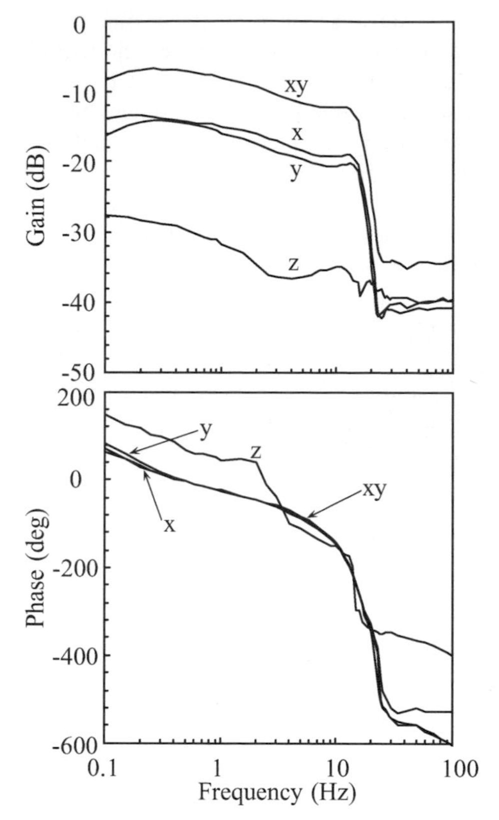


**Figure 9.27.** Frequency characteristics [input: 1.5 V sinusoidal waves, output: displacement (mm)]

Experiments have shown that the new device is capable of supporting dynamic micromanipulation strategies where dynamic conditions, such as adhesion and pushing, can be major. Gripping or grasping is possible using this manipulator, particularly for short-duration applications that require soft handling. Although effective operation was observed, issues of control still require improvement.

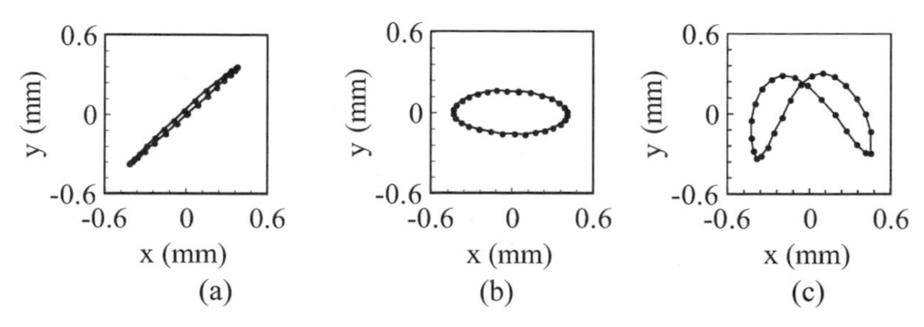


**Figure 9.28.** Lissajou motion of the 3-DOF device in *xy* plane. (a) $V = 1.5$ V, (b) $\phi = 0.5$, $V = 0.9$, 1.5 V, (c) $f = 1$ Hz and 2 Hz.

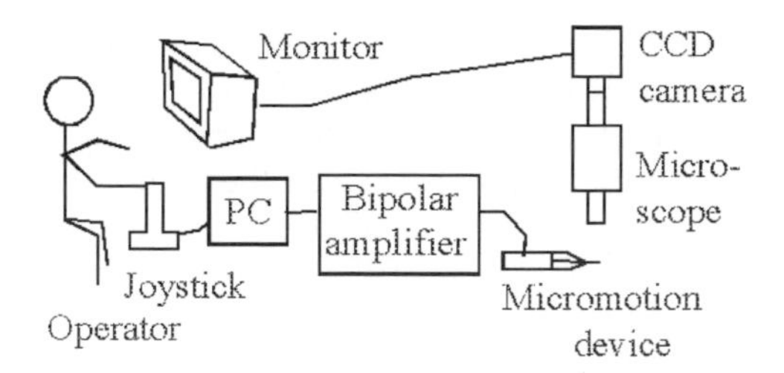


**Figure 9.29.** Experimental setup for telemanipulation

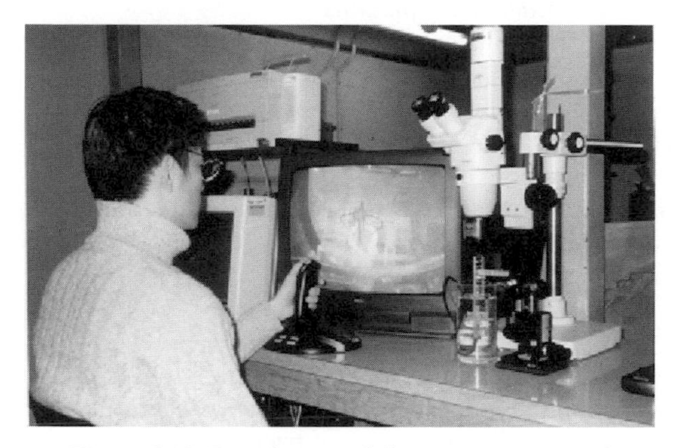

**Figure 9.30.** Experiments of direct operator control

## 9.5 IPMC Sensors

### 9.5.1 Background

'A soft sensing system is extremely important for advanced applications of IPMC actuators, because conventional solid sensors may cancel the flexibility of IPMC. One possible sensor would be IPMC itself. It was reported that IPMC could be a motion or pressure sensor [39–40], because an electric potential would be generated across the composite when the strip is bent suddenly. Although Shahinpoor *et al.*[39] reported that IPMC sensor output was in proportion to the quasi-static displacement of bending, it is easily confirmed that an IPMC sensor does not respond to static deformation and that there are phase differences between the sensor outputs and the displacements. The authors investigated the relationship between the sensor output and dynamic deformations and showed that the velocity of deformation was in proportion to the sensor output voltage [28].

An IPMC sensor has the following advantages in common with an IPMC actuator's features: (1) Soft material does not cause injury to an object and does not interfere with movements of soft actuators, (2) Light weight and simple structure make device size small.

For example, if an IPMC sensor were used in an active catheter system [3,4], the system could detect contacts with blood vessels without causing injury and could avoid a critical operational error. Another example is utilization for a tactile display [14–17], as mentioned above. If IPMC sensors were arranged among the vibrators in the same shape and materials, the display system could control the stimulation by measuring the pressure between the ciliary device and the human skin without a negative influence upon tactile feel.

In this manner, a good combination of IPMC sensors and actuators will extend the ability of EAP applications. The following three arrangements can be considered to combine them:

(1) *Switching the sensor/actuator functions*
Both IPMC sensors and actuators can be operated by the same electrodes, if the electric circuits can be switched in two ways. This arrangement seems useful for a system that has many end effectors like a multilegged walking robot.

(2) *Parallel arrangement*
IPMC sensors can be set in parallel with the actuators in order to detect the motion of their body caused by actuation or external disturbances. It is possible to control the actuators based on sensor feedbacks.

(3) *Patterning both functions on an IPMC film*
If an IPMC film is separated electrically by cutting grooves, both the sensor and the actuator can be unified in the same film. This arrangement is more effective to sense motion than the parallel arrangement because there is less interference with the actuation by the sensor part. Using the patterned IPMC cilium, an active sensing system like an insect's feeler can be realized by comparing motion command and sensor feedback.

### 9.5.2 Basic Characteristics of IPMC Sensor

To investigate the basic characteristics of an IPMC sensor, the relationship between the sensor output and the displacements of vibration were measured, when the IPMC strips were bent by free vibrations. Multiple vibrations with different frequencies were measured by changing the length of an IPMC strip from 8 to 15 mm. The experimental setup is shown in Figure 9.31. A cantilevered IPMC strip was fixed by electrodes. The tip of the strip was hit by the rotary hummer periodically (about 1 Hz) and made a free vibration. An Au-Nafion composite type IPMC [41], which contained the sodium ion, was applied as a sample. The direct-current components of the sensor output are not stable due to the hysteretic influence of previous motion. To avoid the influence, the alternating current components were extracted by a high-pass filter (cutoff frequency: 0.5 Hz), which was set ahead of the amplifier.

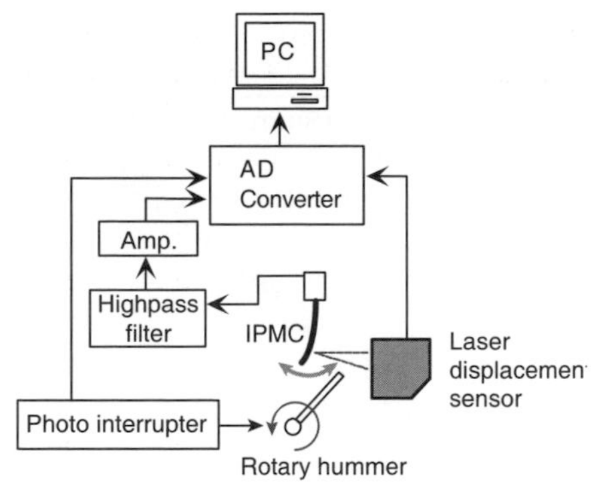


**Figure 9.31.** Experimental setup

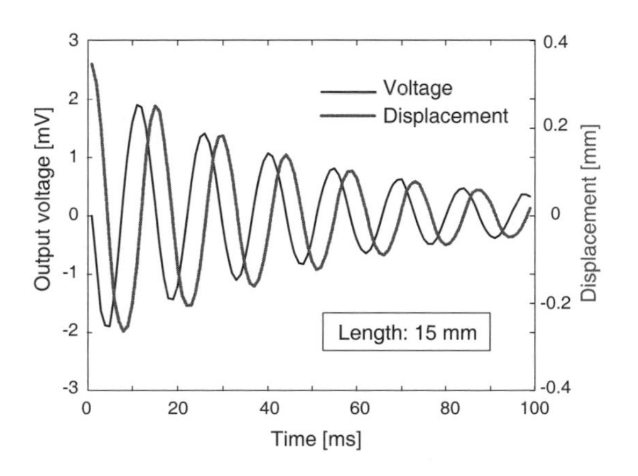


**Figure 9.32.** Displacements vs. sensor output

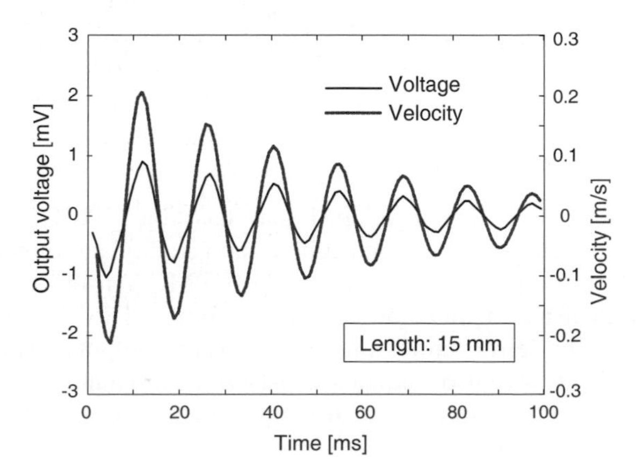


**Figure 9.33.** Velocities vs. sensor output

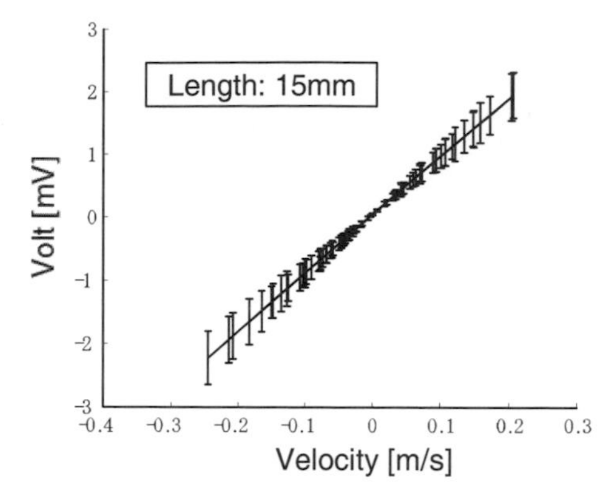


**Figure 9.34.** The relationship between velocities and sensor outputs

Figure 9.32 shows the example of the relationship between the displacement and the sensor output for the length of 15 mm. The displacements were measured at points 5 mm inside from the free ends.

The outputs were generated in the same frequencies as each free vibration, and the mutual relationship between the amplitude of vibration and that of output was sufficiently estimated from the result of measurement. However, it was confirmed that there the sensor outputs had a phasedelay of approximately 90° toward the displacements. These results suggest that the sensor generates voltages in response to the physical value delayed on the displacement by 90°, that is, the velocity that is given by the differentiation of the displacement.

Figure 9.33 shows the results of the relationship, corresponding to Figure 9.32, between sensor outputs and velocities, which were calculated by the difference of the displacements at each sampling time (1 ms). This figure shows clearly that the phases of the velocity are synchronized exactly with that of the output. Figure 9.33 shows the relationship of the velocities and the sensor outputs on the 15 mm length of IPMC. These results showed that an excellent linear relationship exists between the sensor output and the velocity of bending motion despite the length of the IPMC.

### 9.5.3 Three-DOF Tactile Sensor

A 3-DOF tactile sensor was developed that has four IPMC sensor modules combined in a cross shape and can detect both the velocity and direction of motion of the center tip. The parallel arrangements of IPMC sensors contribute to the sensing ability to detect a multi degree of freedom and to the improvement of sensing accuracy by error correcting with several outputs. This cross-shape structure of the IPMC was also studied as a 3-DOF manipulator [10]. If the electric circuits could be switched to actuator driving circuits, the 3-DOF tactile sensors would perform as a soft manipulator.

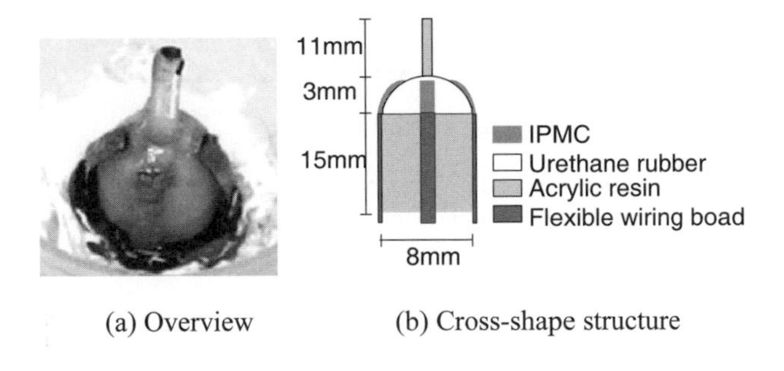


(a) Overview (b) Cross-shape structure

**Figure 9.35.** The structure of the 3-DOF tactile sensor

Figure 9.36 illustrates the structure of the 3-DOF tactile sensor. Four IPMC strips are combined at the center pole in a cross shape. The center pole is also connected to the domed urethane rubber, which has enough softness and durability and can move in multiple directions. This center pole has the function of extending the deformation of the IPMC strip. To make a quantitative vibratory stimulation, the tip of the center pole was connected to an arm module with a low-adhesiveness bond. The sensor outputs were recorded when the arm module made a sinusoidal motion at several frequencies. The displacements of the tip of the center pole were measured by a laser displacement sensor. In addition, to change in the angle of vibration, the sensor rotated 15° at a time from 0° to 180° as shown in Figure 9.36.

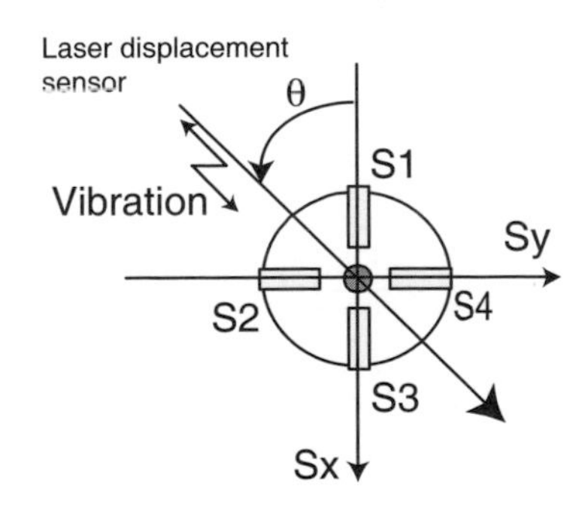


**Figure 9.36.** Rotational angle of vibratory stimuli

The 3-DOF tactile sensor can detect both the velocity and the direction of motion of the center pole by calculating from the four outputs of the IPMC sensors. The four sensors, however, have individual differences in their outputs, because of individual differences in the IPMC sensor itself and structural differences in the manufacturing process. In this study, the four sensor outputs were calibrated by the mean of the peak-to-peak value of sensor outputs when the rotated angle was 0°. and the frequency was 1 Hz on each sensor.

The direction of motion can be estimated by the relationship of the four sensors. As shown in Figure 9.36, consider two axes of *Sx* and *Sy*, and consider the four sensor outputs are *S1*, *S2*, *S3, and S4*. Supposing $V_X$ and $V_Y$ are the components of the the velocity on *Sx* and *Sy*, they can be expressed by the four sensor output as follows

$$V_X = k(S3 - S1) \tag{9.9}$$

$$V_Y = k(S4 - S2) \tag{9.10}$$

where, $k$ is the proportionality constant. Hence, the angle of the motion can be estimated by the relation of $V_X$ and $V_Y$ as follows:

$$V_Y = V_X \tan\theta \tag{9.11}$$

Figure 9.37 shows the comparison between the estimated angle and the theoretical angle by plotting the value of Equation (9.11) and calculating the regression line by the least-squares method for the vibration angles from 0° to 45°. These experimental results show that the estimated angles are in approximate agreement with the theoretical angles.

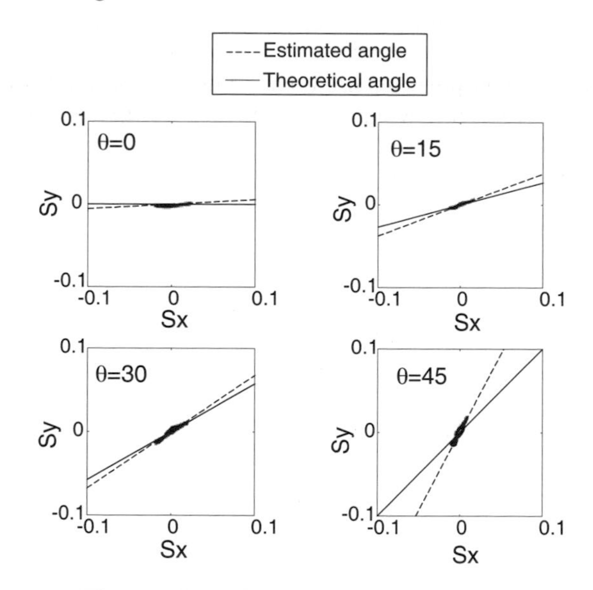


**Figure 9.37.** Estimated directions of motions

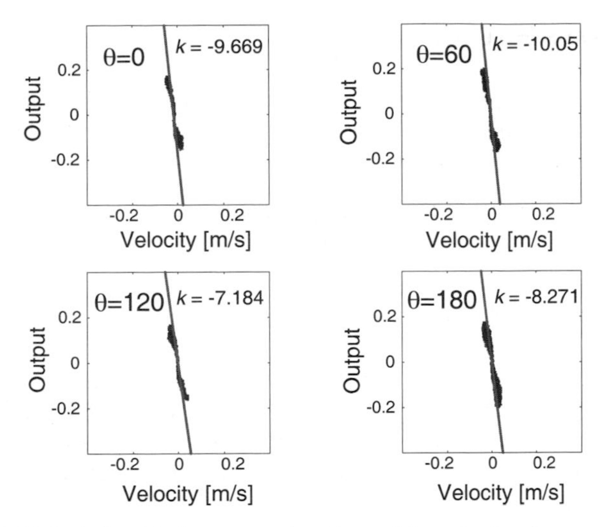


**Figure 9.38.** Relationship between velocities and the calculated sensor outputs

The velocity of the center pole can also be estimated by the vectors $V_X$ and $V_Y$. The velocity estimation is calculated separately according to the condition of the angle estimation as follows:

When $0 < \theta < 90$ :

$$V = k[(S3 - S1)\cos\theta + (S4 - S2)\sin\theta] \tag{9.12}$$

When $90 < \theta < 180$ :

$$V = k[-(S3 - S1)\cos\theta + (S4 - S2)\sin\theta] \tag{9.13}$$

Figure 9.38 shows the relationship between the calculated output and the actual velocity calculated from the displacement of the tip of the center pole, where the frequency of vibration is 1 Hz. The proportionality constants $k$ given by the least-squares method are also shown in the figure. The mean and the standard deviation of the proportionality constant $k$ is calculated as follows

$$k = -8.889 \pm 2.185 \tag{9.14}$$

The velocity of the tip of the center pole can be estimated in realtime by using Equations (9.12) and (9.13), the proportionality constant $k$, and the estimated angle $\theta$.

### 9.5.4 Patterned Sensor on an IPMC Film

If an IPMC film is separated electrically by cutting grooves, both the sensor and the actuator can be unified in the same film. This arrangement is more effective to sense motion than the parallel arrangement because there is less interference with the actuation by the sensor part.

The authors investigated the possibility of a patterned IPMC strip that had both the actuator and the sensor functions [28]. The strip could sense a velocity of bending motion made by the actuator part. As shown in Figure 9.39 an IPMC strip gave a groove to the depth to be isolated using a cutter for acrylic resin board.

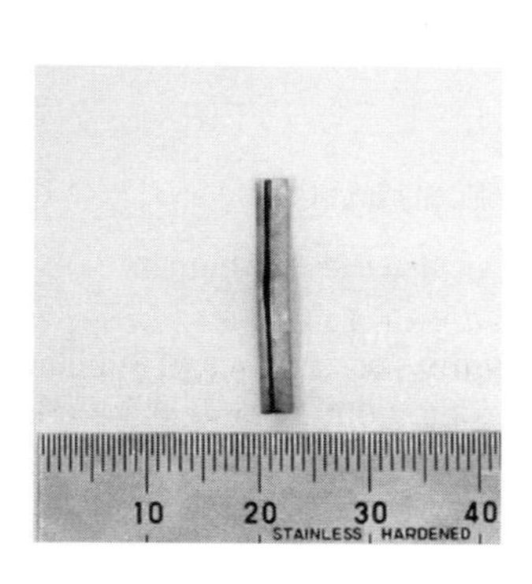


Figure 9.39. Patterned IPMC

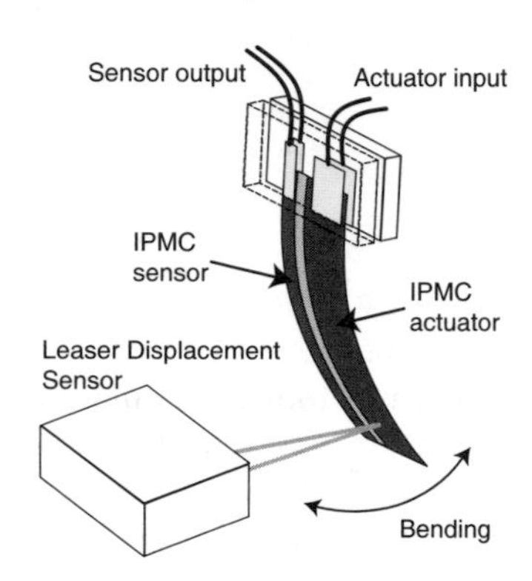


Figure 9.40. Experimental IPMC

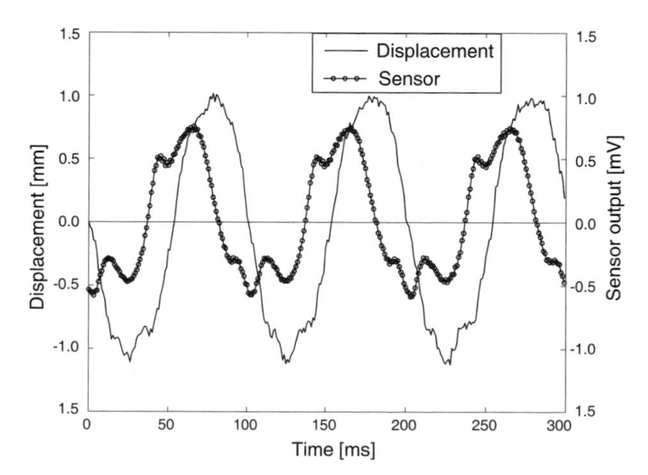


Figure 9.41. Displacement vs. sensor output

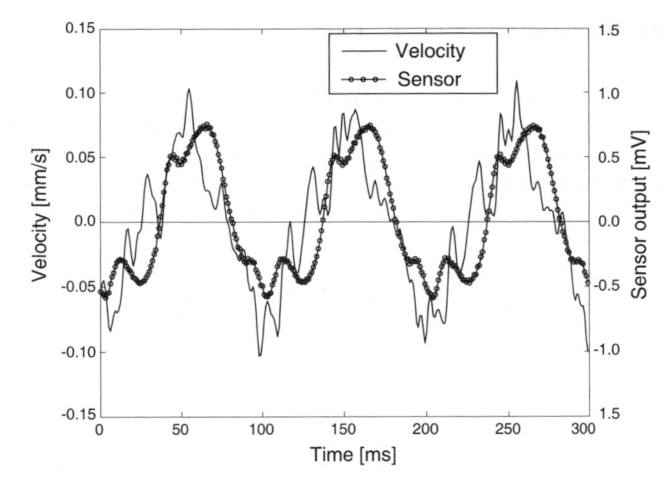


**Figure 9.42.** Velocity vs. sensor output

The size of the strip is 3 × 20 [mm]. The strip is separated into the two sections, the sensor part is 1 mm wide, and the actuator part is 2 mm wide. The experimental setup is shown in Figure 9.40. Two couples of electrodes were arranged for the actuator and the sensor. The actuator part was driven by a sinusoidal input at a frequency of 10 Hz and in an amplitude of 1.5 V. Displacement of the center of the tip was measured by a laser displacement sensor.

Figure 9.41 shows the relationship between the displacement and the sensor voltage. Figure 9.42 also shows the relationship between the velocity and the sensor voltage. It is clear that the latter agrees more with the sensor output, again. The results demonstrate that a patterned IPMC sensor can detect the velocity of the motion made by the actuator part.

This patterning is a preliminary test to investigate the ability of patterned IPMC. Recently, a patterning technique using laser machining, which can cut a groove of 50 μm wide and about 20 μm deep, was developed by the RIKEN Bio-mimetic Control Research Center team [42]. They have developed a multi-DOF robot using the patterned IPMC actuator. If this technique is utilized for the IPMC sensor, an active sensing system like an insect' s feeler can be realized by comparing a motion command and sensor feedback.

## 9.6. Conclusions

In this paper, we described several robotic applications developed using IPMC materials, which the authors have been developed as attractive soft actuators and sensors. We introduced following unique devices as applications of IPCM actuators: (1) haptic interface for virtual tactile displays, (2) distributed actuation devices, and (3) a soft micromanipulation device with three degrees of freedom.

We also focused on aspects of sensor function of IPMC materials. The following applications are described: (1)a three-DOF tactile sensor and (2)a patterned sensor on an IPMC film.

## 9.7 References

[1] Oguro K., Y Kawami, and H. Takenaka, "Bending of an ion-conducting polymer film-electrode composite by an electric stimulus at low voltage," J. of Micromachine Society, Vol. 5, pp. 27-30, 1992.
[2] Shahinpoor M., Conceptual Design, Kinematics and Dynamics of Swimming Robotic Structures using Ionic Polymeric Gel Muscles, Smart Materials and Structures, Vol. 1, No.1, pp.91-94, 1992.
[3] Guo S., T. Fukuda, K. Kosuge, F. Arai, K. Oguro, and M. Negoro, "Micro catheter system with active guide wire," Proc. IEEE International Conference on Robotics and Automation, pp. 79-84, 1995.
[4] Onishi Z., S. Sewa, K. Asaka, N. Fujiwara, and K. Oguro, Bending response of polymer electolete acutator, Proc. SPIE SS-EAPD, pp.121--128, 1999.
[5] Tadokoro S., T. Murakami, S. Fuji, R. Kanno, M. Hattori, and T. Takamori, "An elliptic friction drive element using an ICPF (ionic conducting polymer gel film) actuator," IEEE Control Systems, Vol. 17, No. 3, pp. 60-68, 1997.
[6] Tadokoro S., S. Fuji, M. Fushimi, R. Kanno, T. Kimura, T. Takamori, and K. Oguro, "Development of a distributed actuation device consisting of soft gel actuator elements," Proc. IEEE International Conference on Robotics and Automation, pp. 2155-2160, 1998.
[7] Tadokoro S., S. Fuji, T. Takamori, and K. Oguro, Distributed actuation devices using soft gel actuators, Distributed Manipulation, Kluwer Academic Press, pp. 217-235, 1999.
[8] Guo S., T. Fukuda, N. Kato, and K. Oguro, "Development of underwater microrobot using ICPF actuator," Proc. IEEE International Conference on Robotics and Automation, pp. 1829-1835, 1998.
[9] Tadokoro T., S. Yamagami, M. Ozawa, T. Kimura, T. Takamori, and K. Oguro, "Multi-DOF device for soft micromanipulation consisting of soft gel actuator elements," Proc. IEEE International Conference on Robotics and Automation, pp. 2177-2182, 1999.
[10] Tadokoro S., S. Yamagami, T. Kimura, T. Takamori, and K. Oguro, "Development of a multi-degree-of-freedom micro motion device consisting of soft gel actuators," J. of Robotics and Mechatronics, 2000.
[11] Guo S., S. Hata, K. Sugimoto, T. Fukuda, and K. Oguro, "Development of a new type of capsule micropump," Proc. IEEE International Conference on Robotics and Automation, pp. 2171-2176, 1999.
[12] Bar-Cohen Y., S.P. Leary, K. Oguro, S. Tadokoro, J.S. Harrison, J.G.Smith, and J. Su, "Challenges to the application of IPMC as actuators of planetary mechanisms," Proc. SPIE 7th International Symposium on Smart Structures, Conference on Electro-Active Polymer Actuators and Devices, pp. 140-146, 2000.
[13] Fukuhara M., S. Tadokoro, Y. Bar-Cohen, K. Oguro, and T. Takamori, "A CAE approach in application of Nafion-Pt composite (ICPF) actuators: Analysis for surface wipers of NASA MUSES-CN nanorovers," Proc. SPIE 7th International Symposium on Smart Structures, Conference on Electro-Active Polymer Actuators and Devices, pp. 262-272, 2000.
[14] Konyo M., S. Tadokoro, T. Takamori, and K. Oguro, "Artificial tactile feel display using soft gel actuators," Proc. IEEE International Conference on Robotics and Automation, pp. 3416-3421, 2000.
[15] Konyo M., S. Tadokoro, M. Hira, and T. Takamori, "Quantitative Evaluation of Artificial Tactile Feel Display Integrated with Visual Information", Proc. IEEE International Conference on Intelligent Robotics and Systems, pp. 3060-3065, 2002.

[16] Konyo M., K. Akazawa, S. Tadokoro, and T. Takamori, Wearable Haptic Interface Using ICPF Actuators for Tactile Feel Display in Response to Hand Movements, Journal of Robotics and Mechatronics, Vol. 15, No. 2, pp. 219-226, 2003.
[17] Konyo M., A. Yoshida, S. Tadokoro, and N. Saiwaki, "A tactile synthesis method using multiple frequency vibration for representing virtual touch", IEEE/RSJ International Conference on Intelligent Robots and Systems, pp. 1121-1127, 2005.
[18] Kanno R., A. Kurata, M. Hattori, S. Tadokoro, and T. Takamori, "Characteristics and modeling of ICPF actuator," Proc. Japan-USA Symposium on Flexible Automation, pp. 692-698, 1994.
[19] Kanno R., S. Tadokoro, T. Takamori, M. Hattori, and K. Oguro, "Linear approximate dynamic model of an ICPF (ionic conducting polymer gel film) actuator," Proc. IEEE International Conference on Robotics and Automation, pp. 219-225, 1996.
[20] Kanno R., S. Tadokoro, M. Hattori, T. Takamori, and K. Oguro, "Modeling of ICPF (ionic conducting polymer gel film) actuator, Part 1: Fundamental characteristics and black-box modeling," Trans. of the Japan Society of Mechanical Engineers, Vol. C-62, No. 598, pp. 213-219, 1996(in Japanese).
[21] Kanno R., S. Tadokoro, M. Hattori, T. Takamori, and K. Oguro, "Modeling of ICPF (ionic conducting polymer gel film) actuator, Part 2: Electrical characteristics and linear approximate model," Trans. of the Japan Society of Mechanical Engineers, Vol. C-62, No. 601, pp. 3529-3535, 1996 (in Japanese).
[22] Kanno R., S. Tadokoro, T. Takamori, and K. Oguro, "Modeling of ICPF actuator, Part 3: Considerations of a stress generation function and an approximately linear actuator model," Trans. of the Japan Society of Mechanical Engineers, Vol. C-63, No. 611, pp. 2345-2350, 1997 (in Japanese).
[23] Firoozbakhsh K., M. Shahinpoor, and M. Shavandi, "Mathematical modeling of ionic-interactions and deformation in ionic polymer-metal composite artificial muscles," Proc. SPIE Smart Structure and Material Conference, Proc. SPIE Vol. 3323, pp. 577-587, 1998.
[24] Shahinpoor M., "Active polyelectrolyte gels as electrically controllable artificial muscles and intelligent network structures, Structronic Systems: Smart Structures, Devices and Systems, Part II: Systems and Control," World Scientific, pp. 31-85, 1998.
[25] Tadokoro S., S. Yamagami, T. Takamori, and K. Oguro, "Modeling of Nafion-Pt composite actuators (ICPF) by ionic motion," Proc. SPIE 7th International Symposium on Smart Structures, Conference on Electro-Active Polymer Actuators and Devices, pp. 92-102, 2000.
[26] Tadokoro S., S. Yamagami, T. Takamori, and K. Oguro, "An actuator model of ICPF for robotic applications on the basis of physicochemical hypotheses," Proc. IEEE International Conference on Robotics and Automation, pp. 1340-1346, 2000.
[27] Nemat-Nasser S. and J.Y. Li, "Electromechanical response of ionic polymer metal composites," Proc. SPIE Smart Structures and Materials 2000, Conference on Electro-Active Polymer Actuators and Devices, Vol. 3987, pp. 82-91, 2000.
[28] Konyo M., Y. Konishi, S. Tadokoro, and T. Kishima, Development of Velocity Sensor Using Ionic Polymer-Metal Composites, Proc. SPIE International Symposium on Smart Structures, Conference on Electro-Active Polymer Actuators and Devices, 2003.
[29] Benali-Khoudja M., M. Hafez, J.M. Alexandre, and A. Kheddar, Tactile interfaces: a state-of-the-art survey, 35th International Symposium on Robotics, pp.23-26, 2004.
[30] Shinoda H, N. Asamura, and N. Tomori, A tactile feeling display based on selective stimulation to skin receptors, Proc. IEEE ICRA, pp.435-441,1998.
[31] Kajimoto H, M. Inami, N. Kawakami, and S. Tachi, Smart Touch: Augmentation of Skin Sensation with Electrocutaneous Display, Proc. of the 11th International

Symposium on Haptic Interfaces for Virtual Environment and Teleoperator Systems, pp.40-46, 2003.

[32] Vallbo, Å.B. and Johansson, R.S., Properties of cutaneous mechanoreceptors in the human hand related to touch sensation, Human Neurobiology, 3, pp.3-14, 1984.

[33] Maeno T., Structure and Function of Finger Pad and Tactile Receptors, J. Robot Society of Japan, 18, 6, pp.772-775, 2000 (In Japanese).

[34] Talbot W.H., I. Darian-Smith, H.H. Kornhuber, and V.B. Mountcastle, The Sense of Flutter.Vibration: Comparison of the human Capability with Response Patterns of Mechanoreceptive Afferents from the Monkey Hand, J. Neurophysiology, 31, pp.301-335, 1968.

[35] Freeman A.W., and K.O. Johnson, A Model Accounting for Effects of Vibratory Amplitude on Responses of Cutaneous Mechanoreceptors in Macaque Monkey, J. Physiol., 323, pp.43-64, 1982.

[36] Carrozza M. C., P. Dario, A. Menciassi, and A. Fenu, “Manipulating biological and mechanical micro-objects using LIGA-microfabricated end-effectors,” Proc. IEEE International Conference on Robotics and Automation, pp. 1811-1816, 1998.

[37] Ono T., and M. Esashi, “Evanescent-field-controlled nano-pattern transfer and micro-manipulation,” Proc. IEEE International Workshop on Micro Electro Mechanical Systems, pp. 488-493, 1998.

[38] Zhou Y., B.J. Nelson, and B. Vikramaditya, ”Fusing force and vision feedback for micromanipulation,” Proc. IEEE International Conference on Robotics and Automation, pp. 1220-1225, 1998.

[39] Sadeghipour K., R. Salomon, and S. Neogi, Development of a Novel Electrochemically Active Membrane and `Smart' Material Based Vibration Sensor/Damper, Smart Materials and Structures, Vol.1, No.2, pp.172-179, 1992.

[40] Shahinpoor M., Y. Bar-Cohen, J.O. Simpson, and J. Smith, “Ionic polymer-metal composites (IPMC) as biomimetic sensors, Actuators and Artificial Muscles -- A Review,” Field Responsive Polymers, American Chemical Society, 1999.

[41] Fujiwara N., K. Asaka, Y. Nishimura, K. Oguro, and E. Torikai, Preparation and gold-solid polymer electrolyte composites as electric stimuli-responsive materials, Chem. Materials, Vol. 12, pp.1750-1754, 2000.

[42] Nakabo Y., T. Mukai, and K. Asaka, A Two-Dimensional Multi-DOF Robot Manipulator with a Patterned Artificial Muscle, Proc. Robotics Symposia, 2004 (In Japanese).

# 10

# Dynamic Modeling of Segmented IPMC Actuator

W. Yim[1], K. J. Kim[2]

[1] Department of Mechanical Engineering
University of Nevada, Las Vegas, Nevada 89154, USA
wy@me.unlv.edu
[2] Active Materials and Processing Laboratory (AMPL)
Department of Mechanical Engineering
University of Nevada, Reno, Nevada 89557, USA

## 10.1 Configuration of Segmented IPMC Actuator

Herein, we introduce an analytical modeling method for a segmented IPMC actuator which can exhibit varying curvature along the actuator. This segmented IMPC can generate more flexible propulsion compared with a single strip IPMC where only forward propulsion can be generated by a simple bending motion [1,2]. It is well known in biomimetic system research that a simple bending motion has lower efficiency than a snakelike, wavy motion in propulsion [3]. To realize this complex motion, a segmented IPMC can be a possible solution where each segment of the IPMC can be bent individually. As shown in Figure 10.1, the segmented IPMC design consists of a number of independently electroded sections along the length of the actuator. Each segment of the IMPC can be made by carving the surface of the IPMC and monitoring the electric insulation of each segment. Figure 10.2 shows a three-segment actuator consisting of Nafion (ionomeric polymer) passive substrate layer of thickness $h_b$ where two layers of metallic electrode (platinum) of thickness $h_p$ are placed on both sides. The electrodes for each segment are wired independently from the others, and by selectively activating each segment, varying curvature along the length may be obtained. The magnitude of curvature can be controlled by adjusting the voltage level applied across each segment. By controlling the curvature of the actuator along the length, it is possible to usc this actuator as a steerable device in the water. Here, we focus on the development of an analytical model to predict the free deflection of this segmented actuator.

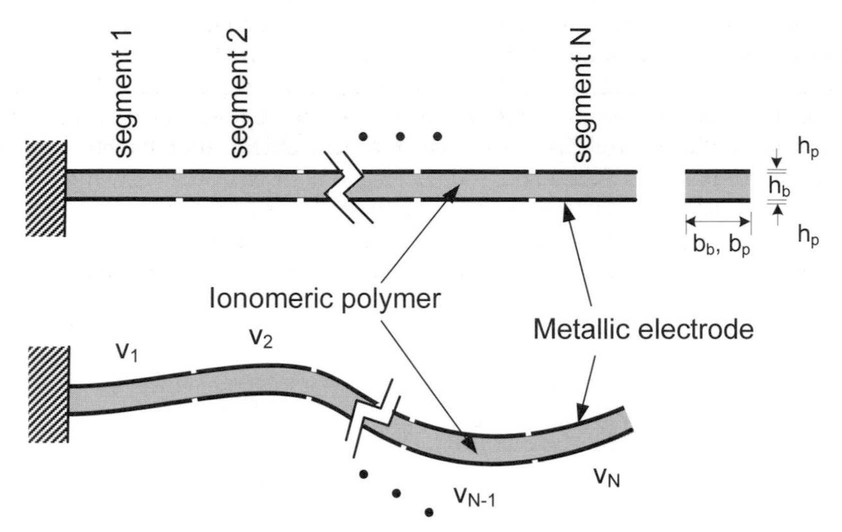


**Figure 10.1.** IPMC with N Segments

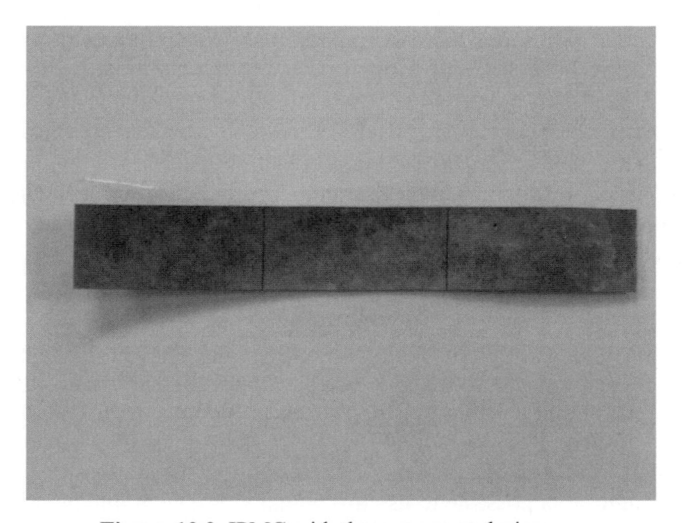

**Figure 10.2.** IPMC with three segment design

## 10.2 RC Model of IPMC

The analytical model is developed based on the clumped RC model of the IPMC [4,7] and a beam bending theory accounting for large deflections. The clumped RC model relates the input voltage applied to the IPMC strip to the charge. It has been shown that the IPMC often exhibits slow relaxation toward a cathode after quick bending towards an anode. However, this relaxation phenomenon associated with the bending curvature of the IPMC strip is ignored in this RC model. The finite-element approach is used to describe the dynamics of the segmented IPMC strip. It is considered as composed of finite elements that can be used to represent a large mechanical deflection of the IPMC using a geometrical approximation of the IPMC shape under no axial and shear loading condition. An energy approach is used to formulate the equations, and the bending moment applied in each segment is assumed to be proportional to the bending curvature determined from the simple first-order model. The modeling steps are described briefly in this section.

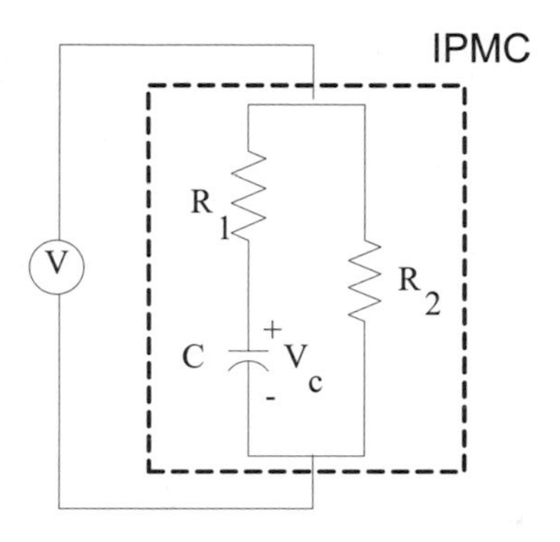


**Figure 10.3.** Clumped RC model for segment *i*

The IPMC has two parallel electrodes and electrolyte between the electrodes. The capacitance formed between two electrodes and internal resistance of electrolyte can be modeled as a simple RC circuit shown in Figure 10.3. For a voltage input to segment $i$, $V_i$, the electric charge, $Q_i$, and the current, $I_i$, in the circuit become

$$\frac{Q_i}{V_i} = \frac{C}{R_1 Cs + 1}$$
$$\frac{I_i}{V_i} = \frac{C(R_1 + R_2)s + 1}{R_1 R_2 Cs + R_2} \tag{10.1}$$

where $s$ is a Laplace complex variable.

Under a step input voltage, the IPMC strip shows a bending towards an anode due to cation migration towards a cathode in the polymer network. This bending moment is modeled using a simple first-order model of

$$\frac{m_i}{Q_i} = \frac{K_u}{\tau_u s + 1} \tag{10.2}$$

where $m_i$ is the bending moment applied to the IPMC segment $i$ and $K_u$, and $\tau_u$ are parameters that can be found using the experimental data. Here, $K_u$ is the gain and $\tau_u$ is the time constant that characterizes the speed of bending moment generation from the electric charge applied across the thickness of the IPMC. The RC model (10.1) and bending moment model (10.2) of the IPMC can be combined into the following linear model that relates the input voltage $V_i$ and bending moment $m_i$ of segment $i$

$$\frac{m_i}{V_i} = \frac{K_u C}{\tau_Q \tau_u s^2 + (\tau_Q + \tau_u)s + 1} = \frac{b_{i0}}{s^2 + a_{i1}s + a_{i0}} \tag{10.3}$$

where $\tau_u = R_1 C$ and $b_{i0}$ and $a_{ij}$ $(j=0,1)$ can be determined from the experimental data and an appropriate system identification techniques.

Equation (10.3) can be further generalized in the following form by including the IPMC relaxation phenomenon commonly observed after a quick bending towards an anode. This generalization can be accomplished by adding one zero to Eq. (10.2) that relates the charge $Q_i$ and the bending moment $m_i$ by a simple lead network of

$$\frac{m_i}{V_i} = \frac{b_{i1}s + b_{i0}}{s^2 + a_{i1}s + a_{i0}} \tag{10.4}$$

## 10.3 Mechanical Model of IPMC

### 10.3.1 Kinematics

Analytical solutions are available for several special cases of geometric nonlinearity in a cantilever beam [8]. Here, the finite-element approach is used to describe the dynamics of the segmented IPMC strip. We assume that the number of segments, $n$, is the same as the number of elements used in the modeling as shown in Figure 10.4. Hence, there are $n+1$ nodes with the nodes of a element ($i$) being node ($i$) and ($i+1$). The displacement of any point on the IPMC is described in terms of nodal displacements and slopes. The lateral displacement at distance $x_i$ can be expressed as follows:

$$v(x_i, t) = N(x_i)\, q_i(t) \tag{10.5}$$

where N is a 1×4 row vector of $N = [N_1(x_i) \quad N_2(x_i) \quad N_3(x_i) \quad N_4(x_i)]$, $q_i(t) = [\eta_i \quad \eta_{i+1}]^T = [v_i(t) \quad \phi_i(t) \quad v_{i+1}(t) \quad \phi_{i+1}(t)]^T$, and $\eta_i$ is a vector associated with nodal coordinates $v_i$ and $\phi_i$ of node $i$ where $v_i$ and $\phi_i$ denote nodal displacement and slope. Hermitian shape functions $N_i(x)$ that can be determined from $v$ and $\phi$ at both ends of the element with a length of $L_i$ become

$$N_1(x) = \frac{1}{L_i^3}\left(2x^3 - 3x^2 L_i + L_i^3\right), \quad N_2(x) = \frac{1}{L_i^3}\left(x^3 L_i - 2x^2 L_i^2 + x L_i^3\right)$$
$$N_3(x) = \frac{1}{L_i^3}\left(-2x^3 + 3x^2 L_i\right), \quad N_4(x) = \frac{1}{L_i^3}\left(x^3 L_i - x^2 L_i^2\right) \tag{10.6}$$

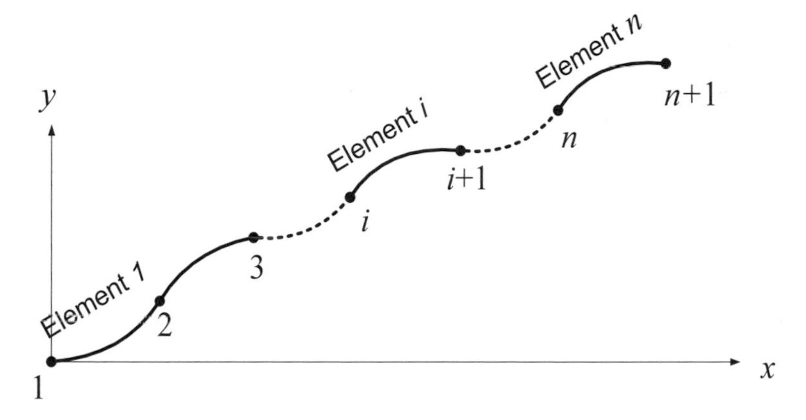


**Figure 10.4.** Finite-element modeling of an IPMC with $n$ elements

The IPMC would not experience axial loading and the axial deformation is ignored, however, its position in the $x$ direction is determined geometrically by the lateral deformation only. As shown in Figure 10.5, an infinitesimal deformation $du$ in the axial direction can be expressed as $dx = ds + du$ where $ds$ is the length of an differential element and can be approximated as

$$ds = \sqrt{(dv)^2 + (dx)^2} \tag{10.7}$$

Using Eq. (10.7), $du/dx$ can be expressed

$$\frac{du}{dx} = 1 - \left[\left(\frac{dv}{dx}\right)^2 + 1\right]^{\frac{1}{2}} \tag{10.8}$$

Noticing that $(a+1)^n = n\,a + 1$ if $a$ is small enough, Eq. (10.8) simplifies to:

$$\frac{du}{dx} = -\frac{1}{2}\left(\frac{dv}{dx}\right)^2 \tag{10.9}$$

Integrating Eq. (10.9) and using (10.5), the axial deformation at distance $x_i$ in element $i$ can be expressed as

$$u\left(x_i,t\right) = -\frac{1}{2}\int_0^{x_i}\left(\frac{dv\left(x_i,t\right)}{dx}\right)^2 dx = q_i^T\left[-\frac{1}{2}\int_0^{x_i} N'(x)^T N'(x)\ dx\right]q_i = q_i^T N_s(x_i)q_i \tag{10.10}$$

where $N'(x) = \dfrac{dN(x)}{dx}$ and $N_s(x_i)$ is a 4×4 matrix defined as

$$N_s(x_i) = -\frac{1}{2}\int_0^{x_i} N'(x)^T N'(x)\ dx \tag{10.11}$$

Also, the axial displacement at node $(i+1)$ can be expressed as

$$u_{i+1}\left(t\right) = u(L_i,t) = q_i^T\left[-\frac{1}{2}\int_0^{L_i} N'(x)^T N'(x)\ dx\right]q_i = q_i^T N_s(L_i)q_i \tag{10.12}$$

From Figure 10.5 the nodal positions of each node can be determined using Eq. (10.12) as

$$
\begin{aligned}
{}^{1}r &= \begin{Bmatrix} 0 \\ 0 \end{Bmatrix} \text{ (fixed boundary condition)} \\
{}^{2}r &= \begin{Bmatrix} L_1 + u_2 \\ N(L_1)q_1(t) \end{Bmatrix} = \begin{Bmatrix} L_1 + q_1{}^T N_s(L_1) q_1 \\ N(L_1)q_1(t) \end{Bmatrix} \\
{}^{3}r &= \begin{Bmatrix} x_2 + L_2 + u_3 \\ N(L_2)q_2(t) \end{Bmatrix} = \begin{Bmatrix} L_1 + q_1{}^T N_s(L_1) q_1 + L_2 + q_2{}^T N_s(L_2) q_2 \\ N(L_2)q_2(t) \end{Bmatrix} \\
&\vdots \\
{}^{i}r &= \begin{Bmatrix} \sum_{j=1}^{i-1} \left( L_j + u_{j+1} \right) \\ N(L_{i-1})q_{i-1}(t) \end{Bmatrix} = \begin{Bmatrix} \sum_{j=1}^{i-1} \left( L_j + q_j{}^T N_s(L_j) q_j \right) \\ N(L_{i-1})q_{i-1}(t) \end{Bmatrix}
\end{aligned}
\tag{10.13}
$$

The position vector, ${}^{i}r_p$, of point $P$ at distance $x_i$ on element $i$ becomes

$$
{}^{i}r_p = \begin{bmatrix} \sum_{j=1}^{i-1} \left( L_j + u_{j+1} \right) + x_i + u(x_i, t) \\ v(x_i, t) \end{bmatrix} = \begin{bmatrix} \sum_{j=1}^{i-1} \left( L_j + q_j{}^T N_s(L_j) q_j \right) + x_i + q_i{}^T N_s(x_i) q_i \\ N(x_i) q_i(t) \end{bmatrix}
\tag{10.14}
$$

Differentiating Eq. (10.14) in time, the velocity of point $P$ can be expressed as follows:

$$
{}^{i}\dot{r}_p = \begin{Bmatrix} \sum_{j=1}^{i-1} \left[ 2 q_j{}^T N_S(L_j) \dot{q}_j \right] + 2 q_i{}^T N_S(x_i) \dot{q}_i \\ N(x_i) \dot{q}_i \end{Bmatrix}
\tag{10.15}
$$

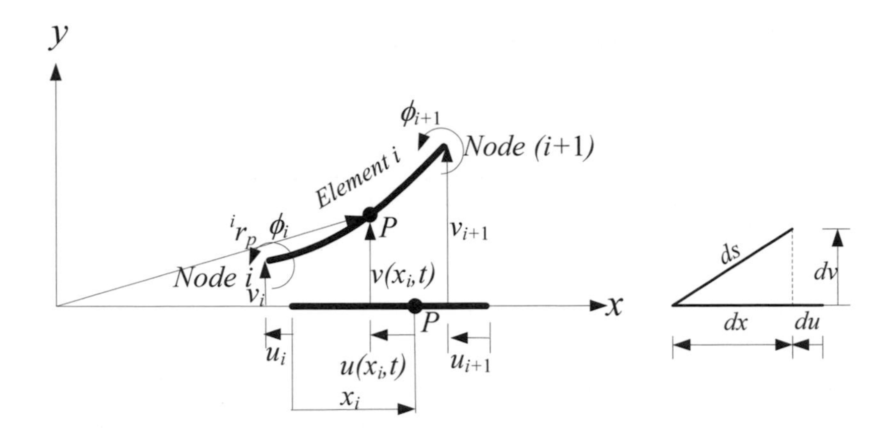


**Figure 10.5.** Deformed element and nodal displacement

### 10.3.3 Energy Formulation

In the kinematic analysis of a beam, axial extension and shearing deformation are ignored, and only lateral velocity contributes to the inertia. This assumption can be justified considering the thin geometry of the IPMC and the pure bending moment assumption. Based on these assumptions, the kinetic energy of the $i$-th element becomes

$$T_i = \frac{\rho}{2}\int_0^{L_i} {}^i\dot{r}_p^T\,{}^i\dot{r}_p dx_i = \frac{1}{2}\dot{\xi}_i^T M_i \dot{\xi}_i \tag{10.16}$$

where $M_i$ is the mass matrix, $\rho$ is the combined density of the IPMC per unit length, and $\xi_i = [\eta_1^T \quad \eta_2^T \quad \cdots \quad \eta_{i+1}^T]^T \in \Re^{2(i+1)}$ is the generalized coordinate. Differentiating $T_i$ of Eq. (10.16) with respect to $\dot{\xi}_i$ leads to

$$\frac{\partial T_i}{\partial \dot{\xi}_i} = \rho\int_0^{L_i} {}^i\dot{r}_p^T \frac{\partial\,{}^i\dot{r}_p}{\partial \dot{\xi}_i} dx_i = \rho\left(\int_0^{L_1} {}^i\dot{r}_p^T \frac{\partial\,{}^i\dot{r}_p}{\partial \dot{\eta}_1} dx_1 \quad \int_0^{L_2} {}^i\dot{r}_p^T \frac{\partial\,{}^i\dot{r}_p}{\partial \dot{\eta}_2} dx_2 \quad \cdots \quad \int_0^{L_i} {}^i\dot{r}_p^T \frac{\partial\,{}^i\dot{r}_p}{\partial \dot{\eta}_{i+1}} dx_i\right) \tag{10.17}$$

also,

$$\frac{\partial T_i}{\partial \dot{\xi}_i} = \begin{pmatrix} \frac{\partial T_i}{\partial \dot{\eta}_1} & \frac{\partial T_i}{\partial \dot{\eta}_2} & \cdots & \frac{\partial T_i}{\partial \dot{\eta}_{i+1}} \end{pmatrix} = \dot{\xi}_i^T M_i \tag{10.18}$$

Noting that $\frac{\partial \dot{r}_p}{\partial \dot{\eta}_i} = \frac{\partial r_p}{\partial \eta_i}$, the mass matrix $M_i$ can be expressed as

$$M_i = \rho \left[ \int_0^{L_i} \left( \frac{\partial\, {}^i r_p}{\partial \xi_i} \right)^T \frac{\partial\, {}^i r_p}{\partial \eta_1} dx_1 \quad \int_0^{L_i} \left( \frac{\partial\, {}^i r_p}{\partial \xi_i} \right)^T \frac{\partial\, {}^i r_p}{\partial \eta_2} dx_2 \quad \cdots \quad \int_0^{L_i} \left( \frac{\partial\, {}^i r_p}{\partial \xi_i} \right)^T \frac{\partial\, {}^i r_p}{\partial \eta_{i+1}} dx_i \right] \tag{10.19}$$

where $M_i \in \Re^{2(i+1)\times 2(i+1)}$. The potential energy of element $i$, including the bending moment, $m_i$, induced by an externally applied voltage $V_i$, can be expressed as

$$U_i = \frac{1}{2} \int_0^{L_i} \frac{1}{EI} \left[ EI \frac{\partial^2 v(x_i,t)}{\partial x_i^2} + m_i \right]^2 dx_i \tag{10.20}$$

where $v(x_i,t)$ is the deflection at point $P$ on element $i$ and $EI$ is the product of Young's modulus of elasticity and the cross-sectional moment of inertia. Note that the potential energy term due to extensional deformation is not included in Eq. (10.20) assuming that any axial deformation is negligible. From Eq. (10.20) the stiffness matrix, $K_i$, of element $i$ is defined as

$$K_i = EI \int_0^{L_i} \left[ \frac{\partial^2 N(x_i)}{\partial x_i^2} \right]^T \left[ \frac{\partial^2 N(x_i)}{\partial x_i^2} \right] dx_i \tag{10.21}$$

Unlike the kinetic energy term shown in Eq. (10.16), the potential energy of element $i$ depends only on the nodal coordinate $q_i = \left[ \eta_i^T \quad \eta_{i+1}^T \right]^T \in \Re^4$. Both $M_i$ and $K_i$ are expanded to the dimension of the generalized coordinate $\xi_e = [\eta_1^T \quad \eta_2^T \quad \cdots \quad \eta_{n+1}^T]^T \in \Re^{2(n+1)}$ for the IPMC with $n$ segments. The expanded matrices become

$$M_{ei} = \begin{bmatrix} M_i & 0_{2(i+1)\times 2(n-i)} \\ 0_{2(n-i)\times 2(i+1)} & 0_{2(n-i)\times 2(n-i)} \end{bmatrix} \tag{10.22}$$

$$K_{ei} = \begin{bmatrix} 0_{2(i-1)\times 2(i-1)} & 0_{2(i-1)\times 4} & 0_{2(i-1)\times 2(n-i)} \\ 0_{4\times 2(i-1)} & K_i & 0_{4\times 2(n-i)} \\ 0_{2(n-i)\times 2(i-1)} & 0_{2(n-i)\times 4} & 0_{2(n-i)\times 2(n-i)} \end{bmatrix}$$

Using Lagrangian dynamics, the equations of motion corresponding to element $i$ can be obtained as

$$M_{ei}\ddot{\xi}_e + K_{ei}\xi_e = B_{ei}m_i(t) \quad i = 1\cdots n \tag{10.23}$$

where $B_{ei} = \begin{bmatrix} 0_{2(i-1)} & 0 & -1 & 0 & 1 & 0_{2(n-i)} \end{bmatrix}^T \in \Re^{2(n+1)}$ is a control input vector for the bending moment input $m_i(t)$ on element $i$. It corresponds to a distributed moment that is replaced by two concentrated moments at the two nodes. Equation (10.23) can be assembled for the entire segments $n$ using $\xi_e = \begin{bmatrix} \eta_2^T & \eta_3^T & \cdots & \eta_{n+1}^T \end{bmatrix}^T = \begin{bmatrix} v_2 & \phi_2 & v_3 & \phi_3 \cdots v_{n+1} & \phi_{n+1} \end{bmatrix}^T \in \Re^{2n}$ and $m = \{m_1\ m_2 \cdots m_n\}^T \in \Re^n$ by noting that $\begin{bmatrix} v_1 & \phi_1 \end{bmatrix}$ is eliminated from the generalized coordinate $\xi_e$ because the first node has zero boundary conditions. This assembled equation becomes

$$\left(\sum_{i=1}^{n} M_{ei}\right)\ddot{\xi}_e + \left(\sum_{i=1}^{n} K_{ei}\right)\xi_e = \begin{bmatrix} B_{e1} & \cdots & B_{en} \end{bmatrix} m$$

or (10.24)

$$M_e\ddot{\xi}_e + K_e\xi_e = B_e m$$

where $M_e \in \Re^{2n\times 2n}$, $K_e \in \Re^{2n\times 2n}$ are the mass and stiffness matrix, respectively, $m \in \Re^n$ is an input moment vector, and $B_e \in \Re^{2n\times n}$ is an input control matrix for $m$. In Eq. (10.24), it can be seen that the material modulus, $E$, can be factored from the stiffness matrix $K_e$. Performing the factorization and transforming Eq. (10.23) into a Laplace domain yields,

$$\left(s^2 M_e + E\overline{K}\right)\xi_e = B_e\mathbf{m} \tag{10.25}$$

where $K_e = E\overline{K}$ and $s$ is the Laplace variable. The viscoelastic property of the IPMC can be included here by replacing $E$ with complex modulus $E^*$. It is well known that the stress-strain relationship of a viscoelastic material includes not only the instantaneous strain, but the strain history as well. This frequency-dependent term of the complex modulus $E^*$ can be modeled by the transfer function *h(s)* represented by the sum of appropriate rational polynomials that depends on the types of viscoeleastic models [5,6].

$$\left(s^2 M_e + E^* \bar{K}\right)\xi_e = \left(s^2 M_e + E(1+h(s))\bar{K}\right)\xi_e = B_e \mathbf{m} \tag{10.26}$$

Equation (10.26) can be transformed back to the time domain using additional variables defined for $h(s)$.

### 10.3.4 State Space Model

Unlike the small deflection model of the IPMC [4] the large deflection model cannot be modeled as a standard linear state space model because the deflection in the axial direction is determined by the lateral deflection of the IPMC, as shown in Eq. (10.10). To realize the combined dynamic model for an entire IPMC length of $n$ segments including linear RC models, Eq. (10.4) can be written as

$$\ddot{m}_i + a_{i1}\dot{m}_i + a_{i0}m_i = b_{i1}\dot{V}_i + b_{i0}V_i, \quad i = 1\cdots n \tag{10.27}$$

By introducing two new variables $z_{i1}$ and $z_{i2}$ for element $i$,

$$\begin{aligned} \dot{z}_{i1} &= z_{i2} \\ \dot{z}_{i2} &= -a_{i1}z_{i2} - a_{i0}z_{i1} + V_i \end{aligned} \tag{10.28}$$

$m_i$ of Eq. (10.27) can be expressed in terms of these new variables $z_{i1}$ and $z_{i2}$ as

$$m_i = b_{i0}z_{i1} + b_{i1}z_{i2}, \quad i = 1\cdots n \tag{10.29}$$

Equation (10.28) can be expanded for an entire IPMC of $n$ segments as

$$\dot{Z} = \begin{bmatrix} & 0_{n\times n} & & & I_{n\times n} & \\ -a_{10} & 0 & 0 & -a_{11} & 0 & 0 \\ 0 & \ddots & 0 & 0 & \ddots & 0 \\ 0 & 0 & -a_{n0} & 0 & 0 & -a_{n1} \end{bmatrix} Z + \begin{bmatrix} 0_{n\times n} \\ I_{n\times n} \end{bmatrix} V \tag{10.30}$$
$$\equiv A_z Z + B_v V$$

where $Z = \{z_{11}\ z_{21} \cdots z_{n1},\ z_{12}\ z_{22} \cdots z_{n2}\}^T \in \Re^{2n}$ and $V = \{V_1\ V_2 \cdots V_n\}^T \in \Re^n$ is an input voltage vector. Equation (10.29) can be also expanded for the input moment vector $u$ using $Z$ as

$$m = \begin{Bmatrix} m_1 \\ \vdots \\ m_n \end{Bmatrix} = \begin{bmatrix} b_{10} & 0 & 0 & b_{11} & 0 & 0 \\ 0 & \ddots & 0 & 0 & \ddots & 0 \\ 0 & 0 & b_{n0} & 0 & 0 & b_{n1} \end{bmatrix} Z \equiv B_u Z \tag{10.31}$$

Now the complete equations of motion for the segmented IPMC actuator can be written as

$$\begin{aligned} M_e\ddot{\xi}_e + E^*\overline{K}\xi_e &= B_e u = B_e B_u Z \\ \dot{Z} &= A_z Z + B_v V \end{aligned} \tag{10.32}$$

Defining the state vector $x = [\xi_e \ \dot{\xi}_e \ Z]^T$, then the state variable representation of the system becomes

$$\dot{x} = f(x) + g(x)V \tag{10.33}$$

where

$$f(x) = \begin{bmatrix} \dot{\xi}_e \\ -M_e^{-1}E^*\overline{K}\xi_e + M_e^{-1}B_e B_u Z \\ A_z Z \end{bmatrix}$$

$$g(x) = \begin{bmatrix} 0 \\ 0 \\ B_v \end{bmatrix}$$

Equation (10.33) can be used for designing control laws for the segmented IPMC strip for various applications. Note that extra state variables must be defined for the viscoelastic property of the IPMC expressed as $E^*$ or *h(s)* in Eq. (10.26) for the state variable of Eq. (10.33).

## 10.4 Computer Simulation

Computer simulation was performed for Eq. (10.33) for $n$=3 for different magnitudes of input voltage in each element. Table 10.1 shows the simulation and material parameters used in this computational study. Figures 10.6 and and 10.7 show the large deflection simulation of the segmented IPMC for the step input voltage of $V=\{3\ 3\ 3\}^T$ volt and $V=\{-3\ 2\ 3\}^T$, respectively. In this simulation, the viscoelastic property of the IPMC is ignored, i.e., $E^*=E$ or $h(s)=0$.

**Table 10.1.** Simulation and material parameters

| | | | |
|---|---|---|---|
| $R_1$ | 160 Ω | A (length of IPMC) | 0.05 m |
| $R_2$ | 700 Ω | b (width of IPMC) | 0.01 m |
| C | 1.0E-3 F | $E_b$ (modulus for Nafion) | 5E7 Pa |
| $\tau_1(=R_1C)$ | 0.16 s | $E_p$ (modulus for electrode(Pt)) | 144E9 Pa |
| $\tau_2$ | 10 s | $h_b$ (thickness of Nafion) | 0.00028 m |
| $\rho_p$ (density of electrode(Pt)) | 21500 kg/m$^3$ | $h_p$ (thickness of electrode) | 0.000002 m |
| $\rho_b$ (density of Nafion) | 2600 kg/m$^3$ | | |

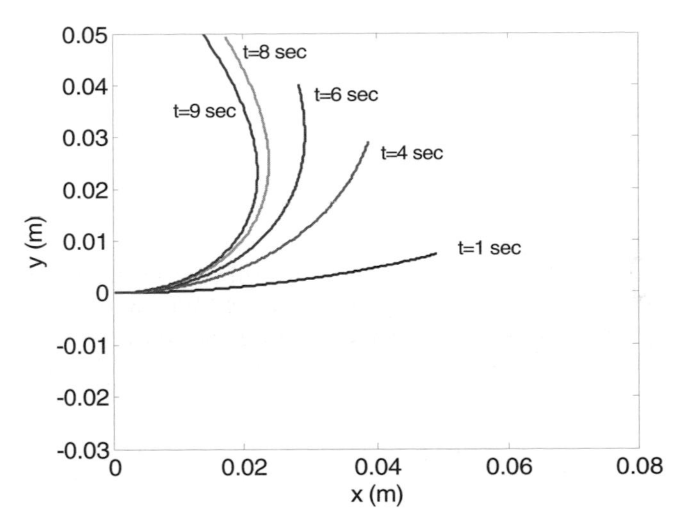


**Figure 10.6.** Deflection of IPMC under $V=[3\ 3\ 3]^T$ (Volt)

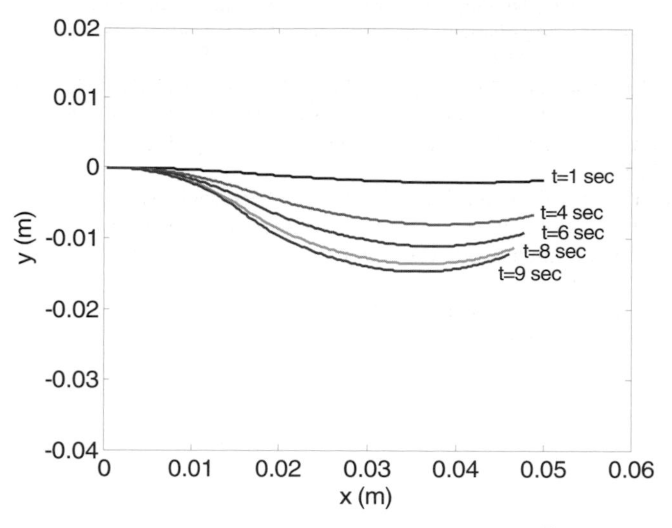


**Figure 10.7.** Deflection of IPMC under $V$=[-3 2 3]$^T$ (Volt)

## 10.5 Conclusions

In this study, we described the analytical framework for deriving the dynamic model of the segmented IPMC. For the electromechanical property of the IPMC, a simple RC model is used and the finite-element approach is used for dynamic modeling of the mechanical portion of the segmented IPMC. The proposed modeling approach can be used for designing various controllers for effective underwater propulsion and other robotics applications where soft and complex motion are required. The proposed segmented IPMC can be a possible solution because each segment of the IPMC can be bent individually. The state space equation developed using a large deflection beam model was simulated for different input voltages, and simulation results show that the proposed model can be effectively used for estimating the motion of the segmented IPMC in various applications.

## 10.6 References

[1] Y. Nakabo, T. Mukai, and K. Asaka, "A multi-DOF robot manipulator with a patterned artificial muscle," The 2nd Conf. on Artificial Muscle, Osaka, 2004.

[2] Y. Bar-Cohen (2002) Electro-active polymers: current capabilities and challenges. Proc. of SPIE Int. Symp. on Smart Structures and Materials, EAPAD

[3] M.J. Lighthill (1960) Note on the swimming of slender fish. J. Fluid Mechanics, 9:305–317.

[4] W. Yim, J. Paquette, S. Heo, and K.J. Kim, "Operation of Ionic Polymer-Metal Composites in Water," SPIE Smart Materials and Structures Conf., #5759-6 (March 2005).

[5] D.F. Golla and P.C. Hughes, “Dynamics of Viscoelastic Structures—A Time-Domain, Finite Element Formulation.” ASME Journal of Applied Mechanics, Vol. 52, pp. 897-906, December, 1985.

[6] D.J. McTavishand and P.C. Hughes, “Finite Element Modeling of Linear Viscoelastic Structures: The GHM Method.” AIAA/ASME/ASCE/AHS/ASC Structures, Structural Dynamics, and Materials Conference, 33rd, Vol. 4, pp. 1753-1763, Dallas, TX, April 13-15, 1992.

[7] X. Bao, Y. Bar-Cohen, and S-S Lih, “Measurements and Macro Models of Ionomeric Polymer Metal Composites (IPMC),” Proceedings of the SPIE Smart Structures and Materials Symposium, EAPAD Conference, San Diego, CA, March 18-21, 2002 (220-227).

[8] J.M. Gere and S.P. Timoshenko, “Mechanics of Materials,” 2nd Edition, Van Nostrand Reinhold Co., New York, 1972.

# Index

## T

Printing: Krips bv, Meppel
Binding: Stürtz, Würzburg

RECEIVED
MAY 22 2007